METHODS IN MOLECULAR BIOLOGY

Series Editor
John M. Walker
School of Life and Medical Sciences
University of Hertfordshire
Hatfield, Hertfordshire, UK

For further volumes:
http://www.springer.com/series/7651

For over 35 years, biological scientists have come to rely on the research protocols and methodologies in the critically acclaimed *Methods in Molecular Biology* series. The series was the first to introduce the step-by-step protocols approach that has become the standard in all biomedical protocol publishing. Each protocol is provided in readily-reproducible step-by-step fashion, opening with an introductory overview, a list of the materials and reagents needed to complete the experiment, and followed by a detailed procedure that is supported with a helpful notes section offering tips and tricks of the trade as well as troubleshooting advice. These hallmark features were introduced by series editor Dr. John Walker and constitute the key ingredient in each and every volume of the *Methods in Molecular Biology* series. Tested and trusted, comprehensive and reliable, all protocols from the series are indexed in PubMed.

Bone Marrow Environment

Methods and Protocols

Edited by

Marion Espéli and Karl Balabanian

Université de Paris, Institut de Recherche Saint Louis, OPALE Carnot Institute, EMiLy, INSERM U1160, Paris, France

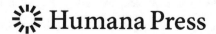 Humana Press

Editors
Marion Espéli
Université de Paris
Institut de Recherche Saint Louis
OPALE Carnot Institute
EMiLy, INSERM U1160
Paris, France

Karl Balabanian
Université de Paris
Institut de Recherche Saint Louis
OPALE Carnot Institute
EMiLy, INSERM U1160
Paris, France

ISSN 1064-3745   ISSN 1940-6029   (electronic)
Methods in Molecular Biology
ISBN 978-1-0716-1427-3   ISBN 978-1-0716-1425-9   (eBook)
https://doi.org/10.1007/978-1-0716-1425-9

This Humana imprint is published by the registered company Springer Science+Business Media, LLC, part of Springer Nature.
The registered company address is: 1 New York Plaza, New York, NY 10004, U.S.A.

Preface

The bone and the marrow (BM) it contains are absolutely essential for the formation and the maintenance of the skeletal and immune-hematological systems. In this complex ecosystem, the fate of hematopoietic, immune, stromal, vascular, and neural cells is intertwined with numerous multi-partner interactions established in a highly dynamic and concerted fashion. Indeed, hematopoietic stem and progenitor cells (HSPCs) co-exist with cells corresponding to intermediate differentiation states as well as with terminally differentiated or effector cells (e.g., memory T cells, plasma cells (PCs)). All these cells of hematopoietic origin interact with numerous and still partly characterized non-hematopoietic stromal cells that act as tissue organizers, orchestrate hematopoietic and immune processes, and are at the origin of skeletal formation and renewal. The higher organ functions of the BM depend on the sophisticated and finely tuned communication established between all the individual cellular components, through direct contact or secretion of soluble factors such as cytokines or growth factors. In particular, highly specialized microenvironments are required for proper development, survival, and function of HSPCs and immune effector cells. This is a long-standing concept in hematology, where HSPCs develop and differentiate in micro-anatomical sites or "niches" within the BM. The notion of niches has become an essential part of how we envision the organization and function of the BM ecosystem. Niches are structures composed of different cell types including mesenchymal stem/stromal cells (MSCs). Although the definition of "niches" may vary depending on the studies, a unifying view could be that they correspond to complex, dynamic microstructures in which several soluble and membrane-anchored mediators are produced allowing the correct positioning of a given cell type and thus facilitating interactions with other cellular actors and access to all the elements necessary for their maintenance or differentiation.

During the last two decades, our understanding of this fascinating organ has dramatically increased in particular thanks to the development and application of dedicated experimental models to unravel the bidirectional interplay established between HSPCs and their environment within the BM. In this timely volume of Methods in Molecular Biology, we intend to bring together classical and cutting-edge protocols to help the scientific community to push forward the spatio-temporal study of the cellular subsets constituting the bone and the marrow in both mouse and human. This volume encompasses protocols to visualize the BM ecosystem, to label, sort, analyze, and culture specific cell subsets as well as techniques allowing the evaluation of the function of some of the cellular elements of the BM.

It is the expertise and dedication of the contributing authors that have made this timely book possible, and we hope it will help new investigators to pursue the characterization of the BM microenvironment in the coming years.

Paris, France

Marion Espéli
Karl Balabanian

Contents

Contributors

ZEINA ABOU NADER • *Université de Paris, Institut de Recherche Saint-Louis, EMiLy, INSERM U1160, Paris, France*

KUTAIBA ALHAJ HUSSEN • *INSERM U976, Université de Paris, École Pratique des Hautes Études/PSL Research University, Institut de Recherche Saint Louis, Paris, France; Service d'Hématologie Biologique, Hôpital Tenon, Hôpitaux Universitaires de l'Est Parisien, Assistance Publique Hôpitaux de Paris, Paris, France*

ADRIENNE ANGINOT • *Université de Paris, Institut de Recherche Saint-Louis, EMiLy, INSERM U1160, Paris, France*

AMINA AOUIDA • *Paris-Saclay University, Gustave Roussy Institute, Institut National de la Santé et de la Recherche Médicale Inserm U1170, Villejuif, France*

MICHEL AURRAND-LIONS • *Aix Marseille University, CNRS, INSERM, Institut Paoli-Calmettes, Cancerology Research Center of Marseille (CRCM), Marseille, France; Equipe Labellisee Ligue Nationale Contre le Cancer, Paris, France*

KARL BALABANIAN • *Université de Paris, Institut de Recherche Saint-Louis, OPALE Carnot Institute, EMiLy, INSERM U1160, Paris, France*

VALERIA BISIO • *Université de Paris, Institut de Recherche Saint-Louis, EMiLy, INSERM U1160, Paris, France*

CLAUDINE BLIN-WAKKACH • *Université Côte d'Azur, CNRS, UMR7370, LP2M, Nice, France*

AMÉLIE BONAUD • *Université de Paris, Institut de Recherche Saint Louis, EMiLy, INSERM U1160, Paris, France*

DOMINIQUE BONNET • *Haematopoietic Stem Cell Lab, The Francis Crick Institute, London, UK*

JÉRÔME BOURGEAIS • *CNRS ERL7001 LNOx "Leukemic Niche & Redox Metabolism", Tours, France, EA7501 GICC, Faculty of Medicine, Tours University, Tours, France; Department of Biological Hematology, Tours University Hospital, Tours, France*

PAUL E. BOURGINE • *Cell, Tissue & Organ Engineering Laboratory, BMC B11, 221 84, Department of Clinical Sciences Lund, Stem Cell Center, Lund University, Lund, Sweden; Wallenberg Centre for Molecular Medicine, Lund University, Lund, Sweden*

STÉPHANE BRUNET • *Université de Paris, CEA/INSERM/AP-HP, Institut de Recherche Saint Louis, UMR976, HIPI, CytoMorpho Lab, Hopital Saint Louis, Paris, France; Université Grenoble-Alpes, CEA/INRA/CNRS, Interdisciplinary Research Institute of Grenoble, UMR5168, LPCV, CytoMorpho Lab, Grenoble, France*

FERNANDO J. CALERO-NIETO • *Department of Haematology, Jeffrey Cheah Biomedical Centre, Puddicombe Way, Wellcome - MRC Cambridge Stem Cell Institute, University of Cambridge, Cambridge, UK*

BRUNO CANQUE • *INSERM U976, Université de Paris, École Pratique des Hautes Études/ PSL Research University, Institut de Recherche Saint Louis, Paris, France*

HUI CHENG • *State Key Laboratory of Experimental Hematology, Institute of Hematology & Blood Diseases Hospital, Collaborative Innovation Center for Cancer Medicine, Tianjin, China*

MARJORIE C. DELAHAYE • *Aix Marseille University, CNRS, INSERM, Institut Paoli-Calmettes, Cancerology Research Center of Marseille (CRCM), Marseille, France; Equipe Labellisee Ligue Nationale Contre le Cancer, Paris, France*

STEVEN J. DUPARD • *Cell, Tissue & Organ Engineering Laboratory, BMC B11, 221 84, Department of Clinical Sciences Lund, Stem Cell Center, Lund University, Lund, Sweden; Wallenberg Centre for Molecular Medicine, Lund University, Lund, Sweden*

ALMUT S. EISELE • *Institut Curie, Université PSL, Sorbonne Université, CNRS UMR168, Laboratoire Physico Chimie Curie, Paris, France*

MARION ESPÉLI • *Université de Paris, Institut de Recherche Saint-Louis, OPALE Carnot Institute, EMiLy, INSERM U1160, Paris, France*

MAXIMILIEN EVRARD • *Department of Microbiology and Immunology, The University of Melbourne and The Peter Doherty Institute for Infection and Immunity, Melbourne, VIC, Australia*

BERTHOLD GÖTTGENS • *Department of Haematology, Jeffrey Cheah Biomedical Centre, Puddicombe Way, Wellcome - MRC Cambridge Stem Cell Institute, University of Cambridge, Cambridge, UK*

JULIEN M. P. GRENIER • *Aix Marseille University, CNRS, INSERM, Institut Paoli-Calmettes, Cancerology Research Center of Marseille (CRCM), Marseille, France; Equipe Labellisee Ligue Nationale Contre le Cancer, Paris, France*

JULIA HALPER • *Université Côte d'Azur, CNRS, UMR7370, LP2M, Nice, France*

MYRIAM L. R. HALTALLI • *Imperial College London, London, UK; The Francis Crick Institute, London, UK; Wellcome—Medical Research Council Cambridge Stem Cell Institute, University of Cambridge, Cambridge, UK*

HOUDA HAOUAS • *Department of Biological and Chemical Engineering, National Institute of Applied Sciences and Technology, Tunis, Tunisia*

ANJA E. HAUSER • *Charité—Universitätsmedizin Berlin, corporate member of Freie Universität Berlin and Humboldt-Universität zu Berlin, Department of Rheumatology and Clinical Immunology, Berlin, Germany; Deutsches Rheuma-Forschungszentrum (DRFZ) Berlin, a Leibniz Institute, Berlin, Germany*

LIANG HE • *Paris-Saclay University, Gustave Roussy Institute, Institut National de la Santé et de la Recherche Médicale Inserm U1170, Villejuif, France*

OLIVIER HÉRAULT • *CNRS ERL7001 LNOx "Leukemic Niche & Redox Metabolism", Tours, France; EA7501 GICC, Faculty of Medicine, Tours University, Tours, France; Department of Biological Hematology, Tours University Hospital, Tours, France*

STEPHAN HOLTKAMP • *Walter-Brendel-Centre of Experimental Medicine, University Hospital, Ludwig-Maximilians-University Munich, BioMedical Centre, Planegg-Martinsried, Munich, Germany*

SEYDOU KEITA • *INSERM U976, Université de Paris, École Pratique des Hautes Études/PSL Research University, Institut de Recherche Saint Louis, Paris, France*

IMMANUEL KWOK • *Singapore Immunology Network (SIgN), A*STAR (Agency for Science, Technology and Research), Biopolis, Singapore; School of Biological Sciences, Nanyang Technological University, Singapore, Singapore*

JACKSON LIANG YAO LI • *Singapore Immunology Network (SIgN), A*STAR (Agency for Science, Technology and Research), Singapore, Singapore*

CRISTINA LO CELSO • *Imperial College London, London, UK; The Francis Crick Institute, London, UK*

FAWZIA LOUACHE • *Paris-Saclay University, Gustave Roussy Institute, Institut National de la Santé et de la Recherche Médicale Inserm U1170, Villejuif, France*

MARIA-BERNADETTE MADEL • *Université Côte d'Azur, CNRS, UMR7370, LP2M, Nice, France; Department of Orthopedic Surgery, Baylor College of Medicine, Houston, TX, USA*

STÉPHANE J. C. MANCINI • *Aix Marseille University, CNRS, INSERM, Institut Paoli-Calmettes, Cancerology Research Center of Marseille (CRCM), Marseille, France; Equipe Labellisee Ligue Nationale Contre le Cancer, Paris, France*

YUKI MATSUSHITA • *University of Michigan School of Dentistry, Ann Arbor, MI, USA*

STEFANO MAZZONI • *Department of Translational Research and New Technology in Medicine, University of Pisa, Pisa, Italy*

AMIRA MEHTAR • *Paris-Saclay University, Gustave Roussy Institute, Institut National de la Santé et de la Recherche Médicale Inserm U1170, Villejuif, France*

SIMÓN MÉNDEZ-FERRER • *Wellcome-MRC Cambridge Stem Cell Institute, Cambridge, UK; Department of Haematology, University of Cambridge, Cambridge, UK; National Health Service Blood and Transplant, Cambridge, UK*

SYED A. MIAN • *Haematopoietic Stem Cell Lab, The Francis Crick Institute, London, UK; Department of Haematology, School of Cancer and Pharmaceutical Sciences, King's College London, London, UK*

YEVIN MUN • *Department of Medical Oncology and Hematology, University of Zurich and University Hospital Zurich, Zurich, Switzerland*

LAI GUAN NG • *Singapore Immunology Network (SIgN), A*STAR (Agency for Science, Technology and Research), Biopolis, Singapore; School of Biological Sciences, Nanyang Technological University, Singapore, Singapore; State Key Laboratory of Experimental Hematology, Institute of Hematology, Chinese Academy of Sciences & Peking Union Medical College, Tianjin, China; Department of Microbiology and Immunology, Yong Loo Lin School of Medicine, National University of Singapore, Singapore, Singapore*

RALUCA A. NIESNER • *Deutsches Rheuma-Forschungszentrum (DRFZ) Berlin, a Leibniz Institute, Berlin, Germany; Veterinary Medicine, Dynamic and Functional in vivo Imaging, Freie Universität Berlin, Berlin, Germany*

CÉSAR NOMBELA-ARRIETA • *Department of Medical Oncology and Hematology, University of Zurich and University Hospital Zurich, Zurich, Switzerland*

NORIAKI ONO • *University of Michigan School of Dentistry, Ann Arbor, MI, USA*

WANIDA ONO • *University of Michigan School of Dentistry, Ann Arbor, MI, USA*

MATTHIEU OPITZ • *Alvéole, 68 Boulevard de Port-Royal, Paris, France*

JULIA P. LEMOS • *Université de Paris, Institut de Recherche Saint-Louis, EMiLy, INSERM U1160, Paris, France*

SIMONE PACINI • *Department of Clinical and Experimental Medicine, Hematology Division, University of Pisa, Pisa, Italy*

FRANCESCA M. PANVINI • *Institute of Life Sciences, Sant'Anna School of Advanced Studies, Pisa, Italy*

LEÏLA PERIÉ • *Institut Curie, Université PSL, Sorbonne Université, CNRS UMR168, Laboratoire Physico Chimie Curie, Paris, France*

VINCENT RONDEAU • *Université de Paris, Institut de Recherche Saint-Louis, EMiLy, INSERM U1160, Paris, France*

CHRISTOPH SCHEIERMANN • *Walter-Brendel-Centre of Experimental Medicine, University Hospital, Ludwig-Maximilians-University Munich, BioMedical Centre, Planegg-Martinsried, Munich, Germany; Centre Médical Universitaire (CMU), Department of Pathology and Immunology, Faculty of Medicine, University of Geneva, Geneva, Switzerland*

CHANGMING SHIH • *Singapore Immunology Network (SIgN), A*STAR (Agency for Science, Technology and Research), Singapore, Singapore*

BENOIT SOUQUET • *Alvéole, 68 Boulevard de Port-Royal, Paris, France; Université de Paris, CEA/INSERM/AP-HP, Institut de Recherche Saint Louis, UMR976, HIPI, CytoMorpho Lab, Hopital Saint Louis, Paris, France; Université Grenoble-Alpes, CEA/INRA/CNRS, Interdisciplinary Research Institute of Grenoble, UMR5168, LPCV, CytoMorpho Lab, Grenoble, France*

KATHERINE H. M. STURGESS • *Department of Haematology, Jeffrey Cheah Biomedical Centre, Puddicombe Way, Wellcome - MRC Cambridge Stem Cell Institute, University of Cambridge, Cambridge, UK*

TAMAR TAK • *Institut Curie, Université PSL, Sorbonne Université, CNRS UMR168, Laboratoire Physico Chimie Curie, Paris, France*

LEONARD TAN • *Singapore Immunology Network (SIgN), A*STAR (Agency for Science, Technology and Research), Singapore, Singapore; Department of Microbiology, Immunology Programme, Yong Loo Lin School of Medicine, National University of Singapore, Singapore, Singapore*

YINGROU TAN • *Singapore Immunology Network (SIgN), A*STAR (Agency for Science, Technology and Research), Singapore, Singapore; National Skin Centre, Singapore, Singapore*

MANUEL THÉRY • *Université de Paris, CEA/INSERM/AP-HP, Institut de Recherche Saint Louis, UMR976, HIPI, CytoMorpho Lab, Hopital Saint Louis, Paris, France; Université Grenoble-Alpes, CEA/INRA/CNRS, Interdisciplinary Research Institute of Grenoble, UMR5168, LPCV, CytoMorpho Lab, Grenoble, France*

CAROLIN ULBRICHT • *Charité—Universitätsmedizin Berlin, corporate member of Freie Universität Berlin and Humboldt-Universität zu Berlin, Department of Rheumatology and Clinical Immunology, Berlin, Germany; Deutsches Rheuma-Forschungszentrum (DRFZ) Berlin, a Leibniz Institute, Berlin, Germany*

BENOIT VIANAY • *Université de Paris, CEA/INSERM/AP-HP, Institut de Recherche Saint Louis, UMR976, HIPI, CytoMorpho Lab, Hopital Saint Louis, Paris, France; Université Grenoble-Alpes, CEA/INRA/CNRS, Interdisciplinary Research Institute of Grenoble, UMR5168, LPCV, CytoMorpho Lab, Grenoble, France*

NICOLA K. WILSON • *Department of Haematology, Jeffrey Cheah Biomedical Centre, Puddicombe Way, Wellcome - MRC Cambridge Stem Cell Institute, University of Cambridge, Cambridge, UK*

Part I

Cell Culture and *In Vitro* Functionnal Assays

Chapter 1

Culture, Expansion and Differentiation of Human Bone Marrow Stromal Cells

Valeria Bisio, Marion Espéli, Karl Balabanian, and Adrienne Anginot

Abstract

Mesenchymal stromal cells (MSC) are a rare, heterogeneous and multipotent population that can be isolated from several tissues. MSC were originally discovered in the bone marrow and studied for their capacity to maintain hematopoietic cells. We will describe here methods to isolate, culture, and bank MSC from human bone marrow. Then, characterization protocols by flow cytometry, clonogenic assays and doubling time evaluation will be developed. Finally, in vitro MSC culture and differentiation into osteoblasts, adipocytes, and chondrocytes will be explained. Thus, this chapter will detail all bases to work on MSC with consensus and clear methods and protocols.

Key words Origin of human samples, Isolation, Expansion, Banking, Characterization, Colony-forming unit—fibroblast, Doubling time, Osteogenic differentiation, Adipogenic differentiation, Chondrogenic differentiation

1 Introduction

The presence of nonhematopoietic cells in the bone marrow (BM) was observed more than 150 years ago by Cohnheim and colleagues but it was only Friedenstein in the sixties, with a series of seminal studies, who demonstrated the adherent and the associated osteogenic potential of this minor BM cell population [1–3]. This population, representing approximately 0.001% to 0.01% of nucleated human BM cells, is commonly known as "human mesenchymal stromal/stem cells" (h-MSC) [4]. Since their discovery, the concept of mesenchymal stromal cells has gained wide popularity; h-MSC have been isolated from many sources, such as peripheral blood, fat, skin, vasculature, muscle and cord blood, but BM is still considered as the most standard source of MSC [5, 6]. However, different methods of isolation and expansion have taken place over the time in various laboratories, allowing for MSC characterization and defining their potential activity. Thus, due to the exponential scientific production, it becomes increasingly difficult to compare

Marion Espéli and Karl Balabanian (eds.), *Bone Marrow Environment: Methods and Protocols*, Methods in Molecular Biology, vol. 2308, https://doi.org/10.1007/978-1-0716-1425-9_1, © Springer Science+Business Media, LLC, part of Springer Nature 2021

and evaluate study outcomes, which hinders progress in the field. To address this issue, the International Society of Cell Therapy (ISCT) [7] clarified the MSC acronym as mesenchymal stromal cells (still debated today) and the essential requirements that define this population.

- A heterogeneous population of cells isolated by adherent culture on plastic substrates that can display different morphologies: a spindle narrow or wide polygonal shape or a more cubic and squamous conformation.

- Expressing the surface antigenic markers CD105 (endoglin), CD73 (Ecto-5′-nuclotidase), CD90 (Thy-1, glycophosphatidy-linositol) and not expressing the markers CD34 (hematopoietic progenitors), CD45 (mature hematopoietic cells), CD11a or CD14 (monocytes, macrophages), CD19 or CD79α (B lymphocytes), and HLA-DR (MSC without stimulation).

- Able to differentiate in vitro into adipocytes, osteoblasts, and chondrocytes.

In the last decades the large number of studies have led to a better understanding of the variety of biological processes engaged by MSC: interaction with various immune cell types (including T, natural killer, and myeloid cells), secretion of soluble factors like growth factors, cytokines and chemokines, inhibition of apoptosis and stimulation of proliferation, and modulation of the immune response [8–10].

All these properties have made MSC a promising candidate in regenerative medicine and cell therapy era. The medical utility of MSC has been investigated in over 950 clinical trials (www. clinicaltrials.gov) where the pathologies treated are most often cardiovascular diseases (myocardial infarction, cardiomyopathy, myocardial ischemia), graft-versus-host disease (GVH), and certain autoimmune, neurodegenerative, and osteoarticular diseases [9]. Nowadays, there are some setbacks in MSC-based medical treatments, one of the highly debated aspects is that the studies showed varying outcomes likely because of the differences in cell isolation, origin and culture conditions, that could result in selection of MSC subpopulations [11]. Moreover, MSC cultured for long periods of time to obtain clinically relevant cell numbers result in important changes in gene expression, clonal selection, thus affecting biological properties, as has recently been shown [12, 13].

If the MSC clinical applications momentarily ended in an empirical clinical use, it has recently emerged the important role of this cell subset in the constitution of the BM niche environment. In human, few works described in depth the organization of the BM as it has been done with mouse models. However, MSC appear to be key components of hematopoietic niches allowing stemness functionalities of normal hematopoietic stem/progenitor cells [14]. This physiological equilibrium can be disrupted in cancer conditions promoting tumor initiation, progression, or therapy resistance. For example, the leukemia BM display "aberrant" MSC with reduced differentiation properties, altered production of soluble factors leading to decreased ability to support hematopoietic stem cells and to negatively influence the immune compartments [10, 15, 16]. Moreover, MSC are also implicated in regulating stemness and chemoresistance of leukemia cells [17]. Isolation of pathological MSC allows to study the pathological niches ex vivo and particularly, the impact of the malignant clone on the MSC biology [18]. These studies open for new investigations exploring more specific and synergic therapeutic approaches targeting the crosstalk between MSC and malignant cells [19, 20].

The expansive and growing MSC *field*, here briefly summarized, is anyway dependent on the ability and efficiency to purify h-MSC from heterogeneous cell populations to avoid errors in culture cell derivation, thus in data interpretations. In this chapter, we present detailed methods, practical tips, and minimal requirement's criteria to efficiently isolate, expand, characterize, and differentiate human BM MSC which can be used as a powerful tool for basic and translational studies.

2 Materials

2.1 Culture Media

1. Complete PBS: phosphate buffered saline (PBS) 1×, without Ca^{2+} or Mg^{2+}, pH 7.4 containing 2 mM EDTA, 100 units/mL penicillin G, 100 μg/mL streptomycin sulfate and 10% fetal bovine serum (FBS).

2. Complete α-MEM FBS medium: α-MEM containing L-glutamine without ribonucleosides or deoxyribonucleosides, 100 units/mL penicillin G, 100 μg/mL streptomycin sulfate and 10% FBS (*see* **Note 1**).

3. Complete α-MEM HS medium: α-MEM containing L-glutamine without ribonucleosides or deoxyribonucleosides, 100 units/mL penicillin G, 100 μg/mL streptomycin sulfate and 10% human serum (HS).

4. Ficoll-Paque with a density of 1.077.

5. 0.5% Trypsin-EDTA, no phenol red.

6. 0.4% Trypan blue.

7. Osteogenic medium (*see* **Note 2**): complete α-MEM medium supplemented with 0.1 μM dexamethasone, 0.05 mM L-acid ascorbic-2-phosphate and 10 mM β-glycerophosphate.

8. Complete DMEM high glucose medium: DMEM containing 4.5 g/L glucose, L-glutamine without ribonucleosides or deoxyribonucleosides, 100 units/mL penicillin G, 100 μg/ mL streptomycin sulfate and 10% FBS (*see* **Note 1**).

9. Adipogenic induction medium (*see* **Note 2**): complete DMEM high glucose supplemented with 1 µM dexamethasone, 0.5 mM 3-isobuthyl-1-methylxanthine (IBMX), 0.2 mM indomethacin, and 0.01 mg/mL insulin.

10. Adipogenic maintenance medium: complete DMEM high glucose containing 0.01 mg/mL insulin.

11. Chondrogenic control medium: complete DMEM high glucose supplemented with 0.1 µM dexamethasone, 1 mM sodium pyruvate, 0.35 mM L-proline, 1× insulin–transferrin–selenium (ITS), 20 µM linoleic acid, 0.05 mM ascorbic acid, and 1.14 mg/mL human albumin.

12. Chondrogenic induction medium (*see* **Notes 2** and **3**): chondrogenic control medium containing 10 ng/mL of human TGFβ3.

13. Dimethylsulfoxide (DMSO).

2.2 Flow Cytometry

1. FACS buffer: PBS 1× supplemented with 2% bovine serum albumin (BSA), 2 mM EDTA.

2. Antibodies and relative isotypic controls (*see* Table 1 for complete list).

3. Flow cytometer apparatus.

2.3 Histology, Dyes and Staining

1. Fix/stain Crystal Violet solution: 0.05% crystal violet solution (w/v), 1%formaldehyde, 1% methanol, PBS 1×.

2. Distilled water.

3. 20 g/L Alizarin Red solution in water (*see* **Note 4**).

4. Alizarin Red extraction buffer: 0.5 N hydrochloric acid, 5% sodium dodecyl sulfate (SDS).

5. Oil Red O stock solution: 0.5% Oil Red O (w/v) in 2-propanol.

6. Oil Red O working solution: Dilute the Oil Red O stock solution in distilled water at 3:2 ratio to obtain a concentration of 0.2% Oil Red O in 40% 2-propanol.

7. Paraffin.

8. EtOH 100° and dilutions (95°, 90°, 80°, 70°).

9. Xylene.

10. Alcian Blue solution: 1% Alcian Blue 8GX in 3% acetic acid solution. Adjust the pH to 2.5 with acetic acid.

11. Nonaqueous mounting medium.

12. Plate reader to measure absorbance from 415 to 520 nm.

13. Optical microscope with camera.

Table 1
MSC-surface markers. Expression pattern of MSC markers. In bold, markers commonly used for MSC characterization. +++, ++, +, and - indicate high, moderate, low, or no expression respectively

CD	Alternative Name	Function	Expression
CD13	ANPEP, aminopeptidase N, AAP, APM, LAP1, P150, PEPN	Aminopeptidase	+++
CD29	Platelet GPIIa, Integrin β1, GP	Cell adhesion	+++
CD44	ECMRII, H-CAM, Pgp-1, Phagocytic glycoprotein I	Cell adhesion and migration	+++
CD45	Leukocyte common antigen B220, protein tyrosine phosphatase	Regulator of cell growth and differentiation	–
CD49b	VLA-2α, Integrin α2, gPla	Cell adhesion	+++
CD49d	VLA-4α, Integrin α4	Cell adhesion and lymphocyte homing	+++
CD49e	VLA-5α, Integrin α5, fibronectin receptor	Cell adhesion	+++
CD54	ICAM-1LFA	Cell adhesion, lymphocyte activation, and migration	++
CD71	TFRC, T9, Transferrin receptor, TFR	Mediates the uptake of transferrin-iron complexes	+++
CD73	Ecto-5′-nuclotidase, NT5E, E5NT, NT5, NTE, eN, eNT	Catalyzes production of extracellular adenosine from AMP	+++
CD90	Thy1	Cell adhesion	+++
CD105	Endoglin, HHT1, ORW, SH-2	May play a role in hematopoiesis and angiogenesis (modulates TGF-beta functions)	+++
CD106	VCAM-1, INCAM-100	Adhesion of lymphocytes, monocytes, eosinophils, and basophils to vascular endothelium	+
CD166	ALCAM, KG-CAM, SC-1, BEN, DM-GRA	Cell adhesion molecule important for intrathymic T-cell development	+++
HLA-DR	MHC Class II	Antigen presentation and immune stimulation	+
HLA-ABC	MHC Class I	Antigen presentation and immune interaction	++

CD cluster of differentiation, *MSC* mesenchymal stromal cell

3   Methods

Carry out all procedures at room temperature and in sterile conditions, unless otherwise specified. All the culture media and buffers should be prewarmed at room temperature before use.

3.1 Isolation of Human BM-MSC

3.1.1 Isolation of BM-MSC from Bone Fragments

1. MSC from donors can be isolated from fresh residues of hip replacement surgery or amputation fragments. *See* **Note 5** for ethical approval and **Note 6** for the anatomical localization.

2. Transfer bone fragments in a recipient you can tightly close. Fill the recipient to 2/3 with complete PBS. Shake energetically. Let the fragments fall down.

3. Transfer the supernatant into 50 mL conical tubes.

4. Refill the recipient with complete PBS to 2/3 again, shake and transfer the supernatant into 50 mL conical tubes.

5. Spin the tubes 10 min at $420 \times g$ at RT.

6. Discard the supernatant and pool the cell pellets into one 50 mL conical tube. Rinse the tubes with complete PBS.

7. Filter the solution with a 70 μm sterile filter in a new tube.

8. Count the cells with a hemocytometer and Trypan Blue or other method (*see* **Note 7**).

9. Plate the cells at 0.2×10^6 mononucleated cells/cm^2 in T150 or T300 cm^2 flasks, depending on the desired amplification, in complete α-MEM FBS medium. More adapted medium can be used to expand MSC at this step (*see* **Note 8**).

10. Incubate the cells at 37 °C with 5% humidified CO_2 for 72 h (*see* **Notes 9** and **10**) to allow adherent cells to attach and change the medium (*see* **Note 11**).

11. After 4 more days, change again the medium then once a week.

12. Examine the cells regularly by phase microscopy until they reach 80% of confluence (*see* **Note 12**). The adherent MSC are defined as passage zero (P0) cells and need to be stocked and expanded by passages. *See* Subheading 3.4.

3.1.2 Isolation of MSC from BM Aspirates

This protocol allows isolation of MSC either from healthy donors or patients with hematological disorders (e.g., myelodysplastic syndromes or acute myeloid leukemia *see* **Note 13**). With this protocol, only MSC can be isolated. A Red Blood Cells (RBC) lysis described in **Note 14** can also be applied if desired to eliminate RBC.

1. Dilute sample twice in complete α-MEM HS medium.

2. Count the cells with a hemocytometer or Trypan Blue or other methods (*see* **Note 7**).

3. Plate the cells at 4×10^6 cells/cm^2 in complete α-MEM HS medium (for AML samples, *see* **Note 15**) and refer to Subheading 3.1.1, **steps 10–12** for cell maintenance.

3.1.3 Isolation of Both BM-MSC and Hematopoietic Cells

This protocol (*see* **Note 16**) can be used to isolate BM-MSC and other cell types (hematopoietic cells) from bone fragments or BM aspirates of healthy donors or patients with hematological disorders.

1. Dilute samples 1:3 with PBS 1× with 2 mM EDTA to obtain 25 mL or multiple.

2. Delicately drop the cell suspension on the surface of 15 mL of room temperature Ficoll already in a 50 mL conical tube (*see* **Note 17**).

3. Centrifuge 20 min at $400 \times g$ at 18 °C without break, *see* **Note 18**.

4. Collect the resulting mononucleated cells (MNC) located at the Ficoll interface and place the cells into a clean 50 mL conical tube.

5. Washed twice with a final volume of 50 mL with PBS 1× at $300 \times g$ for 10 min at 18 °C.

6. Remove the supernatant and resuspend with complete α-MEM FBS medium or complete α-MEM HS medium.

7. Count the cells with a hemocytometer and Trypan Blue or other method (*see* **Note 7**).

8. Plate the cells at 0.4×10^6 cells/cm^2 in complete α-MEM FBS medium (*see* **Note 8**) and refer to Subheading 3.1.1, **steps 10–12** for cell maintenance.

3.2 Amplification and Banking of BM-MSC

1. Passing procedure: remove the media and rinse the flask with PBS 1×. Remove the PBS and add room temperature 1× Trypsin-EDTA solution to the flask, *see* **Note 19** for the volume of Trypsin-EDTA. Distribute the trypsin across the surface area of the flask. Incubate the flask for 5 min at 37 °C. Examine the cells by phase microscopy, 80–90% of the cells have rounded up or become detached, gently beat on the sides of the flask to dislodge any remaining attached cells.

2. Add cell media to the flask. Rock the flask and forth to swirl the media around flask, transfer the cell suspension into a clean conical tube. If necessary, rinse the flask with PBS 1× and combine with the cell suspension.

3. Centrifuge at $300 \times g$ for 10 min at room temperature.

4. Remove the supernatant and resuspend the cells with 1–2 mL of media.

5. Count the cells (*see* **Notes 7 and 20**).

6. Seed cells at a density of 4×10^3 viable cells/cm^2 at this passage (P0 to P1) in complete α-MEM FBS medium (*see* **Note 21**). For later passages, plate at 3×10^3 viable cells/cm^2 (*see* **Note 22** for MSC isolated from patients) in complete α-MEM FBS medium. MSC can be successfully expanded and used through passage 5 without significant loss of stem cell phenotype and differentiation capacities (*see* **Note 23**). The use of MSC after these passages is strongly discouraged.

7. MSC can be frozen from the end of P0, P1 or P2. For this, resuspend up to 1×10^6 cells directly in serum 10% DMSO and freeze first in -80 °C in a cell freezing container and then transfer in liquid nitrogen or in a -150 °C freezer to be kept for a longer time.

3.3 Evaluation of MSC Clonogenicity

1. To evaluate clonogenic capacities of human MSC over time, CFU-F can be achieved at P0 or at later passages. For P0 CFU-F, plate in triplicate 3.3×10^5 total mononucleated cells freshly isolated from bone samples from healthy donors (*see* Subheading 3.1) in 25 cm^2 flask in 7 mL of complete α-MEM FBS medium. For CFU-F at later passages, plate 200 cells in 25 cm^2 flask in triplicate in complete α-MEM FBS medium. Place the flasks in the incubator at 37 °C with 5% humidified CO_2.

2. Change the medium once a week and observe the cells to monitor the growth of the colonies.

3. After 3 weeks (time may be adapted to the growth of the colonies), remove the culture medium and directly fix and stain the cells with the fix/stain crystal violet solution for 20 min at room temperature.

4. Remove the fix/stain crystal violet solution (*see* **Note 24**).

5. Wash delicately the flasks with tap water then with distilled water to avoid traces of limestone.

6. Allow the flasks to dry, take pictures, and quantify the number of colonies with more than 50 cells (Fig. 1).

3.4 Phenotypical Characterization of MSC Population

1. During the first passage, around $10–100 \times 10^3$ MSC are transferred in two 5 mL polystyrene tubes.

2. Spin the cells at $300 \times g$.

3. Resuspend MSC in 50 μL of FACS buffer with isotypic controls or previously titrated anti-human CD45, anti-human CD73, anti-human CD90 and anti-human CD105, each conjugated to a different fluorochrome, during 30 min, at 4 °C in the dark (*see* **Note 25**).

4. Wash with FACS buffer.

A
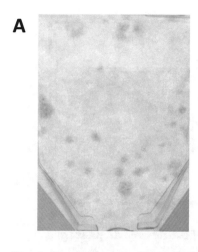

B
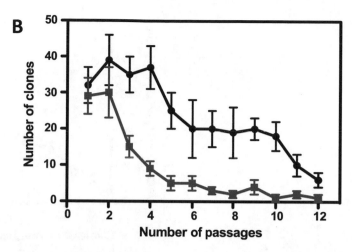

Fig. 1 Clonogenic capacities of MSC over time. MSC clonogenicity is evaluated by CFU-F assays. (**a**) represents a typical picture of a good density of colonies and (**b**) shows the decrease of MSC clonogenicity over passages (passage 1 to passage 12). Clonogenicity can be compared over time between different MSC such as high (black line) or low (red line) clonogenic MSC

5. Resuspend the cell in staining buffer and analyze by flow cytometry.

6. Consider MSC as CD45$^-$, CD73$^+$, CD90$^+$, CD105$^+$. For other markers, *see* Table 1.

3.5 Evaluation of MSC Expansion Capacities

1. When making the first passage (P1), plate the cells at 3×10^3 cells/cm^2 in a 25 cm^2 flask in complete α-MEM FBS medium.

2. Change the medium once a week.

3. Observe the cells regularly until reaching 80% of confluence.

4. When 80% confluence is reached, detach the cells by trypsin procedure (*see* Subheading 3.2) and quantify the number of cells. Note also the number of days necessary to reach 80% of confluence at this passage.

5. Plate again 3×10^3 cells/cm^2. You are at P2.

6. Change medium and make a passage each time 80% of confluence is reached.

7. Repeat steps until cell growing arrest.

8. Determine the population doubling (PD) by the following formula where Ni is the cell number initially seeded and Nf the cell number at confluence obtained at each passage (*see* **Note 26**):

$$PD = \frac{\log(Nf) - \log(Ni)}{\log(2)}$$

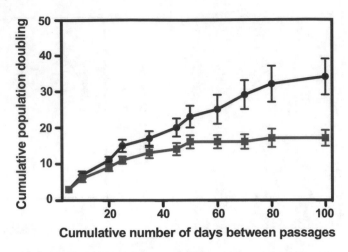

Fig. 2 Expansion capacity of MSC over passages. The graph represents the cumulative population doubling depending on the cumulative number of days between passages. In this situation, for example, we compared two different conditions, control (black line) and treated (red line) MSC. Doing that, we observed that the expansion capacity of treated MSC is significantly reduced compared to control MSC over time. A single measure of doubling time at early passages (<20 days) could not have brought a true information. Only this experiment and this representation are truly informative

9. Calculate the cumulative population doubling (CPD) by making the sum of all previous population doubling numbers and represent this data in y axis and number of days in x axis (Fig. 2). The slope of the curve will indicate the expansion capacities of MSC (*see* **Note 27**).

3.6 Osteogenic Differentiation

1. For this assay, use MSC after P2, plate them at 3×10^3 cells/ cm^2 in complete α-MEM FBS medium, and then put them in the incubator at 37 °C with 5% humidified CO_2. Osteoblast differentiation can be processed in 6-, 12-, or 24-well plates. Prepare 3 wells for induction condition and 3 wells for control medium. If large number of differentiated cells are needed, flasks can also be used.

2. After 24 h of adhesion, remove the medium, add osteogenic medium into the induced wells and add complete α-MEM FBS medium into the control wells.

3. Observe the cells (*see* **Note 28**) and change the media twice a week.

4. After 21 days, wash the cells with PBS 1× then EtOH70° and stain the cultures with Alizarin Red solution during 15 min to reveal the mineralization. *See* **Note 29** for other stainings and Fig. 3a.

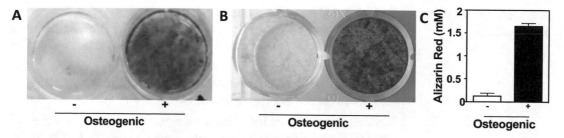

Fig. 3 In vitro osteogenic differentiation of MSC. MSC are cultured in complete (−) or osteogenic (+) α-MEM FBS medium. Alkaline phosphatase (**a**) and Alizarin Red (**b**) stainings are performed after 14 and 21 days respectively. Alizarin Red dye is extracted and quantified by spectrometry to compare the osteogenic differentiation efficiency between induced and noninduced control wells (**c**)

Table 2

Table of dilutions and related concentrations to perform the standard curve. Linearity range is shown in colored boxes

	1	2	3	4	5	6	7	8	9	10	11	12
Alizarin red	pure	1/2	1/4	1/8	1/16	1/32	1/64	1/128	1/256	1/512	1/1024	1/2048
Concentration (g/L)	20	10	5	2,5	1,25	0,625	0,3125	0,1563	0,0781	0,0391	0,0195	0,0098
Concentration (mg/L)						625	312,50	156,25	78,13	39,06	19,53	9,77

5. Wash the wells with distilled water until the disappearance of dissolved stain in the supernatant.

6. Let the wells dry at room temperature and take pictures (Fig. 3b).

7. To quantify the mineral deposit, dissolve Alizarin Red staining using Alizarin Red extraction solution. Adjust the volume regarding the intensity of the staining (*see* **Note 30**).

8. Make a standard curve starting from a known concentration of Alizarin Red (20 g/L) and two by two dilutions until 9.77 mg/mL (12 tubes).

9. Dilute extracted Alizarin Red from the culture wells in the Alizarin Red extraction buffer to be in the linearity range of optical density (OD) quantification (*see* Table 2).

10. Deposit 200 μL of each point of the standard curve and the samples into a flat-bottom 96-well plate. Remove bubbles from the wells.

11. Read the absorbance at 415 nm.

12. Calculate the concentrations using equation of the tendency curve multiplied by the dilution applied on each sample (Fig. 3c).

3.7 Adipogenic Differentiation

1. Use MSC after P2. Plate the cells at 21.5×10^3 cells/cm^2 in complete α-MEM FBS medium and put in the incubator at 37 °C with 5% humidified CO_2. Prepare 3 wells for induction condition and 3 wells for control condition.

2. Let the cells reaching the confluence during maximum 1 week without changing the medium.

3. Alternate three cycles of induction and maintenance: 4 days of induction with adipogenic induction medium followed by 1 day of maintenance with adipogenic maintenance medium. The strict observation of cycle timing is mandatory for an optimal differentiation. Use only adipogenic maintenance medium for the control wells. The fourth and last cycle is made only with adipogenic maintenance medium allowing the loading of lipid droplets within the cells.

4. At the end of the four cycles, wash adipogenic monolayer cultures with PBS 1× and fix 15 min with formaldehyde 4%.

5. Stain with Oil Red O working solution during 30 min allowing lipid droplet detection.

6. Wash 3 to 5 times with distilled water and take pictures (Fig. 4). For other stainings, *see* **Note 31**.

7. To quantify Oil Red O amount, extract the dye with 100% 2-propanol (0.263 mL/cm^2) for 10 min at room temperature [21].

8. Transfer 100 μL to a 96-flat bottom well plate. Make in duplicate.

9. Dilute twice by adding 100 μL of 2-propanol.

10. Read absorbance at 510 nm. Express the result by OD.

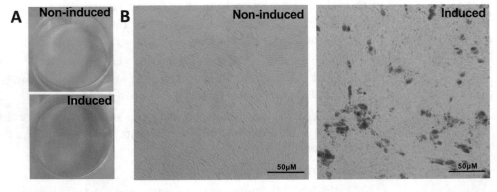

Fig. 4 In vitro adipogenic differentiation of MSC. MSC are cultured in maintenance medium (noninduced) or adipogenic induction medium (induced). Oil Red O staining is performed 21 days later to evaluate the differentiation efficiency. Macroscopic observation (**a**) shows a pale pink/red color in induced wells compared to the white color of noninduced control wells. Microscopic observation shows adipocytes whose cytoplasm is filled with lipid droplets stained in red compared to totally unstained control cells

3.8 Chondrocyte Differentiation

1. Use MSC after P2 and seed 2.5×10^5 cells in a 15 mL conical tube in complete α-MEM FBS medium. Prepare 6 tubes: 3 for induction conditions and 3 for controls.

2. Centrifuge the cells are at $500 \times g$ for 5 min without brake to form small pellets.

3. Remove the supernatants and replace it with 500 μL of chondrogenic control medium.

4. Centrifuge again like in **step 2**.

5. Refill the 3 control tubes with new chondrogenic control medium and the 3 induced tubes with chondrogenic induction medium.

6. Perform a last centrifugation to correctly pellet the cells before incubate at 37 °C with 5% humidified CO_2. Open the cap of a half of a turn to allow air exchange.

7. Change the medium two times a week.

8. After 3 weeks, pellets can be seen by naked eye (*see* **Note 32**). Perform a 4% paraformaldehyde (*see* **Note 33**) fixation and then a graded series of ethanol treatment prior to embed in paraffin (*see* **Note 34**).

9. Realize Paraffin sections of 5 μm thickness including the pellet.

10. Deparaffinize the sections using xylene (*see* **Note 35**) and stained with Alcian Blue solution (*see* **Note 36**).

11. Dehydrate with EtOH through EtOH 70° to EtOH 95° baths, 1 min each.

12. Dehydrate with EtOH 100°, 2 baths, 3 min each.

13. Clear with xylene, 2 baths, 3 min each.

14. Mount with nonaqueous mounting medium.

15. Take pictures using an optical microscope (Fig. 5).

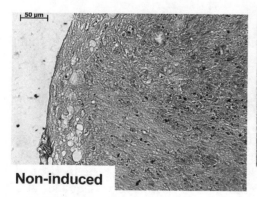
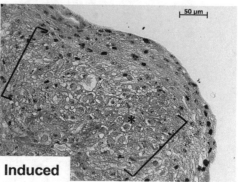

Non-induced   **Induced**

Fig. 5 In vitro chondrogenic differentiation of MSC. MSC are cultured in control (noninduced) or chondrogenic induction (induced) medium. Three weeks later, the pellets are paraffin embedded, cut and stained with Alcian Blue. Pictures shows a large area of differentiated chondrocyte in the induced condition (in the brackets). Hypertrophic chondrocytes (*) are present inside their chondroplast (white space around chondrocyte). In the noninduced condition, only matrix deposits are present

4 Notes

1. FBS is a critical component for these cells. Several batches of FBS should be tested to compare MSC growth, expansion, morphology, CFU and particularly, differentiation potential before starting large experiments. Ten percent of FBS is recommended for the MSC cultures, additional FBS may over-exhaust MSC.

2. Differentiation can be performed with homemade medium, described in this chapter, or with commercial differentiation medium that nowadays are robust tools but still expensive for routine analysis.

3. As the chondrocyte differentiation is a tricky procedure, it is recommended, at least for the first experiments, to use a commercial medium instead of homemade medium.

4. To make Alizarin Red working solution, Alizarin Red is dissolved in water under agitation and pH is adjusted between 4.1 and 4.3 with hydrochloric acid. Filter stain through a paper filter and store tightly capped at room temperature (RT) protected from light for up to 3 months.

5. To work on human samples, an authorization by a competent organization called Institutional Review Board (IRB) should be obtained. It frames research activities while guaranteeing welfare, rights and privacy of human subjects.

6. The anatomical location is an important factor to take into account particularly in terms of MSC differentiation capacity [22].

7. Number of the cells can be checked using Trypan Blue. Dilution factor 1:3, Trypan Blue in PBS 1×. Count cells in three large squares and calculate number of cells/mL using the formula:

$$\text{Number of cells/mL} = \frac{\text{Total cells counted}}{\text{number of squares counted}} \times \text{dilution} \times 10^4$$

   Total number of cells = number of cells/mL × Volume (mL)

8. For MSC initial isolation, mononuclear cells can be plated in α-MEM medium supplemented with 1% of heparin and 5% of human platelet lysate (always add first heparin then platelet lysate). Platelet lysate should be thawed very quickly at 37 °C. Make the complete medium just before use. Instead of platelet lysate you can use 10% human serum (in this case addition of heparin is not required).

9. This delay can be shortened to 48 h if you need to recover hematopoietic cells. For that, aspirate the supernatant, wash once with PBS 1× and perform a density gradient centrifugation (*see* Subheading 3.1.3) to remove erythroid cells and keep only hematopoietic cells.

10. If the nonadherent cells are not removed, hematopoietic cells can adhere, particularly monocytes, and may contaminate the MSC culture.

11. Very few adherent cells can be seen at this point.

12. To preserve stem cell phenotype, do not allow the cells to become confluent. Because MSC are not evenly distributed in the marrow, some aspirates do not have enough MSC to obtain large cultures. If a sample does not grow well or does not present good morphology by the eighth day, discard it.

13. For hematological disorders, samples correspond to a small volume of BM aspirates that are performed to establish myelograms. These samples do not give rise systematically to MSC. We estimate that MSC isolation can be successful every 1/3 to 1/2 patients.

14. Samples can also be treated by Red Blood Cell (RBC) lysis buffer (homemade or commercial) to eliminate erythrocytes from the cell suspension. Then, cells are plated at 4×10^6 cells/cm^2 in complete α-MEM HS serum. Refer to Subheading 3.1.1, **steps 10–12** for cell maintenance.

15. Because of BM invasion by malignant hematopoietic cells particularly for AML patients, the count of the mononucleated cells is not necessary. Thus, in this case, dilute BM samples twice with complete α-MEM with 10% of serum (FBS or Human) and plate in 6-well plate (2 mL per wells). b-FGF (FGF2) can be directly added to culture medium at 10 ng/mL to stimulate the growth (*see* **Note 21**).

16. Density gradient centrifugation causes a significant reduction of BM-mononuclear cells in general (15–30% of the initial content) including some MSC.

17. Do not mix the two phases otherwise the mononuclear cells will not completely separate out during centrifugation.

18. Leave off the brake to avoid to disturb the Ficoll–cell suspension interface.

19. Add 3 mL trypsin/EDTA to 10 cm diameter dish or 5 mL trypsin/EDTA to T75 flask.

20. Typical number of cells from a T-75 flask is between 1 and 3×10^6 total cells, with an average viability usually greater than 90%.

21. The human serum or the platelet lysate used for MSC isolation is, from this moment, removed. All the MSC are cultured from now with complete α-MEM FBS medium.

22. For expansion of pathological MSC, a plating density of 5×10^3 viable cell/cm^2 is recommended.

23. To punctually boost MSC growth speed, add 10 ng/mL b-FGF (FGF2) in the culture medium.

24. You can save the fix/stain crystal violet solution and reuse it at least three times.

25. These 4 markers are mandatory to assess the correct expansion of homogeneous MSC. For the staining, different clones distributed by several companies and conjugated to different fluorochromes can be used. The combination of fluorochromes depends on the flow cytometer device available on site. Many other makers can also be added to distinguish MSC subpopulations (see Table 1) but the consensus staining for MSC characterization relies on the four markers described in Subheading 3.4 and in bold in Table 1 [7].

26. Population doubling represents the total number of times the cells in the population have doubled since their primary isolation in vitro.

27. Doubling time can also be calculated at specific passages using the formula below or using the web site http://www.doubling-time.com/compute.php [23]:

$$\text{Doubling time} = \frac{\text{duration} \times \log(2)}{\log(\text{Nf}) - \log(\text{Ni})}$$

28. Mineralization can be observed with an optical microscope from 14 days after induction for normal cells. Mineralization is visualized as black deposition of minerals on confluent cells.

29. Other colorations can be performed to evaluate osteoblast differentiation including alkaline phosphatase to verify the confluence of differentiating cells (early differentiation staining, Fig. 3a) or von Kossa's staining that detect calcium deposits rather that mineralized nodules detected by Alizarin Red staining (late differentiation stage).

30. Between 250 μL and 1 mL can be added per well of 24-well plate.

31. Fluorescent stainings like Nile Red or Bodipy [24] can also be made and observed by fluorescent microscopy or flow cytometry.

32. Size of the pellet, directly visible by eyes, is an indicator of cell differentiation efficiency. Pellets in control condition are often smaller than the one in the induced conditions.

33. To facilitate the location of the pellet in the paraffin block during the sectioning, put a drop of eosin during fixation. This will stain in red the pellet without hinder the specific staining.

34. To simplify these different steps, place one pellet in a nylon biopsy bag inside an embedding cassette to undergo all the dehydration and impregnation steps. Thus, you will not "lose" the pellet and you will be able to use automatons.

35. Toluene is formally not recommended for deparaffinization because of its well-known toxicity. Other less toxic solvents like Histolemon or Histosol can also be used.

36. Many stainings can be performed to observe chondrocyte differentiation. To observe the pellet histologically, hematoxylin can be used first. Safranin O (proteoglycans) or anti-collagen II staining can also be performed.

Acknowledgments

This work was supported by ANR grants (PRC 17-CE14-0019 and JCJC ANR-19-CE15-001901), the INCa agency under the program PRT-K 2017, and the Association Saint Louis pour la Recherche sur les Leucémies.

References

1. Friedenstein AJ, Chailakhjan RK, Lalykina KS (1970) The development of fibroblast colonies in monolayer cultures of Guinea-pig bone marrow and spleen cells. Cell Tissue Kinet 3:393–403

2. Hernigou P (2015) Bone transplantation and tissue engineering, part IV. Mesenchymal stem cells: history in orthopedic surgery from Cohnheim and Goujon to the Nobel prize of Yamanaka. Int Orthop 39:807–817

3. Cohnheim J (1867) Über entzündung und eiterung. Path Anat Physiol Klin Med 40:1

4. Caplan AI (1991) Mesenchymal stem cells. J Orthop Res 9:641–650

5. Väänänen HK (2005) Mesenchymal stem cells. Ann Med 37:469–479

6. Pittenger MF, Discher DE, Péault BM et al (2019) Mesenchymal stem cell perspective: cell biology to clinical progress. NPJ Regen Med 4:22

7. Dominici M, Le Blanc K, Mueller I et al (2006) Minimal criteria for defining multipotent mesenchymal stromal cells. The International Society for Cellular Therapy position statement. Cytotherapy 8:315–317

8. Konala VBR, Mamidi MK, Bhonde R et al (2016) The current landscape of the mesenchymal stromal cell secretome: a new paradigm for cell-free regeneration. Cytotherapy 18:13–24

9. Gomez-Salazar M, Gonzalez-Galofre ZN, Casamitjana J et al (2020) Five decades later, are mesenchymal stem cells still relevant? Front Bioeng Biotechnol 8:148

10. Rivera-Cruz CM, Shearer JJ, Figueiredo Neto M et al (2017) The immunomodulatory effects of mesenchymal stem cell polarization within the tumor microenvironment niche. Stem Cells Int 2017:4015039

11. Galipeau J, Sensébé L (2018) Mesenchymal stromal cells: clinical challenges and therapeutic opportunities. Cell Stem Cell 22:824–833

12. Kim M, Rhee J-K, Choi H et al (2017) Passage-dependent accumulation of somatic mutations in mesenchymal stromal cells during in vitro culture revealed by whole genome sequencing. Sci Rep 7:14508

13. Jiang T, Xu G, Wang Q et al (2017) In vitro expansion impaired the stemness of early

passage mesenchymal stem cells for treatment of cartilage defects. Cell Death Dis 8:e2851

14. Malfuson J-V, Boutin L, Clay D et al (2014) SP/drug efflux functionality of hematopoietic progenitors is controlled by mesenchymal niche through VLA-4/CD44 axis. Leukemia 28:853–864

15. Sarhan D, Wang J, Arvindam US et al (2020) Mesenchymal stromal cells shape the MDS microenvironment by inducing suppressive monocytes that dampen NK cell function. JCI Insight 5:e130155

16. Poon Z, Dighe N, Venkatesan SS et al (2019) Bone marrow MSCs in MDS: contribution towards dysfunctional hematopoiesis and potential targets for disease response to hypomethylating therapy. Leukemia 33:1487–1500

17. Boutin L, Arnautou P, Trignol A et al (2020) Mesenchymal stromal cells confer chemoresistance to myeloid leukemia blasts through side population functionality and ABC transporter activation. Haematologica 105:987–9998

18. Martinaud C, Desterke C, Konopacki J et al (2015) Osteogenic potential of mesenchymal stromal cells contributes to primary myelofibrosis. Cancer Res 75:4753–4765

19. Borriello A, Caldarelli I, Bencivenga D et al (2016) Tyrosine kinase inhibitors and mesenchymal stromal cells: effects on self-renewal, commitment and functions. Oncotarget 8:5540–5565

20. Hmadcha A, Martin-Montalvo A, Gauthier BR et al (2020) Therapeutic potential of mesenchymal stem cells for cancer therapy. Front Bioeng Biotechnol 8:43

21. Kraus NA, Ehebauer F, Zapp B et al (2016) Quantitative assessment of adipocyte differentiation in cell culture. Adipocytes 5:351–358

22. Scarpone M, Kuebler D, Chambers A et al (2019) Isolation of clinically relevant concentrations of bone marrow mesenchymal stem cells without centrifugation. J Transl Med 17:10

23. Roth V. 2006. Doubling time computing. http://www.doubling-time.com/compute.php

24. Durandt C, van Vollenstee FA, Dessels C et al (2016) Novel flow cytometric approach for the detection of adipocyte subpopulations during adipogenesis. J Lipid Res 57:729–742

Chapter 2

Differentiation and Phenotyping of Murine Osteoclasts from Bone Marrow Progenitors, Monocytes, and Dendritic Cells

Julia Halper, Maria-Bernadette Madel, and Claudine Blin-Wakkach

Abstract

Bone physiology is dictated by various players, including osteoclasts (OCLs) as bone resorbing cells, osteoblasts (capable of bone formation), osteocytes, or mesenchymal stem cells, to mention the most important players. All these cells are in tight communication with each other and influence the constantly occurring process of bone remodeling to meet changing requirements on the skeletal system. In order to understand these interplays, one must investigate isolated functions of the various cell types. However, OCL research displays a special drawback: due to their giant size, low abundance, and tight attachment on the bone surface, ex vivo isolation of sufficient amounts of mature OCLs is limited or not conceivable in most species including mice. Moreover, OCLs can be obtained from different progenitors in vivo as well as in vitro. Thus, in vitro differentiation of OCLs from various progenitor cells remains essential in the analysis of OCL biology, underlining the importance of reliable gold standard protocols to be applied throughout OCL research. This chapter will deal with in vitro differentiation of OCLs from murine bone marrow cells, as well as isolated monocytes and dendritic cells that have already been validated in numerous studies.

Key words Osteoclasts, Monocytes, Dendritic cells, Osteoimmunology, Bone resorption, Cell culture

1 Introduction

Osteoclasts (OCLs) are giant, long-lived multinucleated cells specialized in bone resorption. They are the only cell type capable of degrading mineralized bone matrix and function in tight balance with the bone-forming osteoblasts (OBLs). This interplay allows to permanently maintain and ensure physiological bone turnover, which is an equilibrium (or coupling) of bone resorption and formation [1] that is also tightly controlled by other cells in the bone environment. Depending on their environmental conditions (healthy or pathological) as well as on the developmental stage, OCLs originate from various precursor cell types from the monocytic lineage [2–6] that fuse together and build the big OCL syncytia (Fig. 1a). The driving cytokine for precursor differentiation and fusion is receptor activator of NF-κB ligand (RANK-L),

Marion Espéli and Karl Balabanian (eds.), *Bone Marrow Environment: Methods and Protocols*, Methods in Molecular Biology, vol. 2308, https://doi.org/10.1007/978-1-0716-1425-9_2, © Springer Science+Business Media, LLC, part of Springer Nature 2021

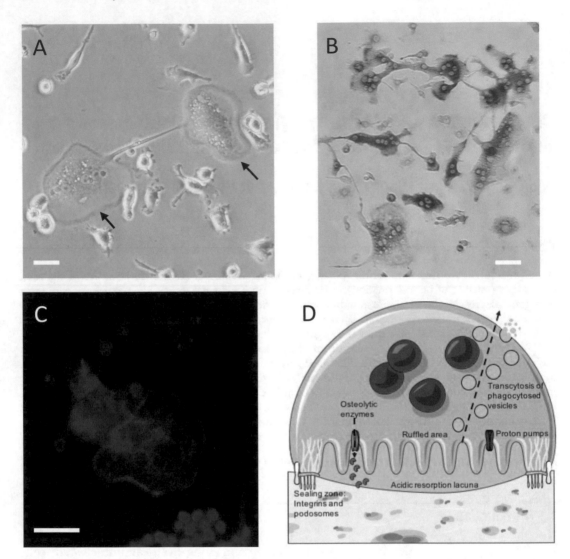

Fig. 1 Major characteristics of mature OCLs. (**a**) Representative light microscopy image of mature OCLs in culture (indicated with arrows), surrounded by precursor cells that did not fuse; Scale bar = 100 μm; (**b**) TRAcP positive OCLs, as indicated by the purple color that develops upon substrate procession; Scale bar = 200 μm; (**c**) Confocal image of immunofluorescence staining of the podosomes on mature OCLs; blue = DAPI, red = Phalloidin (Actin), Scale bar = 50 μm; (**d**) Schematic depiction of OCL attachment to the bone. Expression of integrins and podosomes form the sealing zone and the intracellular actin ring (stained in **c**), to ensure delimitation of the highly acidic resorption lacuna. OCL polarization forms the basal ruffled area containing high numbers of proton pumps to maintain the acidic environment necessary for bone resorption. Degradation products are phagocytosed and vesicles are transported to the apical extracellular space through transcytosis

which was only discovered in the late 1990s [7–9]. Before this date, in vitro OCL differentiation was possible only in coculture between OCL precursors and OBLs [10], one of the main producers of RANK-L. Since its discovery, a milestone in bone research, investigations on OCL function have extended far beyond the classical bone-centered view. In addition to in vivo experimentation, the possibility to efficiently differentiate OCLs in vitro opened broad perspectives for the analysis of the mechanisms involved in OCL differentiation and function and for their interaction with other bone marrow (BM) cells, mainly immune cells. With these emerging discoveries of OCL intersection with the immune system, the new field of osteoimmunology has arisen [11]. Initially, it explored the modulation of OCL differentiation by immune cells, in particular T cells, through their cytokine production [12–15]. Nowadays, it becomes evident that, in turn, OCLs actively participate in immune responses by reacting to stimuli from immune cells, as well as being cytokine producers that drive inflammation and participate in a vicious circle between inflammation and bone destruction as observed in chronic inflammatory diseases [2, 16, 17], which are frequently associated with osteoporosis. Furthermore, it was recently shown that OCLs are highly heterogeneous in their phenotype and function [2, 16, 18]. These studies suggest that depending on the precursor cells that fuse, OCL functions can become extremely diverse in terms of bone resorption capacity and participation in inflammatory processes [2, 18]. The latter being due to their capability of antigen uptake and processing and further presentation and activation of T cells [16, 19]. However, a lot of questions remain to be answered in terms of OCL functional differences as well as OCL precursor characteristics making in vitro analyses on OCLs essential.

Per definition, a mature OCL is multinucleated (with at least three nuclei, Fig. 1a), expresses specific markers, the main one being the tartrate-resistant acid phosphatase (TRAcP, Fig. 1b) and possess the capacity to degrade the bone matrix. The latter is achieved by tight attachment onto the bone surface via adhesion structures (podosomes, Fig. 1c, d) and expression of different integrins. This sealing zone delimits a specialized resorption lacuna between the bone surface and the OCL ruffled border membrane [20]. Subsequently, due to high expression of various osteolytic enzymes (anhydrases, phosphatases) and proton pumps (ATPases) in this ruffled area, massive acidification of the resorption lacunae enables bone degradation [21]. Beside bone resorption, OCLs are professional antigen-presenting cells [16, 22] and are capable to phagocytose degradation products comparable to giant macrophages [2]. Those phagocytic vesicles are trafficked toward the apical site and cargo is released into the extracellular space [2, 23]. Importantly, the typical OCL signature to allow characterization includes various genes involved in this degradation process.

As mentioned above, OCLs develop from fusion of precursors from the monocytic/macrophagic lineage [2, 5] when stimulated with RANK-L and macrophage colony-stimulating factor (M-CSF). Nevertheless, recent insights into the field of osteoimmunology revealed functionally different OCL phenotypes, depending on the type of precursor cells [2, 4, 6, 16, 24, 25]. It was shown that OCL differentiation, phenotype, and function are divergent between physiological and pathological conditions frequently accompanied by secondary osteoporosis. In chronic inflammatory diseases, for example, those affecting the joints (rheumatoid arthritis) or the gut (inflammatory bowel diseases), dendritic cells (DCs) as well as inflammatory monocytes participate as OCL precursors [2, 16, 17, 26]. Moreover, OCL progenitors in fetal stages have an erythromyeloid origin, contrasting with those present in adults that originate from BM hematopoietic stem cells [6, 24].

To study OCL biology in physiology or pathology, in vivo approaches include investigations on general bone status (bone mineral density, robustness on mechanical demands, measures of trabecular and cortical bone architecture and volume) [27]. However, in order to perform functional analysis on pure OCL populations, OCLs need to be isolated from the bone environment, a crucial procedure that harbors its pitfall: due to the sparseness of OCLs present in the bone in vivo as well as their tight attachment onto the bone surface, ex vivo isolation and subsequent cultivation approaches are not possible in most of species [28, 29]. Therefore, in vitro studies of differentiated primary OCL progenitors are indispensable. Possible sources for the myeloid precursor cells are BM, spleen, or whole blood, the latter is typically used for human OCL generation [17, 30]. OCL differentiation is a dynamic process that involves fusion of precursors, and therefore needs a reliable experimental procedure to adjust OCL research toward reproducible standards. The following protocols have been developed and constantly adapted over the past years by various laboratories and are efficient and reliable gold standard protocols used in OCL research.

Herein, we aim to cover murine OCL differentiation from widely used OCL progenitors. The first one is from total BM cells that reflect the mix of various precursor cells present in vivo. Furthermore, we will describe OCL differentiation from BM monocytic CD11b+ cells (MN-OCLs) and BM-derived CD11c+ dendritic cells (DC-OCLs), as both are well known to give rise to OCLs in different conditions with opposing immune function [2, 16, 18, 25].

2   Materials

2.1   Isolation of BM Cells

1. Mice.

2. Sterile surgery equipment (scissors, tweezers), 1 mL syringe with 25 G needle.

3. $1 \times$ PBS (without Mg^{2+} and Ca^{2+}).

2.2 Cell Culture (Necessarily Sterile)	1. 1× PBS (without Mg^{2+} and Ca^{2+}) to wash cells.
	2. Red blood cell lysis buffer to deplete erythrocytes.
	3. PSE: PBS 1× + 1% filtered fetal bovine serum + 2 mM EDTA.
	4. Medium for OCLs (OCL medium): αMEM medium with ribonucleosides, deoxyribonucleosides, and L-glutamine, 5% filtered serum (characterized fetal bovine serum (*see* **Note 1**), 1% penicillin–streptomycin, 50 mM β-mercaptoethanol. Keep at 4 °C and bring to 37 °C the required volume of medium just before the experiment.
	5. Medium for DCs (DC medium): RPMI medium with L-glutamine, 5% filtered serum (characterized fetal bovine serum, *see* **Note 1**), 1% penicillin–streptomycin, 50 mM β-mercaptoethanol. Keep at 4 °C and bring to 37 °C the required volume of medium just before the experiment.
	6. Cytokines: mRANK-L, mM-CSF, mIL-4, mGM-CSF. Keep at −80 °C (*see* **Note 2**).
	7. Cell culture plates (24, 48 or 96-well plates, *see* **Note 3**).
2.3 Magnetic Separation (See Note 4)	1. Magnet and magnetic stand, Anti-biotin MicroBeads, Cell Separation Columns (the size is depending on the quantity of cells), cell strainer/sterile filters (40 μm).
	2. Antibodies: biotinylated anti-CD11b antibody, biotinylated anti-CD11c antibody.
2.4 Phenotypic Analysis	1. TRAcP Staining: Acid Phosphatase/Leukocyte (TRAP) Kit.
	2. Resorption assay: Osteo assay plates, 2% Alizarin Red S solution (in distilled water and ammonium hydroxide), 1% Toluidine Blue (in distilled water and sodium borate).
	3. Detachment: Accutase solution, 1x PBS, OCL medium (from Subheading 2.2). Nuclear staining: 5 μg/μL Hoechst 33342.

3   Methods

3.1 Collection of Murine BM Cells	1. Dissect murine legs with scissors and tweezers to remove skin, fur, and muscle tissue. Take out skeletal parts by cutting them from the hip pan (pelvis, Fig. 2a) (*see* **Note 5**).
	2. Keep legs in sterile 1× PBS at 4 °C until further use (*see* **Note 6**).
	3. Separate femur and tibia at the tibiofemoral joint, remove fibula and patella. Cut off bone necks, just as far as necessary to access the bone cavity and to introduce the needle into the medullary cavity (Fig. 2b). Flush out the BM with a 25 G needle adjacent to a 1 mL syringe containing 1 mL of sterile

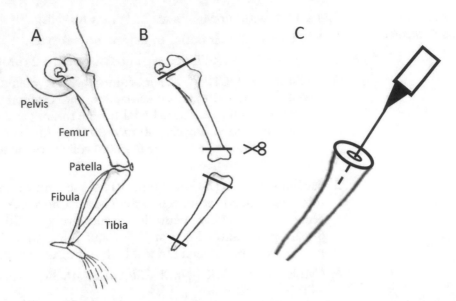

Fig. 2 Schematic anatomical depiction of bones for OCL isolation. (**a**) Murine hindlimb bones after dissection; (**b**) Separated femur and tibia with indications where to cut off bone necks in order to access the medullary cavity with the needle; (**c**) Indication of how deep the needle should be inserted into the medullary cavity, in order to flush out the BM efficiently

1× PBS in a 15 mL falcon tube. Insert the needle only as superficial as necessary in order to efficiently rinse out cells throughout the whole length of the bone (Fig. 2c). Importantly, flush multiple times from both apical and basal ends of the bones to collect as many cells as possible, until the bone color changes from its red appearance (due to erythrocytes present in the BM) into white/beige, indicating the empty, mineralized bone. Homogenize the cells using a 1 mL pipette and centrifuge for 5 min at 340 × g at room temperature.

4. Remove supernatant, resuspend the cell pellet in 500 µL of red blood cell lysis buffer per leg (femur and tibia) and incubate at room temperature for a maximum of 10 min.

5. Make up to 5 mL with 1× PBS to wash the cells and centrifuge again (5 min at 340 × g, room temperature).

6. Remove the supernatant and resuspend the cells in 5 mL 1× PSE per leg used. Count the cells by staining dead cells with trypan blue (1:10 diluted, in 1:1 ratio with the cell suspension) in a counting chamber before continuing with one of the 3 following protocols.

3.2 Differentiation of OCLs from Total BM Cells

1. Total BM cells can be directly differentiated into OCLs within 4–5 days of culture in culture plates at a density of 2.5 × 10^5 cells/cm^2 in 300 µL (*see* **Notes 3** and **7**).

Table 1
Number of cells and volume of OCL medium per culture plate required for OCL differentiation from total BM cells

Plate	Number of cells/well	Medium volume/well
24-well plate	5×10^5	500 μL
48-well plate	2.5×10^5	300 μL
96-well plate	1.25×10^5	150 μL

2. According to Table 1, collect the volume of PBS containing the total number of BM cells from Subheading 3.1, **step 6**) required for the experiment. Centrifuge the cells for 5 min at $340 \times g$ at room temperature. Remove supernatant and resuspend in the total volume required for the experiment (Table 1) in OCL medium supplemented with 30 ng/mL RANK-L and 25 ng/mL M-CSF. Seed cells at required density in the appropriate plates (Table 1, *see* **Note 2**).

3. After 3 days of culture, change the medium to boost the differentiation and to remove dead cells. Multinucleated cells appear around day 4 (*see* **Note 8**). Fig. 1a displays a mature OCL and the good time point to start with OCL analysis. For subsequent detachment and analysis of mature OCLs, refer to Subheading 3.5.

3.3 OCL Differentiation from CD11b⁺ Monocytic Cells (See Note 4)

1. Collect total BM cells from Subheading 3.1, **step 6**) and centrifuge for 5 min at $340 \times g$ at room temperature. Resuspend up to 1×10^7 per 100 μL cold PSE and incubated with 0.5 μg of biotinylated anti-CD11b antibody at 4 °C for 15 min (*see* **Note 9**).

2. Wash antibody excess by adding 2 mL of PSE and centrifuging for 5 min at $340 \times g$.

3. Resuspend cells in 80 μL cold PSE (per maximum of 10^7 cells, *see* **Note 10**) and add 20 μL of anti-biotin microbeads (per maximum of 10^7 cells) (*see* **Note 10**), mix well with the pipette. Incubate at 4 °C for 15 min and wash cells with 2 mL PSE. Centrifuge and resuspend in 500 μL ice-cold PSE.

4. Place the MS columns in the magnetic stand (*see* **Notes 4, 10, and 11**) and balance with 500 μL of ice-cold PSE. Then introduce the cells in the column in 500 μL of ice-cold PSE and let the cells enter into the column (*see* **Note 11**).

5. Wash the column with 500 μL PSE for three times (*see* **Note 12**), before detaching the column from the magnetic stand and placing it into a 15 mL falcon tube. Add 1 mL ice-cold PSE to the column and immediately push the piston thoroughly to

Table 2
Number of cells and volume of OCL medium per culture plate required for OCL differentiation from CD11b⁺ BM cells

Plate	Number of cells/well	Medium volume/well
24-well plate	2×10^5	500 µL
48-well plate	1×10^5	300 µL
96-well plate	0.5×10^5	150 µL

release the attached CD11b⁺ monocytes from the column. Count cells by using trypan blue (diluted 1:10 in 1:1 ratio with the cell suspension, *see* **Note 13**).

6. If large amounts of cells are needed, use LS columns instead of the MS ones (*see* **Notes 10** and **14**). Balance the LS column with 3 mL of ice-cold PSE before introducing the cells and perform each of the three washing steps with 3 mL ice-cold PSE as well (*see* **Notes 11** and **12**). Conduct the final piston flush in a volume of 5 mL ice-cold PSE, collect and count cells using trypan blue as in Subheading 3.3, **step 5**) (*see* **Note 13**).

7. According to Table 2, collect the volume of PSE containing the total number of CD11b⁺ cells (from **steps 5** or **6**) required for the experiment. Centrifuge the cells for 5 min at $340 \times g$ at room temperature. Remove supernatant and resuspend in the total volume required for the experiment (Table 2) in OCL medium supplemented with 30 ng/mL RANK-L and 25 ng/mL M-CSF (*see* **Note 2**). Seed the number of cells required for the experiment in the appropriate plates according to Table 2.

8. Change media after 3 days of culture (OCL medium supplemented with 30 ng/mL RANK-L and 25 ng/mL M-CSF, *see* **Note 2**) and follow differentiation. Multinucleated cells appear after about 4 days (*see* **Note 8**).

3.4 OCL Differentiation from BM-Derived CD11c⁺ DCs

1. Seed total BM cells (*see* **Note 15**) from Subheading 3.1, **step 6**) in 24-well plates at 5×10^5 cells per well in 500 µL of DC medium supplemented with 10 ng/mL both IL-4 and GM-CSF (*see* **Note 2**).

2. After 3 days of culture, add fresh DC medium (supplemented with 10 ng/mL both IL-4 and GM-CSF, *see* **Note 2**) to the wells (300 µL per well of a 24-well plate). The DCs in culture do not attach to the well but form nonadherent clusters that lay on the bottom of the wells. Once these clusters are clearly visible, usually after 5–6 days of differentiation (2–3 days after medium addition), harvest them by gently pipetting up and down. Do not scrape the cells or use enzymatic detachment for the cells as this will lead to high contamination by cells other than DCs.

Table 3
Number of cells and volume of OCL medium per culture plate required for OCL differentiation from CD11c$^+$ BM-derived DCs cells

Plate	Number of cells/well	Medium volume/well
24-well plate	2×10^4	500 μL
48-well plate	1×10^4	300 μL
96-well plate	0.5×10^4	150 μL

3. Isolate CD11c$^+$ DCs from these cells using a magnetic separation system (*see* **Note 4**) as described in Subheading 3.3, **step 2**) to Subheading 3.3, **step 6**), and 1 μg of biotinylated anti-CD11c antibodies (instead of CD11b$^+$ antibodies in Subheading 3.3). After separation and counting of cells, seed the appropriate number of cells according to Table 3 in OCL medium supplemented with 30 ng/mL RANK-L and 25 ng/mL M-CSF (*see* **Notes 2** and **16**).

4. Change the medium after 3 or 4 days and tightly follow differentiation, which will take approximately 6–7 days in total after seeding (*see* **Note 8**).

3.5 Validating the OCL Phenotype (See Note 17)

3.5.1 TRAcP Staining

1. Carry out quantificational analysis of fully differentiated and fixed OCLs by staining the OCL marker enzyme TRAcP upon substrate addition. Perform this fast and efficient procedure directly inside the culture plates (Acid Phosphatase, Leucocyte (TRAP) Kit), following the manufacturer's recommendations. TRAcP positive cells appear pink/purple in color.

2. On a light microscope, enumerate OCLs as TRAcP-expressing cells having 3 or more nuclei applying the criteria of a bona fide OCL (Figs. 1b and 3a) (*see* **Notes 17** and **18**).

3. Alternatively, fix and stain detached cells (for OCL detachment *see* next paragraph and **Note 20**) in suspension for endogenous TRAcP expression (ELF 97 Endogenous Phosphatase Detection Kit) following the manufacturer's recommendations. Analyze the cells by flow cytometry as described [16, 18](*see* **Note 19**).

3.5.2 Detachment of OCLs (See Notes 20–22)

1. Use Accutase enzymatic solution for this procedure in order to preserve OCL survival as well as integrin expression and other surface markers (*see* **Note 21**).

2. Briefly, wash wells with PBS and incubate with 300 μL Accutase solution per well (of a 24-well plate, ensure the well is covered with solution) for around 30 min at 37 °C, until cells appear spheroidal in shape (*see* **Note 21**). Collect the Accutase containing the detached cells in 2 mL of OCL medium to inactivate the enzymatic reaction.

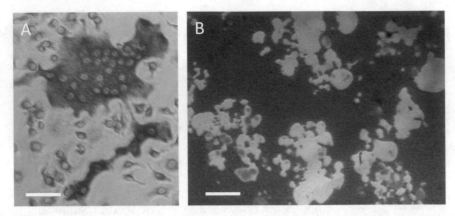

Fig. 3 Validation of the presence of bona fide mature OCLs. (**a**) Zoom into the well after TRAcP staining to verify bona fide OCLs. All cells that appear purple express TRAcP, but only the ones with 3 nuclei and more account for OCLs; Scale Bar = 200 μm; (**b**) Resorption pits of OCLs seeded on an calcium phosphate-coated Osteoassay plate; residual matrix is stained with 2% Alizarin Red After removal of the cells; Scale bar = 100 μm

3. Thoroughly wash the wells with PBS by pipetting up and down multiple times, in order to remove all residual, tightly attached OCLs (detailed protocol can be found in Madel et al. [28]; *see* **Notes 21** and **22**).

3.5.3 Resorption Assays

1. In order to investigate matrix dissolution activity, seed OCLs (or progenitor cells) onto calcium phosphate coated osteoassay plates or dentin slices in OCL medium with 30 μg/mL RANK-L for a set time (overnight when starting from mature OCLs, up to 7 days when starting from progenitors).

2. Visualize OCL dissolution capacity on osteoassay plates by eliminating the cells by overnight incubation in water and counterstaining of residual mineralized area with 2% Alizarin Red S solution (Fig. 3b). Further analyze dissolution capacity by microscopical quantification of resorbed areas.

3. Alternatively, seed OCLs on dentin slices to quantify their bone resorption capacity. Therefore, counterstain with 1% Toluidine blue after incubation on dentin and after detaching the cells using ultrasound [31] (*see* **Note 23**).

3.5.4 Analysis of Multinucleation on OCLs

1. To analyze or sort OCLs using FACS, it is recommended to stain the nuclei, in order to efficiently distinguish mature, bona fide OCLs with three or more nuclei (*see* **Note 17**) from their precursors (having 1–2 nuclei) or possible cell doublets/aggregates. A reliable procedure was recently set up by Madel et al., that uses the DNA intercalating agent Hoechst 33342 to ensure further cultivation of sorted OCLs due to nontoxic intercalation properties [28].

2. Incubate OCLs with Hoechst 33342 for around 30 min at 37 °C and subsequently wash with PSE. To perform further staining or other experimental procedures (*see* **Note 19**), strictly carry out all subsequent steps on ice and in the dark, in order to prevent passive export of H33342 as much as possible (detailed protocol can be found in Madel et al. [28]).

4   Notes

1. Using characterized serum allows to reduce the percentage of serum in the medium to 5%, in contrast to 10% uncharacterized serum. Furthermore, reproducibility of OCL differentiation is facilitated when characterized medium is used.

2. Cytokines are one of the most important things for successful OCL differentiation and can vary a lot between suppliers. The indicated concentrations are validated for the indicated suppliers: mRANKL (R&D), mM-CSF (R&D), mIL-4 (Peprotech), mGM-CSF (Peprotech). When using different ones, the concentrations and/or differentiation times require adjustment.

3. Cell culture plates might influence differentiation process; we recommend the use of Corning plates for a maximal efficiency.

4. Magnetic separation procedure will be explained throughout this chapter for the use of the Miltenyi MACS separation system provided by Miltenyi Biotec (other systems/suppliers may be used as well).

5. Importantly, bones must be cut out in their total entity to prevent opening of the BM cavity and possible contamination of the BM cells for further cultivation. If necessary, remove remaining muscles carefully in order to avoid breaking the bones.

6. All subsequent steps are necessarily carried out in a sterile environment under a culture hood (biosafety cabinet) and using sterile equipment.

7. Differentiation is much less efficient when using wells with a surface larger than the one of 24-well plates.

8. Microscopic evaluation remains crucial, as differentiation efficiency and time might vary from one experiment to another or between laboratories.

9. Keep PSE ice cold throughout procedure.

10. MS columns are suitable for a maximum of 2×10^8 cells with an expected number of labeled cells up to 1×10^7.

11. Column flow through speed can be slightly accelerated if columns are precooled (4 °C).

12. Collect flow through, in case magnetic separation procedure needs to be repeated.

13. Purity of the separated cells will be around 90% but can be increased if the procedure is repeated, however, overall cell yield will be reduced in case of repetitive separation steps.

14. LS columns have a retraction capacity for up to 2×10^9 total cells containing up to 1×10^8 labeled cells. Adjust the quantity of antibodies and beads according to the cell count.

15. DCs are not represented in the BM in sufficient amounts to start differentiation directly. Thus, this protocol uses DCs derived in vitro from BM cells.

16. Seeded cell densities differ from total BM cells/monocyte to DC precursors due to a different efficiency of fusion. Please use Tables 1, 2, 3 to use appropriate cell densities.

17. Bona fide OCLs are described as multinucleated cells (with three nuclei or more) expressing TRAcP and capable to resorb bone or mineralized matrix [2]. These characteristics have to be confirmed before concluding on the presence of OCLs.

18. TRAcP staining on OCLs can be performed in later stages of the differentiation, as cells are fixed and stained inside of the wells. Also, later stages usually contain more and bigger OCLs, facilitating the microscopical readout.

19. Hoechst 33342 and ELF staining cannot be carried out at the same time, due to fluorescence signals in the same channel.

20. Some procedures to verify the OCL phenotype do require detachment from culture plates, a critical point as OCLs are heterogeneous in appearance and size.

21. Accutase treatment used to detach OCLs is not an optimal environment, leading to higher numbers of apoptotic OCLs with time of enzymatic treatment. Therefore, incubation with Accutase solution should be limited to an absolute maximum of 40 min at 37 °C. Furthermore, strong attachment increases the risk of spontaneous death of OCLs due to the increased mechanical demands during the detachment procedure.

22. If a subsequent flow cytometric analysis or sorting procedure is applied, differentiation should be stopped as soon as OCLs are present in the culture to avoid the OCLs being too large. With increasing size, the OCLs are attached stronger on the plates, leading to less efficient detachment from the culture plate and therefore to an increase in loss of OCL number for further analysis.

23. Of note, dentin resorption by OCLs is much more difficult to quantify than dissolution of calcium phosphate in the osteoassay plate. Moreover, compared to the osteoassay plate, dentin is not uniform and a lot of replicates (at least 10) are necessary to obtain statistically significant resorption data.

Acknowledgments

The work was supported by the Agence Nationale de la Recherche (ANR-16-CE14-0030) as well as by the French government, managed by the ANR as part of the Investissement d'Avenir UCA^{JEDI} project (ANR-15-IDEX-01), the Fondation Arthritis, the Société Française de Biologie des Tissus Minéralisés (SFBTM), the European Calcified Tissue Society (ECTS), and the American Society of Bone and Mineral Research (ASBMR). M-B. M. is supported by the Fondation pour la Recherche Médicale (FRM, ECO20160736019).

References

1. Charles JF, Aliprantis AO (2014) Osteoclasts: more than 'bone eaters. Trends Mol Med 20:449–459. https://doi.org/10.1016/j.molmed.2014.06.001

2. Madel M-B, Ibáñez L, Wakkach A et al (2019) Immune function and diversity of osteoclasts in normal and pathological conditions. Front Immunol 10:1408. https://doi.org/10.3389/fimmu.2019.01408

3. Jacome-Galarza CE, Lee S-K, Lorenzo JA, Aguila HL (2013) Identification, characterization, and isolation of a common progenitor for osteoclasts, macrophages, and dendritic cells from murine bone marrow and periphery. J Bone Miner Res 28:1203–1213. https://doi.org/10.1002/jbmr.1822

4. Wakkach A, Mansour A, Dacquin R et al (2008) Bone marrow microenvironment controls the in vivo differentiation of murine dendritic cells into osteoclasts. Blood 112:5074–5083. https://doi.org/10.1182/blood-2008-01-132787

5. Walker DG (1975) Control of bone resorption by hematopoietic tissue. The induction and reversal of congenital osteopetrosis in mice through use of bone marrow and splenic transplants. J Exp Med 142:651–663. https://doi.org/10.1084/jem.142.3.651

6. Yahara Y, Barrientos T, Tang YJ et al (2020) Erythromyeloid progenitors give rise to a population of osteoclasts that contribute to bone homeostasis and repair. Nat Cell Biol 22:49–59. https://doi.org/10.1038/s41556-019-0437-8

7. Horwood NJ, Elliott J, Martin TJ et al (1998) Osteotropic agents regulate the expression of osteoclast differentiation factor and osteoprotegerin in osteoblastic stromal cells. Endocrinology 139:4743–4746. https://doi.org/10.1210/endo.139.11.6433

8. Simonet WS, Lacey DL, Dunstan CR et al (1997) Osteoprotegerin: a novel secreted protein involved in the regulation of bone density. Cell 89:309–319. https://doi.org/10.1016/s0092-8674(00)80209-3

9. Yasuda H, Shima N, Nakagawa N et al (1998) Osteoclast differentiation factor is a ligand for osteoprotegerin/osteoclastogenesis-inhibitory factor and is identical to TRANCE/RANKL. Proc Natl Acad Sci U S A 95:3597–3602

10. Takahashi N, Yamana H, Yoshiki S et al (1988) Osteoclast-like cell formation and its regulation by Osteotropic hormones in mouse bone marrow cultures. Endocrinology 122:1373–1382. https://doi.org/10.1210/endo-122-4-1373

11. Arron JR, Choi Y (2000) Bone versus immune system. Nature 408:535–536. https://doi.org/10.1038/35046196

12. Ciucci T, Ibáñez L, Boucoiran A et al (2015) Bone marrow Th17 TNFα cells induce osteoclast differentiation, and link bone destruction to IBD. Gut 64:1072–1081. https://doi.org/10.1136/gutjnl-2014-306947

13. Kong Y-Y, Feige U, Sarosi I et al (1999) Activated T cells regulate bone loss and joint destruction in adjuvant arthritis through osteoprotegerin ligand. Nature 402:304–309. https://doi.org/10.1038/46303

14. Sato K, Suematsu A, Okamoto K et al (2006) Th17 functions as an osteoclastogenic helper T cell subset that links T cell activation and bone destruction. J Exp Med 203:2673–2682. https://doi.org/10.1084/jem.20061775

15. Zaiss MM, Axmann R, Zwerina J et al (2007) Treg cells suppress osteoclast formation: a new link between the immune system and bone. Arthritis Rheum 56:4104–4112. https://doi.org/10.1002/art.23138

16. Ibáñez L, Abou-Ezzi G, Ciucci T et al (2016) Inflammatory osteoclasts prime TNF-α-producing CD4+ T cells and express CX3CR1. J Bone Miner Res 31:1899–1908. https://doi.org/10.1002/jbmr.2868

17. de Vries TJ, el Bakkali I, Kamradt T et al (2019) What are the peripheral blood determinants for increased osteoclast formation in the various inflammatory diseases associated with bone loss? Front Immunol 10:505. https://doi.org/10.3389/fimmu.2019.00505

18. Madel M-B, Ibáñez L, Ciucci T et al (2020) Dissecting the phenotypic and functional heterogeneity of mouse inflammatory osteoclasts by the expression of Cx3cr1. eLife 9:e54493. https://doi.org/10.7554/eLife.54493

19. Kiesel JR, Buchwald ZS, Aurora R (2009) Cross-presentation by osteoclasts induces FoxP3 in CD8+ T cells. J Immunol 182:5477–5487. https://doi.org/10.4049/jimmunol.0803897

20. Vaananen HK, Zhao H, Mulari M et al (2000) The cell biology of osteoclast function. J Cell Sci 113:377–381

21. Cappariello A, Maurizi A, Veeriah V et al (2014) The great beauty of the osteoclast. Arch Biochem Biophys 558:70–78. https://doi.org/10.1016/j.abb.2014.06.017

22. Li H, Hong S, Qian J et al (2010) Cross talk between the bone and immune systems: osteoclasts function as antigen-presenting cells and activate CD4+ and CD8+ T cells. Blood 116:210–217. https://doi.org/10.1182/blood-2009-11-255026

23. Salo J, Lehenkari P, Mulari M et al (1997) Removal of osteoclast bone resorption products by transcytosis. Science 276:270–273. https://doi.org/10.1126/science.276.5310.270

24. Jacome-Galarza CE, Percin GI, Muller JT et al (2019) Developmental origin, functional maintenance and genetic rescue of osteoclasts. Nature 568:541–545. https://doi.org/10.1038/s41586-019-1105-7

25. Rivollier A, Mazzorana M, Tebib J et al (2004) Immature dendritic cell transdifferentiation into osteoclasts: a novel pathway sustained by the rheumatoid arthritis microenvironment. Blood 104:4029–4037. https://doi.org/10.1182/blood-2004-01-0041

26. Ammari M, Presumey J, Ponsolles C et al (2018) Delivery of miR-146a to Ly6Chigh monocytes inhibits pathogenic bone erosion in inflammatory arthritis. Theranostics 8:5972–5985. https://doi.org/10.7150/thno.29313

27. Osterhoff G, Morgan EF, Shefelbine SJ et al (2016) Bone mechanical properties and changes with osteoporosis. Injury 47:S11–S20. https://doi.org/10.1016/S0020-1383(16)47003-8

28. Madel M-B, Ibáñez L, Rouleau M et al (2018) A novel reliable and efficient procedure for purification of mature osteoclasts allowing functional assays in mouse cells. Front Immunol 9:2567. https://doi.org/10.3389/fimmu.2018.02567

29. Marino S, Logan JG, Mellis D et al (2014) Generation and culture of osteoclasts. BoneKEy Rep 3:570. https://doi.org/10.1038/bonekey.2014.65

30. Abdallah D, Jourdain M-L, Braux J et al (2018) An optimized method to generate human active osteoclasts from peripheral blood monocytes. Front Immunol 9:632. https://doi.org/10.3389/fimmu.2018.00632

31. Vesprey A, Yang W (2016) Pit assay to measure the bone resorptive activity of bone marrow-derived osteoclasts. BIO-Protoc 6:e1836. https://doi.org/10.21769/BioProtoc.1836

Chapter 3

Culture, Expansion and Differentiation of Mouse Bone-Derived Mesenchymal Stromal Cells

Zeina Abou Nader, Marion Espéli, Karl Balabanian, and Julia P. Lemos

Abstract

Mesenchymal stem/stromal cells (MSCs) are multipotent adult cells that are present in several tissues including the bone marrow (BM), in which they can differentiate in a variety of cell types such as osteoblasts, chondrocytes and adipocytes. The isolation of MSCs has been carried out by many studies that aim to control their differentiation into cartilaginous and bone cells in vitro in order to use this technology in the repair of damaged tissues. Here we describe the minimum requirements and an efficient method for isolation, expansion of mouse bone-derived multipotent mesenchymal stromal cells and their differentiation into osteoblasts, responsible for the bone matrix synthesis and mineralization.

Key words Bone, Mouse, Mesenchymal stem/stromal cells, Isolation, Expansion, Osteogenesis

1 Introduction

Stem cells are undifferentiated or partially differentiated cells capable of self-renewing—maintaining their undifferentiated state and the stem cell pool throughout life—and with the unique ability to develop into specialized cell types [1]. In both humans and mice, there are several main categories of stem cells: (1) pluripotent stem cells, such as the embryonic stem cells, (2) induced pluripotent stem cells and (3) somatic stem cells, commonly called "adult" stem cells, which are undifferentiated cells, found throughout the body after development, in close proximity to vessels and capable to respond to specific signs of the microenvironment to repair damaged tissues [2].

Mesenchymal stem/stromal cells (MSCs) are rare and heterogeneous multipotent adult cells, first isolated from the bone marrow (BM) stroma, but also present in many fetal and post-natal tissues, such as fat, skin, brain, umbilical cord and cord blood, among others [3–17]. MSCs have the ability to renew and differentiate into various connective tissue lines, including osteoblasts (bone-making cells), chondrocytes (cartilage-making cells) and

Marion Espéli and Karl Balabanian (eds.), *Bone Marrow Environment: Methods and Protocols*, Methods in Molecular Biology, vol. 2308, https://doi.org/10.1007/978-1-0716-1425-9_3, © Springer Science+Business Media, LLC, part of Springer Nature 2021

adipocytes (fat-making cells), besides tendon, muscle and the marrow stroma [18] under specific stimulations. These cells were first described by Friedenstein and colleagues, who found that MSCs adhere to culture plates, resemble fibroblasts in vitro and form spindle-shaped colonies [19]. Within the BM, the MSCs and derivatives also provide microenvironmental support for hematopoietic stem and progenitor cells (HSPC) maintenance and differentiation [20].

Recently, MSCs have attracted the attention of several researchers, as they are of great interest to be used in the treatment of many human diseases, such as inflammatory, neurodegenerative and autoimmune disorders [21–23]. Several methods have been used for MSC isolation, mostly based on the ability of the MSCs to selectively adhere to plastic surfaces. After isolation, they have been used in preclinical and clinical applications in regenerative medicine [24–26], such as bone repair [27]. The main endogenous source of MSCs for bone healing is derived from the marrow cavity, the periosteum and the endosteum [27].

Moreover, the isolation and the in vitro expansion of MSCs are of great importance for the study and understanding of the hematopoietic and stromal cell niche interactions that govern both hematopoiesis and progenitor specification within the BM [28].

Here we describe the minimum requirements and an efficient method for isolation, expansion and differentiation of mouse multipotent MSCs from mineralized bones. The bone is a great source of MSCs; in comparison with the marrow cavity, it contains high numbers of skeletal stromal cells including osteoprogenitors that contrast with the low representation of hematopoietic cells, which represent the main type of culture contamination especially in mouse [29, 30]. We will detail how bone fraction can be separated from the marrow one and then processed for enriching in stromal cells. Subsequently, in vitro bone-derived MSC culture and expansion will be addressed and, finally, their differentiation into osteoblasts and the quantification of the mineralization will be explained.

2   Materials

Use only sterile material and prepare all the solutions using sterile reagents and under laminar flow hood to provide an aseptic working area. Unless specified otherwise, prepare and store all reagents at 4 °C.

2.1   Plastic and Equipment

1. 2 mL and 0.5 mL polypropylene microtubes.
2. 18 Gauge needle.
3. Surgical material for mouse.
4. Scalpel.

5. Cell culture dishes.

6. 15 mL and 50 mL conical tubes.

7. 70 μm nylon cell strainer.

8. 25 cm^2 and 75 cm^2 plastic flasks.

9. 24- and 96-well plates (*see* **Note 1**).

10. Sterile plastic pipettes, micropipettes and tips.

11. KOVA Glasstic Slide 10 with Grids or another hemocytometer.

12. 200 mL glass beaker.

13. 200 mL and 300 mL volumetric flasks.

14. Amber glass bottle.

15. Filter paper.

16. Funnel.

17. 5 mL polypropylene tubes.

18. Hood with vertical laminar flow.

19. Heated orbital shaker.

20. 37 °C, 5% CO$_2$ humidified incubator.

21. Centrifuge.

22. Optical microscope.

23. Absorbance plate reader.

24. Magnetic stir bar and a magnetic stirrer.

25. Analytical balance.

26. pH meter.

2.2 Reagents

1. 1× Phosphate-buffered saline (PBS): Dilute 10× PBS at 1:10 in distillated water.

2. PBS–2% FBS containing 1× PBS supplemented with 2% of fetal bovine serum (FBS) (*see* **Note 2**).

3. Collagenase Type 1 solution: Resuspend the Collagenase 1 powder according to the manufacturer's instructions. Prepare aliquots and store at −20 °C protected from light. Prior to use, thaw on ice and dilute the collagenase type 1 in PBS–2% FBS at 2.5 mg/mL. Prepare enough volume to have 5 mL of the solution per mouse (*see* **Note 3**).

4. 0.1% Trypan Blue: Dilute 0.4% Trypan Blue at 1:4 in 1× PBS.

5. Trypsin solution containing 1× trypsin and 0.5% EDTA.

6. Complete α-MEM Medium containing α-Minimum Essential Medium (α-MEM) GlutaMAX™ Supplement with nucleosides, supplemented with 10% FBS (*see* **Note 4**), 100 U/mL penicillin-Streptomycin (PS), and 50 μM 2-mercaptoethanol. Mix components (*see* **Note 5**).

7. Osteogenic differentiation medium containing complete α-MEM medium supplemented with 50 μg/mL of L-ascorbic acid and 2.15 mg/mL of β-Glycerophosphate. Mix well (*see* **Note 6**).

8. 4% Formaldehyde solution containing 37% formaldehyde solution diluted in 1× PBS.

9. 2 N HCl solution: In a 15 mL conical tube, add 5 mL of distilled water and 1 mL of 12 N HCl. Mix well. This solution will allow to adjust the pH of the Alizarin Red S solution (*see* **Note 7**).

10. 10% Ammonia. This solution is to adjust the pH of the Alizarin Red S solution.

11. Alizarin Red S solution containing Alizarin Red S at 20 g/L diluted in distillated water. (*see* **Note 8**).

12. 0.5 N HCl solution: In a 300 mL volumetric flask, add 287.5 mL of distilled water and 12.5 mL of 2 N HCl. Mix well (*see* **Note 9**).

13. 0.5 N HCl–5% SDS solution: For 200 mL, put in a 200 mL volumetric flask 150 mL of 0.5 N HCl and 50 mL of 20% SDS. Mix well (*see* **Note 10**).

3   Methods

Carry out all procedures at room temperature (unless indicated otherwise) and under laminar flow hood. Use only sterile material.

3.1   Isolation of Bone-Derived MSCs

3.1.1   BM Flush

1. Remove the mouse femur, tibia and hip bones (*see* **Note 11**).

2. Using a scalpel, clean the bones entirely, removing all the skin and the muscles.

3. After cleaning, cut one of the epiphyses of each bone to enable the flush of the BM.

4. Prepare two polypropylene microtubes of 2 mL and two of 0.5 mL per animal (Fig. 1a) (*see* **Note 12**).

5. Make holes in the bottom of the 0.5 mL polypropylene microtubes with an 18 Gauge needle and place it within each major one (name each tube appropriately) (Fig. 1b).

6. Place 1 femur, 1 tibia and 1 hip bone within the 0.5 mL microtube, taking care to place the bone hole down and close the 0.5 mL tube (Fig. 1c).

7. Centrifuge at $17,000 \times g$ for 1 min (*see* **Note 13**).

8. Save the empty compact bones in the 0.5 mL microtube for the next step.

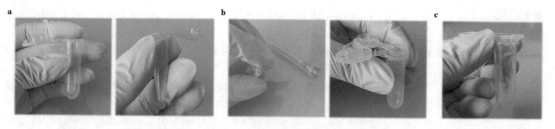

Fig. 1 Tube preparation for bone and marrow fraction separation. (**a**) Take 2- and 0.5-mL polypropylene tubes. (**b**) With an 18 Gauge needle, make a hole in the bottom of the 0.5 mL tube and place it inside the 2 mL one. (**c**) Place the bones inside the 0.5 mL tube with the cut epiphysis down

Fig. 2 Mouse bone cutting prior to the digestion with Collagenase type 1. (**a**) Marrow-flushed compact bones placed in a small cell culture dish. (**b**) After addition of 1 mL of the collagenase type 1, cut the bones in very small pieces until they acquire a gelatinous consistency

3.1.2 Digestion of Compact Bones

1. Perform the bone digestion with 5 mL per animal of Collagenase type 1 solution: take 1 mL of the Collagenase type 1 solution and plate it in a 35 mm diameter cell culture dish for cutting the bones in small pieces (Fig. 2). After cutting, place all the pieces and the entire volume in a 50 mL conical tube and complete with 4 mL of collagenase type 1 solution (*see* **Note 14**).

2. Incubate for 45 min at 37 °C under orbital agitation.

3. Prepare another 50 mL conical centrifuge tube per animal and place a 70 μm nylon cell strainer on each tube.

4. Filter the solution to remove bone debris, placing the bone pieces within the cell strainer.

5. Wash with 10 mL of PBS–2% FBS and filter again using a new 70 μm nylon cell strainer.

6. Centrifuge for 10 min at 300 × g at room temperature and resuspend the cells in 1 mL of PBS–2% FBS.

3.2 Culture and Expansion of Bone-Derived MSCs

3.2.1 Cell Counting and Seeding

1. Count the cells using Trypan Blue at a dilution of 1:10: Take 10 µL of the cell suspension into a well of a 96 well-plate and add 90 µL of 0.1% Trypan blue. Mix gently by pipetting up and down several times.

2. Transfer 10 µL of Trypan Blue–cell suspension on a chamber slide. Observe the chamber slide under an optic microscope with a 20× objective and count live nonblue cells (*see* **Note 15**).

3. Seed the cells in 25 cm^2 plastic flasks at a density of 100,000 cells/cm^2 in 5 mL of complete α-MEM medium.

4. Put the flask, lying down, at 37 °C in a humidified incubator containing 5% CO_2 (*see* **Note 16**).

5. After 3 days, check the flask under an optical microscope. MSCs should be attached to the bottom of the flask showing an elongated fibroblast-like morphology; nonadherent hematopoietic cells would appear round (Fig. 3a).

6. Remove the nonadherent cells and debris by changing the medium: Discard the culture medium from the flask by aspiration, wash the cells twice by adding and aspirating 5 mL of 1× PBS. After washing, add 5 mL of fresh complete α-MEM medium to the flask and place it in the incubator (*see* **Note 17**).

3.2.2 Cell Passaging and Splitting

Within approximately 7 days, the culture reaches 70–90% confluence. So, perform the cell passaging and subculture the cells into a new flask to keep cells alive and growing under suitable cultured conditions for extended periods of time.

1. Rinse the flask twice with 5 mL of 1× PBS and add 1 mL of trypsin solution. Make sure the solution covers all the surface and incubate the cells at 37 °C for 5 min (*see* **Note 18**).

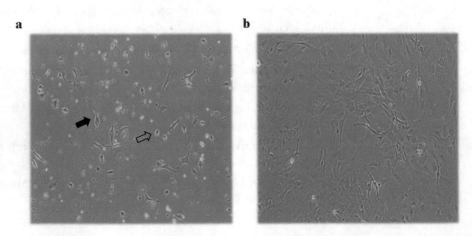

Fig. 3 Culture expansion of bone-derived mouse MSCs. (**a**) At day 3 of culture, few MSCs with elongated fibroblast-like morphology (black full arrow) and round nonadherent hematopoietic cells (black empty arrow) can be observed. (**b**) At passage 2, after 14 days, the culture is more confluent with many expanded MSCs and almost none contaminating hematopoietic cells

2. Rapidly observe the cells under an optical microscope to ensure that all the cells have been detached (nonadherent cells will appear round and floating) (*see* **Note 19**).

3. Immediately, neutralize the trypsin solution by adding 7 mL of complete α-MEM medium and gently pipet up and down the cell suspension several times.

4. Collect the cell suspension in a 15 mL conical tube, centrifuge at $300 \times g$ for 10 min at room temperature, remove the supernatant by aspiration and resuspend the pellet with 2 mL of complete α-MEM medium.

5. Count the cells using trypan blue and plate the cell suspension in 75 cm^2 flasks at a density of 3000 cells/cm^2 in 15 mL of complete α-MEM medium.

6. Put the flask, lying down, at 37 °C in a humidified incubator containing 5% CO_2.

7. Change the culture medium each 4 days and perform the cell passaging once a week. At passage 2, the culture should have more isolated MSCs and fewer hematopoietic cells (Fig. 3b) (*see* **Note 20**).

3.3 Osteoblastic Differentiation

3.3.1 Osteoblastic Differentiation Culture

1. Harvest the MSCs by trypsinization as described in the previous section (*see* **Note 21**).

2. Count the cells as previously described and reseed in a 24-well plate at a density of 3000 cells/cm^2 in 500 μL of complete α-MEM medium. Consider three wells for the osteogenic differentiation induction (osteogenic medium) and two wells for the negative controls (α-MEM medium). Incubate the cells at 37 °C in a humidified incubator containing 5% CO_2 for 24 h.

3. After 24 h, remove gently the medium from all the wells and add 500 μL of osteogenic medium. In the negative control wells, add 500 μL of complete α-MEM medium.

4. Incubate the cells at 37 °C in a 5% CO_2 humidified incubator for 3 weeks, changing the medium every 2 days (*see* **Note 22**).

5. Observe the cells under a light microscope each 2 or 3 days. After 14 days, differentiating cells will exhibit morphological changes and black or dark brown spots will start to appear indicating mineral deposition (Fig. 4a, b). At day 21 and 28, the culture shows more mineralized nodules (Fig. 4c, d).

3.3.2 Alizarin Red S Staining and Quantification

After 28 days of osteogenic differentiation, use Alizarin Red S staining to evaluate and quantify the bone matrix mineralization.

1. Discard the differentiation medium, wash the cells gently three times with 500 μL of 1× PBS and fix the cells with 200 μL of 4% formaldehyde for 20 min at room temperature (*see* **Note 23**).

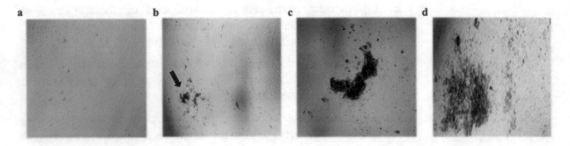

Fig. 4 Osteogenic differentiation of bone-derived mouse MSCs. (**a**) Negative control MSCs grown in classical α-MEM medium at day 14 of culture. (**b**) After 14 days of osteogenic differentiation induction, mineralized areas (black arrow) start to be observed. At day 21 (**c**) and day 28 (**d**) of osteogenic differentiation, more mineralized nodules can be visualized

2. Wash the cells three times with 1 mL of distilled water per well.

3. Add 1 mL of Alizarin Red S solution per well and incubate for 20 min at room temperature.

4. Remove the dye after the incubation: fill each well with 2 mL of distilled water, agitate the plate manually and horizontally to remove any unbound Alizarin Red S dye and remove the water by inverting the plate. Repeat the wash step at least four times or until the dissolved stain disappears in the supernatant.

5. After the last wash, invert the plate on a paper towel to dry the wells.

6. After drying, Alizarin Red S-stained wells would appear red and negative control wells should not show any staining (Fig. 5a).

7. In order to quantify the intensity of Alizarin Red S staining, add 1 mL of 0.5 N HCl–5% SDS solution to extract the fixed dye. Agitate manually for 10 s.

8. Collect all the liquid of each well into 5 mL polypropylene tubes already identified.

9. Prepare Alizarin Red S standards: Take 13 polypropylene tubes of 5 mL and label them from "1" to "13." In the tube "1," add 300 μL of Red Alizarin S solution and in the other 12 tubes, add 150 μL of 0.5 N HCl–5% SDS solution.

10. Transfer 150 μL of the Alizarin Red S solution from tube "1" to "tube "2." Mix slowly using a micropipette fixed to 150 μL to obtain the 10 g/L concentration of Alizarin Red S solution.

11. Repeat **step 10** for tubes "3" to "12" to serially dilute the Alizarin Red S solution at 1:2. Tube "13" serves as a blank solution and contains only 150 μL of 0.5 N HCl–5% SDS solution (Table 1).

12. Transfer 150 μL per well of the samples, the blank and the standard solutions to a flat-bottom 96 well-plate. Remove bubbles from the wells. Read the absorbance at 415 nm using an absorbance plate reader (*see* **Note 24**).

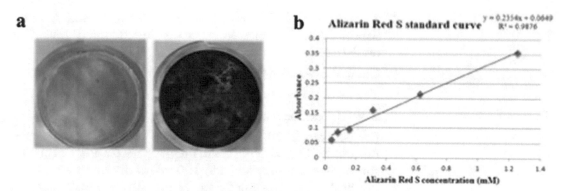

Fig. 5 Alizarin Red S staining and quantification of osteogenic differentiation. (**a**) Negative control well showing any Alizarin Red S staining (left) and a positive Alizarin Red S staining well after 28 days of osteoblastic differentiation induction (right). (**b**) Example of a linear standard curve, with the R^2 value and the equation $y = ax + b$, in which a is the slope and b is the y-intercept

Table 1
Twofold serial dilution of Alizarin Red S solution in 0.5 N HCl–5% SDS

Tube	1	2	3	4	5	6	7	8	9	10	11	12	13
Alizarin Red S solution (20g/L)	Pure (300µL)	1/2	1/4	1/8	1/16	1/32	1/64	1/128	1/256	1/512	1/1024	1/2048	0
0.5 N HCl/ 5 % SDS	0	150 µL	150 µL	150 µL	150 µL	150 µL	150 µL	150 µL	150 µL	150 µL	150 µL	150 µL	150 µL
Final concentration (g/L)	20	10	5	2.5	1.25	0.625	0.3125	0.156	0.078	0.039	0.0195	0.0097	0

13. In order to quantify the concentration of the samples, create a linear standard curve using the absorbance value of the Alizarin Red S standard solutions. Calculate the Alizarin Red S concentration in the samples using the equation of the standard curve and consider the factor of dilution if applied (Fig. 5b).

4   Notes

1. Use 96-well plate with a transparent flat bottom to read the absorbance.

2. Prior to the use of FBS, inactivate the complement proteins by heating the serum for 30 min at 56 °C. After inactivation, it can be aliquoted into 50 mL conical tubes and stored at −20 °C.

3. It is preferable to prepare the enzyme solution in the day of use. In this case, it can be kept at room temperature until the use. Otherwise, store at 4 °C to avoid multiple freeze/thaw cycles.

4. A specific batch of FBS suitable for MSC culture and differentiation must be selected, such as the Sigma-Aldrich, cat no. F7524.

5. Complete α-MEM can be stored for 2–3 weeks at 4 °C.

6. The medium can be kept for up to 4 weeks at 4 °C.

7. 2 N HCl can be stored for 1 month at room temperature.

8. For 200 mL of 20 g/L Alizarin red S solution, weigh 4 g of Alizarin Red S powder in a 250 mL glass beaker. Put a magnetic stir bar in the beaker and add 150 mL of distilled water. Place the beaker on a magnetic stirrer for 40 min, make sure all powder is dissolved. Mix manually and adjust the pH to 4.1–4.3 with 2 N HCl or 10% ammonia. Transfer the solution to a 200 mL volumetric flask and complete to 200 mL with distilled water. Filter the solution using a filter paper and a funnel: Place the funnel on top of a 250 mL glass beaker, put a filter paper in the funnel and pour slowly the Alizarin Red S solution in the funnel. Stock the filtrated solution in an amber glass bottle in a dark place for 3 months at room temperature.

9. 0.5 N HCl should be prepared extemporaneously and maintained at room temperature.

10. 0.5 N HCl–5% SDS is prepared extemporaneously and maintained at room temperature.

11. If sample transportation is necessary between the laboratory facilities, place the bones in a 24-well plate with enough sterile PBS–2% FBS to cover it in order to preserve the organs until the next step.

12. If BM cells (marrow fraction) are dedicated for other assays, add 200 μL of PBS–2% FBS in the 2 mL microtubes.

13. After centrifugation, the marrow and the bone fractions will be separated. The BM will be in the bottom of the 2 mL microtube and the empty compact bones will be in the 0.5 mL microtubes. If the BM has not gone down well, cut the bone better and repeat the process.

14. Be careful while cutting; the bones tend to escape from the dish. Consider holding the cell culture lid creating a barrier to protect and avoid loss/contamination of the bone pieces.

15. The total number of bone cells can vary according to the mouse age, gender and genotype. Young adult wild-type mice should present minimum 10^6 cells, after femur, tibia and hip bone removal. However, some genotypes and conditions can affect the bone metabolism, thus modifying cell count.

16. If a culture flask with gas-permeable filter in the cap is not used, open the cap of a half of a turn to allow air exchange.

17. Medium must be changed gently to avoid detachment of adherent cells.

18. Trypsin is toxic to the cells; the incubation should not exceed 5 min.

19. Gently tap the side of the flask a couple of times to release any remaining attached cells, be careful to not splash the medium into the cap of the flask.

20. The culture expansion of bone MSCs should not exceed five passages to preserve the stemness and multipotency of the cells.

21. Use preferentially the cells at passages 3–4, in which the cells are more active with high multilineage potential and contain fewer hematopoietic cells.

22. Dispense media onto the side of the well rather than directly onto the cells as to not disturb the cell monolayer.

23. Formaldehyde is flammable, irritant and a potential carcinogen. Avoid inhalation and contact. Prepare and use the solution in a certified chemical hood and wear safety glasses and gloves.

24. If necessary, dilute the extracted Alizarin Red S from the culture wells with 0.5 N HCl–5% SDS solution in order to be in the linearity domain of optical density (OD) quantification.

Acknowledgments

This work was supported by People Programme (Marie Curie Actions) of the European Union's Seventh Framework Programme (FP7/2007-2013) under REA grant agreement n. PCOFUND-GA-2013-609102, through the PRESTIGE cofinancing program coordinated by Campus France; by the ANR grant 17-CE14-0019; the INCa agency under the program PRT-K 2017 and by the French Ministry.

References

1. Wei X, Yang X, Han ZP et al (2013) Mesenchymal stem cells: a new trend for cell therapy. Acta Pharmacol Sin 34:747–754

2. Ullah I, Subbarao RB, Rho GJ (2015) Human mesenchymal stem cells—current trends and future prospective. Biosci Rep 35:1–18

3. Friedenstein AJ, Deriglasova UF, Kulagina NN et al (1974) Precursors for fibroblasts in different populations of hematopoietic cells as detected by the in vitro colony assay method. Exp Hematol 2:83–92

4. Erices A, Conget P, Minguell JJ (2000) Mesenchymal progenitor cells in the human umbilical cord. Br J Haematol 109:235–242

5. Riekstina U, Muceniece R, Cakstina I et al (2008) Characterization of human skin-derived mesenchymal stem cell proliferation rate in different growth conditions. Cytotechnology 58:153–162

6. Zebardast N, Lickorish D, Davies JE (2010) Human umbilical cord perivascular cells (HUCPVC): a mesenchymal cell source for dermal wound healing. Organogenesis 6:197–203

7. Gronthos S, Zannettino ACW (2011) Methods for the purification and characterization of human adipose-derived stem cells. Methods Mol Biol 702:109–120

8. Paul G, Özen I, Christophersen NS et al (2012) The adult human brain harbors multipotent perivascular mesenchymal stem cells. PLoS One 7:1–11

9. Appaix F, Nissou M-F, van der Sanden B et al (2014) Brain mesenchymal stem cells: the other stem cells of the brain? World J Stem Cells 6:134–143

10. Carapagnoli C, Roberts IAG, Kumar S et al (2001) Identification of mesenchymal stem cells in human first trimester fetal blood, liver and bone marrow. Blood 98:2396–2402

11. Zuk PA, Zhu M, Mizuno H et al (2001) Multilineage cells from human adipose tissue: implications for cell-based therapies. Tissue Eng 7:211–228

12. Romanov YA, Svintsitskaya VA, Smirnov VN (2003) Searching for alternative sources of postnatal human mesenchymal stem cells: candidate MSC-like cells from umbilical cord. Stem Cells 21:105–110

13. Alsalameh S, Amin R, Gemba T et al (2004) Identification of mesenchymal progenitor cells in Normal and osteoarthritic human articular cartilage. Arthritis Rheum 50:1522–1532

14. Bieback K, Kern S, Kluter H et al (2004) Critical parameters for the isolation of mesenchymal stem cells from umbilical cord blood. Stem Cells 22:625–634

15. Sarugaser R, Lickorish D, Baksh D et al (2005) Human umbilical cord perivascular (HUCPV) cells: a source of mesenchymal progenitors. Stem Cells 23:220–229

16. Hiraoka K, Grogan S, Olee T et al (2006) Mesenchymal progenitor cells in adult human articular cartilage. Biorheology 43:447–454

17. Manca MF, Zwart I, Beo J et al (2008) Characterization of mesenchymal stromal cells derived from full-term umbilical cord blood. Cytotherapy 10:54–68

18. Caplan AI (1991) Mesenchymal stem cells. Orthop Res Soc 9:641–650

19. Friedenstein AJ, Gorskaja UF, Kulagina NN (1976) Fibroblast precursors in normal and irradiated mouse hematopoietic organs. Exp Hematol 4:267–274

20. Pontikoglou C, Deschaseaux F, Sensebé L et al (2011) Bone marrow mesenchymal stem cells: biological properties and their role in hematopoiesis and hematopoietic stem cell transplantation. Stem Cell Rev Rep 7:569–589

21. Joyce N, Annett G, Wirthlin L et al (2010) Mesenchymal stem cells for the treatment of neurodegenerative disease. Regen Med 5:933–946

22. Figueroa FE, Carrión F, Villanueva S et al (2012) Mesenchymal stem cell treatment for autoimmune diseases: a critical review. Biol Res 45:269–277

23. Wang LT, Ting CH, Yen ML et al (2016) Human mesenchymal stem cells (MSCs) for treatment towards immune- and inflammation-mediated diseases: review of current clinical trials. J Biomed Sci 23:1–13

24. Le Blanc K, Ringdén O (2007) Immunomodulation by mesenchymal stem cells and clinical experience. J Intern Med 262:509–525

25. Kim N, Cho SG (2016) Overcoming immunoregulatory plasticity of mesenchymal stem cells for accelerated clinical applications. Int J Hematol 103:129–137

26. Han Y, Li X, Zhang Y et al (2019) Mesenchymal stem cells for regenerative medicine. Cell 8:886

27. Knight MN, Hankenson KD (2013) Mesenchymal stem cells in bone regeneration. Adv Wound Care 2:306–316

28. Aqmasheh S, Shamsasanjan K, Akbarzadehlaleh P et al (2017) Effects of mesenchymal stem cell derivatives on hematopoiesis and hematopoietic stem cells. Adv Pharm Bull 7:165–177

29. Zhu H, Guo ZK, Jiang XX et al (2010) A protocol for isolation and culture of mesenchymal stem cells from mouse compact bone. Nat Protoc 5:550–560

30. Short B, Wagey R (2013) Isolation and culture of mesenchymal stem cells from mouse compact bone. Methods Mol Biol 946:335–347

Chapter 4

Feeder-Free Differentiation Assay for Mouse Hematopoietic Stem and Progenitor Cells

Vincent Rondeau, Marion Espéli, and Karl Balabanian

Abstract

Hematopoietic stem cells (HSCs) are responsible for replenishing immune cells and reside in bone marrow (BM) niches, which provide all cellular and molecular components required for their lifelong maintenance and differentiation. Although HSCs have been extensively analyzed and characterized, their ex vivo expansion, which constitutes a promising approach for therapeutic development in regenerative medicine, remains challenging. Here, we describe an original in vitro system allowing to quantify by flow cytometry the differentiation of mouse HSCs into lineage-primed multipotent hematopoietic progenitors (MPPs) in a cytokine-supplemented feeder-free medium.

Key words Hematopoietic stem and progenitor cells, Fluorescence-activated cell sorting (FACS), In vitro feeder-free differentiation, Cytokines, Cell fate

1 Introduction

Blood production is a tightly regulated process that starts with hematopoietic stem cells (HSCs) which are unique in their capacity to self-renew and replenish the entire blood system through production of a series of increasingly committed progenitor cells within the bone marrow (BM) microenvironment [1]. In mice, long-term (LT)-HSCs form a rare, quiescent population that displays long-term reconstitution capacity [2]. They differentiate through short-term (ST)-HSCs that are characterized by shorter reconstitution ability [3] and then hematopoietic multipotent progenitors (MPPs). The major divergence of lymphoid and myeloid lineages occurs at the MPP stage. Indeed, common lymphoid progenitors and common myeloid progenitors are generated from phenotypically and functionally distinct subpopulations of lineage-biased MPPs, that is, MPP2 (Lin$^-$Sca-l$^+$ c-Kit$^+$[LSK] CD48$^+$CD150$^+$Flt3$^-$) and MPP3 (LSK CD48$^+$CD150$^-$Flt3$^-$)

Marion Espéli and Karl Balabanian (eds.), *Bone Marrow Environment: Methods and Protocols*, Methods in Molecular Biology, vol. 2308, https://doi.org/10.1007/978-1-0716-1425-9_4, © Springer Science+Business Media, LLC, part of Springer Nature 2021

are reported as distinct myeloid-biased MPP subsets that operate together with lymphoid-primed MPP4 (LSK $CD48^{+/-}CD150^-Flt3^+$) to control blood leukocyte production [4, 5].

Hematopoietic stem and progenitor cells (HSPCs) are thought to reside in (peri)-vascular niches, which regulate their lifelong maintenance and differentiation by secreting cytokines, growth factors and glycoproteins among others. These regulatory molecules induce in HSCs a complex signaling network balancing self-renewal and differentiation. Although HSCs have been extensively analyzed and characterized, the expansion and differentiation of these cells in vitro remains challenging for the scientific community. In vitro culture of HSCs promotes their differentiation, suggesting that both intrinsic and extrinsic signals are required to control their identity and fate. The most frequently used models involve cocultures of HSCs with feeders such as fibroblastic or stromal cell lines (e.g., MEF, MS5, OP9). Although these methods allow for the study of "natural" interactions and cross talk between distinct cell populations, the specific and accurate control of the cultures as well as the reproducibility and the standardization remain difficult. To our knowledge, an in vitro system composed of a natural feeder source that permits to selectively expand MPPs and concomitantly to drive their lymphoid-myeloid commitment has never been reported. Consequently, for all these reasons, the number of histo-compatible donor HS(P)Cs is still a limiting factor for effective BM transplantation.

We recently developed an in vitro feeder-free system permitting to rapidly and selectively expand MPPs including lineage-biased subsets [6, 7]. By contrast to published protocols that mostly rely on clonogenic assays, our protocol is adapted to quantify by flow cytometry the differentiation of mouse HSCs into MPPs in a cytokine-supplemented feeder-free medium. Currently, several phenotypic definitions are used to describe the spectrum of HSPC populations in the mouse BM compartment, likely leading to overlapping definitions. For instance, some studies encompass LT-HSCs (LSK $Flt3^-CD34^-CD48^-CD150^+$) and ST-HSCs (LSK $Flt3^-CD34^+CD48^-CD150^+$) in a population named SLAMs (LSK $Flt3^-CD48^-CD150^+$), which subsequently differentiate into MPPs (LSK $Flt3^-CD48^-CD150^-$) [8]. These populations overlap at least partially with LT-HSCs (LSK $Flt3^-CD48^-CD150^+$) and ST-HSCs (LSK $Flt3^-CD48^-CD150^-$) populations defined by others [7]. Despite this lack of consensus, our protocol allows to assess by flow cytometry the capacities of both SLAMs and ST-HSCs to differentiate into lineage-primed MPPs.

2   Materials

2.1   Plastic and Equipment

1. Dissection tool set (forceps, scissors, and scalpels).
2. 0.5 mL and 2 mL Eppendorf tubes.
3. 15 mL and 50 mL conical tubes.
4. 18 Gauge needle.
5. 70μm nylon cell strainer.
6. 5 mL FACS tubes (*see* **Note 1**).
7. U-bottomed 96-well culture plate.
8. Centrifuge for 2 mL and 15 mL or 50 mL tubes.
9. Cell Separation Magnet.
10. Water bath.
11. Flow cytometer apparatus including a cell sorter and a cell analyzer allowing the simultaneous analysis of at least 10 different fluorescences.
12. Software platforms for FACS data analyses such as FACSDive (BD Biosciences) or FlowJo (Treestar, Inc.).

2.2   Reagents

1. Complete Phosphate-buffered saline (PBS) containing 1× PBS, 100 units/mL penicillin G, 100μg/mL streptomycin sulfate and 2% of Fetal Bovine Serum (FBS) (*see* **Note 2**).
2. Complete DMEM FBS medium containing DMEM with 100 units/mL penicillin G, 100μg/mL streptomycin sulfate and 10% FBS.
3. Complete differentiation medium containing DMEM with 100 units/mL penicillin G, 100μg/mL streptomycin sulfate, 10% FBS, 50 ng/mL murine Stem Cell Factor (SCF), 10 ng/mL murine IL-3, 10 ng/mL murine IL-6, 5 ng/mL murine IL-7, and 5 ng/mL murine Flt3-L.
4. 0.1% Trypan Blue: Dilute 0.4% Trypan Blue at 1:4 in 1× PBS.

2.3   Antibodies and Dilutions

1. FACS buffer containing PBS 1×, 3% bovine serum albumin (BSA) and 2 mM EDTA.
2. Mouse Hematopoietic Progenitor (Stem) Cell Enrichment Set or another kit allowing isolation of stem and progenitor cells.
3. Antibodies (*see* Table 1) are diluted in complete PBS (*see* **Note 3**).

Table 1
List of antibodies used for flow-cytometry analyses. Combinations of fluorochromes and antibodies concentrations presented here are indicative

Antibody	Clone	Fluorochrome	Final concentration (μg/mL)
CD3	17A2	APC	1
CD45R (B220)	RA3-6B2	APC	1
CD11b	M1/70	APC	1
Gr-1	RB6-8C5	APC	1
TER-119	TER-119	APC	1
c-Kit	2B8	PerCP-Cyanine5.5	4
Sca-1	E13-161.7	PE-Cyanine7	2
CD48	HM48-1	Pacific Blue	2
CD150	TC15-12F12.2	PE	2
CD135	A2F10	PE-TexasRed	4
Live/Dead		eFluor506	

3 Methods

3.1 Marrow Sample Isolation

1. Kill mice by cervical dislocation or CO_2 inhalation in accordance with local ethical guidelines and protocols (*see* **Note 4**).

2. Harvest carefully the two hip bones, two femurs, and two tibias. Remove all muscles with a scalpel and place the bones in sufficient volume of PBS to submerge them (*see* **Note 5**).

3. Carefully remove the cartilage at one extremity of the femurs and tibias with a scalpel in order to allow the release of the marrow fraction after centrifugation (*see* **Note 6**).

4. Take a 0.5 mL Eppendorf tube and bore a hole in the bottom with an 18 Gauge needle. Place maximum one complete leg (hip bone, femur, and tibia) in it with the open extremity of the bones down.

5. Place the tube containing the bones in a 2 mL Eppendorf collection tube previously filled with 200μL of complete PBS.

6. Centrifuge at $15,000 \times g$ during 1 min.

7. Discard the 0.5 mL Eppendorf tube containing the bone fraction and continue with the marrow fraction which form a pellet in the bottom of the 2 mL Eppendorf tube (*see* **Note 7** and Fig. 1).

8. Add 800μL of complete PBS to obtain a volume of 1 mL and resuspend the pellet (*see* **Note 8**).

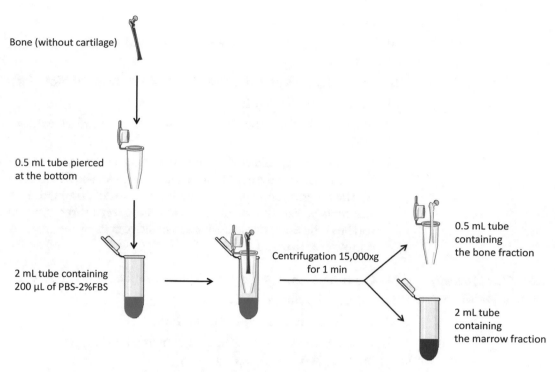

Bone (without cartilage)

0.5 mL tube pierced
at the bottom

2 mL tube containing
200 µL of PBS-2%FBS

Centrifugation 15,000xg
for 1 min

0.5 mL tube
containing
the bone fraction

2 mL tube
containing
the marrow fraction

Fig. 1 Schematic representation of the experimental procedures used to mechanically separate marrow and bone fractions of mouse long bones. The procedure is described in Subheading 3.1

9. Take a 50 mL Falcon and place a 70µm cell strainer on the top of it. Humidify the strainer with 1 mL of complete PBS.

10. Transfer cell suspension into the cell strainer.

11. Wash the cell strainer with 10 mL of complete PBS and discard it.

12. Centrifuge for 5 min at 300 × g at 4 °C.

13. Discard the supernatant.

14. Resuspend cells in 10 mL of complete PBS.

15. Count the cells with a hemocytometer and 0.1% Trypan Blue or other method (*see* **Note 9**).

3.2 Depletion of Lineage-Positive Cells

1. Centrifuge the marrow fraction 5 min at 300 × g at 4 °C.

2. Discard the supernatant.

3. Resuspend cells at 1×10^8/mL in complete PBS.

4. Add 3µL of biotin cocktail (containing anti-CD3, anti-B220, anti-TER-119, anti-CD11b, and anti-Gr-1 antibodies) *per* 1×10^6 cells.

5. Incubate for 30 min at 4 °C.

6. Wash with 10 mL of complete PBS.

7. Centrifuge for 5 min at 300 × g at 4 °C.

8. Discard the supernatant.

9. Resuspend in 1 mL of complete PBS and transfer in a 5 mL FACS tube.

10. Add 3μL of streptavidin beads *per* 1×10^6 cells (*see* **Note 10**).

11. Incubate for 30 min at 4 °C.

12. Place the FACS tube on a Cell Separation Magnet without washing.

13. Wait for 5 min, collect delicately the supernatant by pipetting and place it in a 15 mL collection tube. Remove the FACS tube off the magnet, resuspend the pellet in 1 mL of complete PBS and replace it in the magnet. Repeat these steps three times in total to efficiently deplete Lineage-positive cells of the marrow sample (*see* **Note 11**).

3.3 Flow Cytometry Staining and Sorting

1. Centrifuge Lineage-negative isolate 5 min at 300 × *g* at 4 °C.

2. Discard the supernatant.

3. Resuspend in 1 mL of complete PBS.

4. Count the cells with a hemocytometer and Trypan Blue or other method (*see* **Note 9**).

5. Centrifuge for 5 min at 300 × *g* at 4 °C.

6. Discard the supernatant.

7. If not ahead, prepare 100μL of antibody mix *per* sample containing anti-CD3, anti-B220, anti-CD11-b, anti-Gr-1, anti-TER-119, anti-Sca-1, anti-CD48, anti-CD150, anti-c-Kit, anti-CD135 antibodies, diluted in complete PBS to reach the appropriate concentrations, and 0.25μL of Live/Dead (*see* **Notes 3**, **12** and **13**). This cocktail of antibodies has been designed to allow the concomitant sorting of SLAMs (LSK Flt3⁻CD48⁻CD150⁺) and ST-HSCs (LSK Flt3⁻CD48⁻CD150⁻).

8. Resuspend the cells in 100μL of antibody mix.

9. Incubate for 45 min at 4 °C in the dark.

10. Wash with 1 mL of complete PBS.

11. Centrifuge for 5 min at 300 × *g* at 4 °C.

12. Discard the supernatant.

13. Resuspend in 1 mL of FACS buffer.

14. Prepare the complete differentiation medium as described in the material part and warm it at 37 °C in a water bath (*see* **Note 14**).

15. Take a U-bottom 96-well culture plate and fill the wells with 200μL of complete differentiation medium.

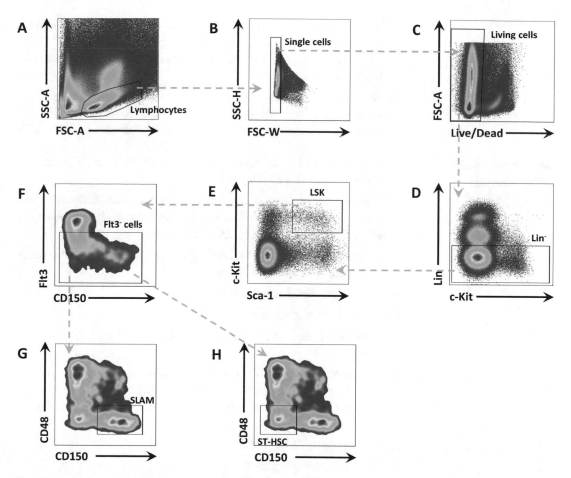

Fig. 2 Representative density-plots depict the flow-cytometric gating strategy used for mouse BM LSK SLAMs or ST-HSCs sorting. The gating strategy (panels **A** to **H**) is described in Subheading 3.3

16. Pass the cell suspension to the cell sorter according to manufacturer's instructions and sort 450 SLAMs or ST-HSCs per well (*see* **Notes 15** and **16**). *See* Fig. 2 for the gating strategy described step by step below.

17. Gate on lymphocyte population, excluding dead cells, debris and monocytes/macrophage neutrophils on the basis of size/scatter specificity (Fig. 2a).

18. Gate out cell doublets (Fig. 2b).

19. Gate on lived cells by excluding dead cells positive for the staining with a viability dye (Fig. 2c).

20. Gate on Lineage-negative (Lin⁻) cells (Fig. 2d).

21. Gate on LSK (Lin Sca-1⁺c-Kit⁺) cells (Fig. 2e).

22. Make a dot plot between Flt3 and CD150 and gate on Flt3⁻ cells (Fig. 2f).

23. Make a dot plot between CD48 and CD150 and sort SLAMs ($CD48^-CD150^+$, Fig. 2g) or ST-HSCs ($CD48^-CD150^-$, Fig. 2h) directly in the culture plate containing complete differentiation medium.

24. After sorting, incubate the cells at 37 °C in a humidified atmosphere containing 5% of CO_2. SLAM and ST-HSC cultures could be maintained until 4 days of culture (*see* **Note 17**).

3.4   In Vitro HSPC Differentiation

1. Prepare one FACS tube per well and fill it with 1 mL of complete DMEM FBS medium.

2. Remove the culture plate from the incubator. Resuspend the cells in the well by several up and down pipetting and place the cell suspension into the FACS tube.

3. Wash the well with 200µL of complete DMEM FBS medium and place it in the FACS tube.

4. Centrifuge for 5 min at $300 \times g$ at 4 °C.

5. Discard the supernatant.

6. Resuspend the pellet with 1 mL of complete DMEM FBS medium.

7. Incubate the cells at 37 °C in a humidified atmosphere containing 5% of CO_2 during 1 h. This step allows the cells to reexpress c-Kit and Flt3/CD135 which are receptors for SCF and Flt3-L respectively.

8. Centrifuge at $300 \times g$ for 5 min at 4 °C.

9. Discard the supernatant.

10. Prepare 50µL of antibody mix per sample containing anti-CD3, anti-B220, anti-CD11-b, anti-Gr-1, anti-TER-119, anti-Sca-1, anti-CD48, anti-CD150, anti-c-Kit, and anti-CD135 antibodies, diluted in complete PBS to reach the appropriate concentration, and 0.25µL of Live/Dead (*see* **Notes 3**, **12**, and **13**). This cocktail of antibodies has been designed to allow the concomitant identification of SLAMs (LSK $Flt3^-CD48^-CD150^+$), MPPs (LSK $Flt3^-CD48^-CD150^-$), MPP2 (LSK $Flt3^-CD48^+CD150^+$), MPP3 (LSK $Flt3^-CD48^+CD150^-$) and MPP4 (LSK $Flt3^+CD48^{+/-}CD150^-$) in SLAM cultures and the concomitant identification of ST-HSCs (LSK $Flt3^-CD48^-CD150^-$), MPP2 (LSK $Flt3^-CD48^+CD150^+$), MPP3 (LSK $Flt3^-CD48^+CD150^-$) and MPP4 (LSK $Flt3^+CD48^{+/-}CD150^-$) in ST-HSC cultures.

11. Resuspend cells in 50µL of antibody mix.

12. Incubate for 45 min at 4 °C in the dark.

13. Wash with 1 mL of complete PBS.

14. Centrifuge for 5 min at $300 \times g$ at 4 °C.

After 4 days of SLAM culture :

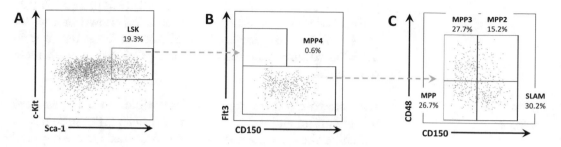

After 4 days of ST-HSC culture :

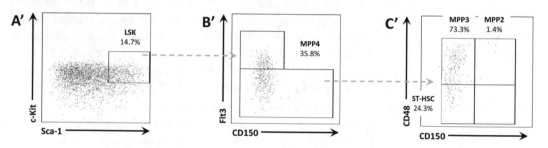

Fig. 3 Representative dot-plots depict the flow-cytometric gating strategy used for identification of the different MPP subsets generated starting from LSK SLAMs or ST-HSCs after 4 days of culture. The gating strategy (panels **A** to **C** and **A'** to **C'**) is described in Subheading 3.5

15. Discard the supernatant.

16. Resuspend in 300µL of FACS buffer (*see* **Note 18**).

17. Pass the cell suspension to the flow cytometer according to the manufacturer's instructions (*see* **Note 19**).

3.5 Flow Cytometric Analyses

See Fig. 3 for gating strategy described step-by-step below.

1. Gate on lymphocyte population on the basis of size/scatter specificity.

2. Gate out cell doublets.

3. Gate on living cells by excluding dead cells that are positive for the staining with the viability dye.

4. Gate on the Lin⁻ cells.

5. LSK cells: gate on c-Kit⁺Sca-1⁺ cells (Figs. 3a, a′).

6. On this population, make a dot plot between Flt3 and CD150, isolate MPP4 that are Flt3⁺CD150⁻ and continue the gating of the other populations in the fraction of Flt3⁻ cells (Fig. 3b, b′). As shown by the dot plots, MPP4 are generated in ST-HSC cultures but not in SLAM cultures.

7. Make a dot plot between CD48 and CD150 (Fig. 3c and c′). Highlight   SLAMs   (CD48⁻CD150⁺),   MPPs (CD48⁻CD150⁻), MPP2 (CD48⁺CD150⁺) and MPP3

(CD48$^+$CD150$^-$) in SLAM cultures and ST-HSCs (CD48$^-$CD150$^-$), MPP2 (CD48$^+$CD150$^+$) and MPP3 (CD48$^+$CD150$^-$) in ST-HSC cultures. As shown by the dot plots, MPP2 are generated in SLAM cultures but not in ST-HSC cultures whereas MPP3 are produced in both types of cultures.

3.6 Additional Analyses

1. It is noteworthy that our protocol is flexible and can be adapted to the culture and the generation of other HSPC subtypes by enriching the cytokine cocktail with additional factors required for specific cell types.

2. The abovementioned in vitro analyses of HSC differentiation into MPP subsets can be implemented by functional assays. This protocol is compatible with in vitro studies aiming to evaluate cell cycle (e.g., Ki-67/Dapi, CellTraceViolet or BromodeoxyUridine pulse-chase kinetics) and apoptosis (Annexin V/Propidium Iodide) status of HSPCs by flow cytometry.

3. This experimental protocol constitutes a method to expand in vitro HSPCs for which isolation of significant numbers directly from mouse BM remains challenging. Consequently, achievement of experimental procedures requiring a large number of HSPCs is difficult. For instance, in vivo differentiation potential of HSPCs is defined by transplanting these cells into recipient mice that are preconditioned by lethal irradiation and therefore devoid of endogenous hematopoietic system. Injection of thousands of MPPs is required to follow their differentiation in recipient mice due to their absence of self-renewal potential and long-term reconstitution ability. Such a large amount of MPPs can be difficult to obtain from donor mice especially if the latter display alterations of the hematopoietic compartment. Furthermore, study of HSPCs–stroma cross talk is addressed in vitro using coculture systems where a single HSPC subtype is plated with different stromal cells. This leads to multiply the experimental conditions and the number of cells required. Therefore, execution of both abovementioned experimental procedures can be facilitated by achieving a preliminary step of HSPC expansion using our protocol.

4 Notes

1. Use either polypropylene or polystyrene tubes depending on the compatibility with your flow cytometer. If you have the choice, prefer polypropylene tubes that have the advantage of minimizing the binding of cells on the walls of the tubes.

2. Complete PBS is used for all wash steps.

3. Antibodies concentrations noted in Table 1 have been carefully determined according to our own personal settings but are indicative as they can vary depending on the supplier of the antibodies, the fluorochrome to which they are conjugated or the flow cytometer used. Each antibody needs to be properly titrated to determine the optimal concentration in your personal settings.

4. To well studying these very rare cell populations, make the experiment within a few hours after mouse euthanasia.

5. To maintain the bones in a wet environment, put bones into a culture plate or petri dishes.

6. Concerning the hips, separation with femurs during dissection is sufficient to allow the extraction of marrow fraction through the femur articulation.

7. After centrifugation, check that the bones are emptied of marrow fraction. If "red" marrow is still visible inside the diaphysis, ensure that one extremity of the bones is correctly open and turned down and repeat the centrifugation step.

8. If fat aggregates are visible during resuspension, remove them using the tips.

9. Number of cells can be checked using Trypan Blue. If cells are exposed to Trypan Blue for extended periods of time, viable cells may also take up the dye and alter the counting. Count cells in 3 large squares and calculate number of cells/mL using the formula: Number of cell/mL = (Total cells counted/ Number of squared counted) × Dilution × 10^4. Total number of cells = Number of cells/mL × Volume (mL).

10. The buffer containing the beads can induce substantial cell death. To avoid this issue, proceed the following step. Prior to put the beads in contact with the cells, add the required volume of beads on the magnet, wait for 5 min, remove the volume and replace it by the same volume of PBS.

11. When collecting the supernatant, avoid touching the walls of the tubes with the tip to prevent any contamination with Lineage-positive cells.

12. Antibody mix can be prepared maximum 24 h prior to the experiment, ready to use and store at 4 °C protected from light.

13. Always prepare an extra volume (approximately 10%) to anticipate some pipetting imprecisions.

14. Cytokines activity can be impacted by repeated freeze–thaw cycles. Prepare aliquots containing a volume corresponding to maximum two uses and always thaw these aliquots on ice.

15. To remove any potential aggregates, transfer and filter cell suspension into a new FACS tube with cell strainer snap cap (35 μm nylon mesh).

16. If the cell sorter allows it, sort the cells directly into the culture plate. If not, sort the cells in FACS tubes containing complete DMEM FBS medium, centrifuge, resuspend in complete differentiation medium and dispatch the volume in the culture plate.

17. Throughout the time of the experiment, observe the viability and the division of the cells under a microscope. If differentiation occurs normally, you should observe an increase in the number of cells present in the wells of the culture plate.

18. Cells can be kept before analysis with the flow cytometer up to 2 days after the staining although immediate acquisition is warmly recommended. In such cases, resuspend cells in 300µL of PBS-1% PFA during 30 min, wash with 1 mL of complete PBS and resuspend in PBS-3% BSA-2 mM EDTA.

19. As the cell counting is realized using the flow cytometer, it is crucial to record all the events present in the FACS tube. To this end, prefer to pass the sample at the minimum speed allowed by the flow cytometer and do not stop to record the events before all the volume present in the FACS tube has been aspirated.

Acknowledgments

This work was supported by the ANR grant PRC 17-CE14-0019; the INCa agency under the program PRT-K 2017; a PhD fellowship from the Fondation pour la Recherche Médicale, and a fourth year PhD fellowship from the Ligue Nationale Contre le Cancer.

References

1. Doulatov S, Notta F, Laurenti E et al (2012) Hematopoiesis: a human perspective. Cell Stem Cell 10:120–136

2. Osawa M, Hanada K, Hamada H et al (1996) Long-term lymphohematopoietic reconstitution by a single CD34-low/negative hematopoietic stem cell. Science 273:242–245

3. Zhong RK, Astle CM, Harrison DE (1996) Distinct developmental patterns of short-term and long-term functioning lymphoid and myeloid precursors defined by competitive limiting dilution analysis in vivo. J Immunol 157:138–145

4. Wilson A, Laurenti E, Oser G et al (2008) Hematopoietic stem cells reversibly switch from dormancy to self-renewal during homeostasis and repair. Cell 135:1118–1129

5. Pietras EM, Reynaud D, Kang YA et al (2015) Functionally distinct subsets of lineage-biased multipotent progenitors control blood production in normal and regenerative conditions. Cell Stem Cell 17:35–46

6. Freitas C, Wittner M, Nguyen J et al (2017) Lymphoid differentiation of hematopoietic stem cells requires efficient Cxcr4 desensitization. J Exp Med 214:2023–2040

7. Freitas C, Rondeau V, Balabanian K (2018) New method to obtain lymphoid progenitors. Patent WO/2018/177971 (PCT/EP2018/057580)

8. Yilmaz OH, Kiel MJ, Morrison SJ (2006) SLAM family markers are conserved among hematopoietic stem cells from old and reconstituted mice and markedly increase their purity. Blood 107:924–930

Chapter 5

In Vitro Analysis of Energy Metabolism in Bone-Marrow Mesenchymal Stromal Cells

Jérôme Bourgeais and Olivier Hérault

Abstract

Bone marrow mesenchymal stromal cells (MSCs) play an essential role in the regulation of normal and leukemic hematopoiesis. Their multipotent potential of differentiation also makes them an interesting therapeutic tool. Among factors involved in the regulation of MSCs, energy metabolism plays a key role in their proliferation and differentiation. Seahorse Bioscience introduced extracellular flux technology to the life sciences market in 2006. This methodology allows, in living cells and in real time, the concomitant determination of basal oxygen consumption, glycolysis rates, ATP production, and respiratory capacity in a single experiment. Here we describe the protocol used to study concomitantly the respiratory and glycolytic metabolism of primary MSCs from the determination of oxygen consumption (OCR) and extracellular acidification (ECAR) rates.

Key words Mesenchymal stromal cells, Seahorse, Energy metabolism, Glycolysis, Mitochondrial respiration

---

1 Introduction

In the bone marrow (BM), mesenchymal stromal cells (MSCs) play a key role in normal hematopoiesis but also in leukemogenic processes. The most striking feature of these cells is their ability to differentiate into multiple lineages, including adipocytes, osteoblasts, chondrocytes, fibroblasts, and myoblasts. They also contribute to the synthesis of extracellular matrix, release of soluble factors, and regulation of the oxidative metabolism of hematopoietic cells. All these characteristics contribute to make them one of the most important elements of the hematopoietic microenvironment by preserving BM homeostasis. In addition, their multipotent capacity and expansion potential make them an attractive cell source for tissue engineering applications.

Energy metabolism is intimately linked to the differentiation of MSCs. Osteoblastic [1–3] and adipocytic [4–7] differentiation is characterized by an increase in mitochondrial respiratory

Marion Espéli and Karl Balabanian (eds.), *Bone Marrow Environment: Methods and Protocols*, Methods in Molecular Biology, vol. 2308, https://doi.org/10.1007/978-1-0716-1425-9_5, © Springer Science+Business Media, LLC, part of Springer Nature 2021

metabolism, via the generation of reactive oxygen species (ROS) [8, 9]. Conversely, chondrocyte differentiation is improved by a predominance of glycolytic metabolism [3, 10].

As in cancer-associated fibroblasts, BM MSCs probably play an important role in tumorigenesis in connection with the regulation of energy metabolism. For instance, in osteosarcomas, the cancer-associated microenvironment metabolically supports adjacent cancer cells (reverse Warburg effect) by decreasing the mitochondrial respiration of MSCs in favor of aerobic glycolysis in order to produce lactate recycled by tumor cells via the MCT1 transporter [11]. When considering hematological diseases, the energy metabolism of BM MSCs is poorly known and ongoing studies in our group aim at clarifying this point in acute myeloid leukemia (ClinicalTrials.gov # NCT03918655).

The importance of energy metabolism in the regulation of MSCs and more generally in tumor processes now makes it an emerging hallmark of cancer [12]. The development of new technology/techniques to analyze respiratory and glycolytic metabolism is therefore essential for the acquisition of new knowledge. The study of hematopoietic cells' energy metabolism in their niche is of interest to understand the mechanisms of chemoresistance [13, 14]. Interestingly, mitochondrial transfers from MSCs to leukemia cells via an endocytic pathway system are probably involved in the chemoresistance process [15, 16]. All these hypotheses rely on a good understanding of MSCs metabolism, not readily available yet.

Seahorse XF Analyzers™ are new tools useful to decipher the energy metabolism of MSCs in the normal or leukemic hematopoietic niche. This technology allows to measure in real time the consumption of oxygen directly linked to mitochondrial respiration in living cells, as well as the extracellular acidification correlated with glycolysis. The technology of Seahorse XF Analyzers™ is based on the quantification of measurable changes in two parameters. The dynamic assessment of the concentrations of dissolved oxygen, called Oxygen Consumption Rate (OCR) that is a direct reflect of mitochondrial respiration, while that of free protons or extracellular acidification rate (ECAR) depends on the glycolytic activity. Measures are performed by solid-state sensor probes residing 200µm above a cell monolayer, for periods of 2–5 min and the instrument calculates both OCR and ECAR. The coupling of these probes with a four compounds injection system makes it possible to measure in real time the main parameters of the mitochondrial respiration (basal and maximal respiration, ATP production, proton leak, spare respiratory capacity and nonmitochondrial respiration) as well as glycolytic function (basal glycolysis, glycolytic capacity, glycolytic reserve, and nonglycolytic acidification).

The following protocol aims to provide a clear and complete explanation of Seahorse metabolic analysis of BM MSCs, both as isolated and purified MSCs or in combinations studies with the addition of leukemic cells.

2 Materials

2.1 Bone Marrow MSCs

1. Obtain primary BM MSCs from healthy patients (without *hematological disease*), after receiving informed donor consent and in accordance with the procedure approved by the local ethics committee (*see* **Note 1**). MSCs were cultured, amplified, passaged once, and characterized in vitro as reported elsewhere [17] before use on passage 2.

2. Cell culture incubator at 37 °C with 5% CO_2.

3. Cell culture growth medium: αMEM medium supplemented with 10% fetal bovine serum, 2 mM L-glutamine, 100 U/mL penicillin G, 0.1 mg/mL streptomycin, 25μg/mL Fungizone, 1 ng/mL FGF2.

4. Cell detachment solution: 0.05% Trypsin-EDTA.

5. Seahorse 96-wells XF96 Cell Culture Microplate (Agilent, Santa Clara, USA) (*see* **Note 2**).

2.2 SeaHorse Experiment

1. Seahorse XF Analyzers™ (e.g., XFe96 device from Agilent) .

2. A non-CO_2 37 °C incubator.

3. Cell culture grade sterile water.

4. Reaction culture medium: XF DMEM medium pH 7.4 (Agilent), supplemented with 2 mM XF Glutamine Solution (Agilent).

5. XFe96 FluxPak (Agilent).

6. Seahorse XF Calibrant Solution (Agilent).

7. Port A injection stock solution: XF 1.0 M Glucose Solution (Agilent).

8. Port B injection stock solution: 1 mM oligomycin. Weigh 3.96 mg of oligomycin A and add DMSO to a volume of 5 mL. Make stock aliquots of 50μL and store at −20 °C.

9. Port C injection stock solution: 2,4-dinitrophenol (DNP). Weigh 46.03 mg of DNP and add DMSO to a volume of 25 mL. Make stock aliquots of 500μL and store at −20 °C.

10. Port D injection stock solution (antimycin A): 1 mM antimycin A. Weigh 2.74 mg of antimycin A and add DMSO to a volume of 5 mL. Make stock aliquots of 50μL and store at −20 °C.

11. Port D injection stock solution (rotenone): 1 mM rotenone. Weigh 1.97 mg of rotenone and add ethanol to a volume of 5 mL. Make stock aliquots of 500µL and store at $-20\,°C$.

---

3   Methods

3.1   Day Prior to Assay: Cartridge Calibration and Cell Plate Preparation

1. Hydrate a Seahorse XFe96 sensor cartridge by filling the utility plate with 200µL of cell culture grade sterile water. Replace the sensor cartridge onto the utility plate and incubate overnight in a non-CO_2 37 °C incubator (*see* **Notes 3** and **4**).

2. Preheat overnight 20 mL of Seahorse XF Calibrant to 37 °C for the second calibration step.

3. Recover confluent passage 2 MSCs by trypsinization (*see* **Note 5**) and plate the cells in 80µL of cell culture growth medium at the optimal cell concentration (between 20,000 and 40,000 cells per well) in Seahorse 96-well XFe96 Cell Culture Microplates (*see* **Notes 6–8**).

3.2   Day Prior to Assay: Template Preparation on Seahorse Wave Acquisition Software

1. Build the template on the Seahorse Wave acquisition software. Use an existing template or create one. In the "Group Definitions" tab, define injection strategies as follows.

   (a)   Port A: 100 mM glucose (10×).

   (b)   Port B: 10µM oligomycin (10×).

   (c)   Port C: 1 mM DNP (10×).

   (d)   Port D: 5µM rotenone (10×) and 5µM antimycin A (10×) mixture.

   Define a pretreatment if relevant. In assay media, select "Seahorse XF DMEM Medium, pH 7.4" to allow simultaneous analysis of the glycolytic and respiratory parameters as well as a thorough study of the amount of ATP. Exhaustively incorporate information in the "cell type" tab, in particular concerning the identification, cell concentration and dating of samples.

2. Go to the "Plate map" tab after generating groups and define the position of the different samples in the Seahorse 96-well XF^{c}96 Cell Culture Microplates. Retain the preselected position of the backgrounds in wells A1, A12, H1, and H12.

3. Create the acquisition protocol in the tab "Protocol." Check that the calibrate and equilibrate steps are selected in the initialization phase. For the baseline phase, select four cycles characterized by 3 min of mixing followed by 3 min of measurement. Then, add four injections, each consisting of three cycles (3 min of mixing followed by 3 min of

measurement). Rename the different injection entries with the compound present in the port (glucose, oligomycin, DNP, rotenone–antimycin A).

4. Rename and save the template.

5. Switch on the Seahorse and open the template to allow the device to preheat to 37 °C (*see* **Note 9**).

3.3 Day of Assay: Cartridge Calibration and Cell Plate Preparation

1. Replace water from the utility plate with 200μL of the pre-warmed XF Calibrant solution. Replace the Seahorse XFe96 sensor cartridge onto the utility plate and incubate for 1 h in a non-CO_2 37 °C incubator.

2. Warm 50 mL reaction culture medium to 37 °C for 1 h.

3. Remove the 80μL cell culture growth medium from the Seahorse 96-well XFe96 Cell Culture Microplates, rinse the MSC layer with 100μL of reaction culture medium before adding 180μL of reaction culture medium (*see* **Note 10**). Control cells under microscope (confluence, cell health, morphology, seeding uniformity, and purity). Incubate the Cell Culture Microplates for 1 h in a non-CO_2 37 °C incubator.

3.4 Day of Assay: Cartridge Loading

1. Port A injection solution making: dilute 340μL of XF 1.0 M Glucose Solution (Port A injection stock solution) in 3.06 mL of reaction culture medium to obtain a final concentration of 100 mM (10×). Mix to homogenize the solution. Load all ports A of the Seahorse XFe96 sensor cartridge with 20μL of this solution (*see* **Note 11**).

2. Port B injection solution making: dilute 35μL of 1 mM oligomycin solution (Port B injection stock solution), an ATP synthase inhibitor, in 3.465 mL of reaction culture medium to obtain a final concentration of 10μM (10×). Mix to homogenize the solution. Load all ports B of the Seahorse XFe96 sensor cartridge with 22μL of this solution.

3. Port C injection solution making: dilute 400μL of 2,4-dinitrophénol 10 mM solution (Port C injection stock solution), an uncoupling agent that collapses the proton gradient and disrupts the mitochondrial membrane potential, in 3.6 mL of reaction culture medium to obtain a final concentration of 1 mM (10×). Mix to homogenize the solution. Load all ports C of the Seahorse XFe96 sensor cartridge with 25μL of this solution.

4. Port D injection solution making: dilute 20μL of 1 mM antimycin A, a complex III inhibitor and 20μL of 1 mM rotenone, a complex I inhibitor (Port D injection stock solution) in 3.96 mL of reaction culture medium to obtain a final concentration of 5μM of the two compounds (10×). Mix to homogenize the solution. Load all ports C of the Seahorse XFe96 sensor cartridge with 27μL of this solution.

5. After loading the different injection ports with the appropriate compounds, it is important to visually check their correct loading. The solution should not be confined to one side of the port but occupy the entire well in order to avoid the passage of the air pulse next to the droplet and allow a proper injection.

3.5 Day of Assay: Run the XFe Test

1. Open the template created the day before.

2. Complete the project and plate information in the tab "Run Assay," then start the run.

3. According to the Seahorse Analyzer software instructions, gently insert the loaded sensor cartridge and its utility plate into the device and start the calibration step by clicking on "I'm ready."

4. At the end of the calibration and the incubation time of the Cell Culture Microplates (Subheading 3.3, **step 4**), replace the utility plate with the plate containing the cells, then start the measurement by clicking on "I'm ready."

5. At the end of the acquisition, extract the cartridge and the plate containing the cells from the Seahorse Analyzer. Remove the reaction culture medium of the Cell Culture Microplates by aspiration, taking care not to detach the cell layer and store the plate at -20 °C for normalization of data on the quantity of DNA or protein according to the desired protocol. Check the reproducibility of replicates for OCR and ECAR as well as the absence of signal and variation in empty wells. Then export the data.

3.6 Results Analysis

The OCR and ECAR are automatically determined by the succession of twelve O_2 and H^+ concentration measurements over a specific time interval (3 min). The main parameters of glycolysis and respiration can be calculated from the same data extracted to the Seahorse Wave software or semiautomatically using the report generator files running under Microsoft Excel software. The XF Mito Stress Test Report Generator automatically calculates the main respiratory parameters from OCR data. The XF Glycolysis Stress Test Report Generator makes it possible to determine the main glycolysis parameters from ECAR data. In all cases, the first step is to perform data normalization with the selected technique (DNA or protein content, cell imaging, . . .). Consider the mean of at least three replicate wells and calculate the standard error of the mean (SEM).

3.6.1 Mitochondrial Respiration Determination

1. The keys parameters of mitochondrial respiration are determined through the successive injection of compounds that target the elements of the electron transport chain (ETC) (Fig. 1a and Table 1).

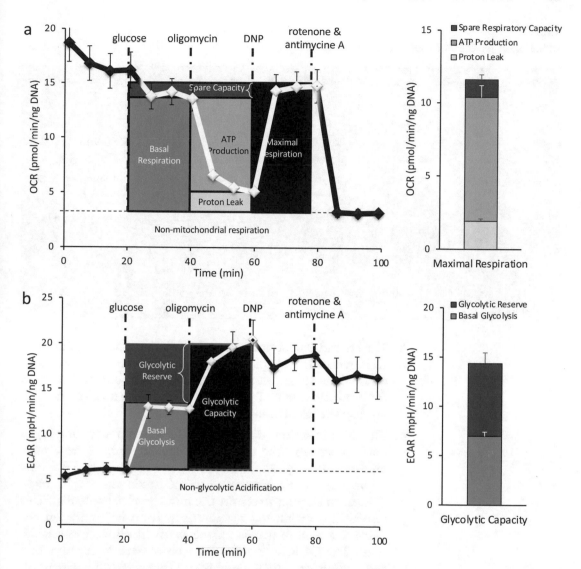

Fig. 1 Definition and determination of the main respiratory and glycolytic parameters. Primary BM MSCs were plated on the Seahorse 96-well Cell Culture microplates and evaluation of mitochondrial respiration and glycolytic function was performed on a Seahorse analyzer by the determination of OCR and ECAR. (**a**) Basal and maximal respiration, ATP-linked and proton leak were determined by measuring OCR (pmol/min/ng DNA) after sequential exposure to ATP synthase inhibitor oligomycin (1μM), uncoupler DNP (100μM) and complex III inhibitor antimycin A (0.5μM) plus complex I inhibitor rotenone (0.5μM). (**b**) Basal glycolysis, glycolytic capacity and glycolytic reserve were determined by measuring ECAR (mpH/min/ng DNA) after sequential exposure to glucose (10 mM) and ATP synthase inhibitor oligomycin (1μM). Data are normalized for DNA concentration from a single biological replicate and presented as mean ± SEM of three independent wells

2. The first injection (glucose) creates an optimal environment for the cells. The resulting OCR is the sum of basal and nonmitochondrial respiration.

Table 1
Calculation for the main respiratory parameters

Respiratory parameter	Parameter calculation
Basal respiration	Last OCR measurement before oligomycin injection minus nonmitochondrial respiration
H+ (proton) leak	Minimum OCR measurement after oligomycin addition but prior to DNP injection minus nonmitochondrial respiration
ATP production	Last OCR measurement before oligomycin injection minus minimum OCR measurement after oligomycin addition but prior to DNP injection
Maximal respiration	Maximum OCR measurement after DNP injection minus nonmitochondrial respiration
Spare respiratory capacity	Maximal respiration minus basal respiration
Nonmitochondrial respiration	Minimum OCR measurement after rotenone–antimycin A mix injection

3. The second compound injected is oligomycin, an inhibitor of ATP synthase (complex V of the mitochondrial respiratory chain). The latter causes a decrease in OCR directly linked to the production of ATP (*see* Table 1 for more details on the ATP production calculation).

4. The third injection (dinitrophenol) adds an uncoupling agent that dissipates the proton gradient from the inter-mitochondrial space. In response, complexes I to IV of the respiratory chain are activated to the maximum of their capacities in an attempt to restore the mitochondrial potential. This results in a major and maximum increase in respiration (*see* Table 1 for more details on the maximal respiration calculation). The OCR obtained is the sum of maximum respiration and nonmitochondrial respiration. From there, the spare respiratory capacity can be deduced, defined as the difference between maximal and basal respiration. The spare capacity is an essential cellular parameter and determines the ability of the cells to adapt to an increased demand for mitochondrial energy (*see* Table 1 for more details on spare capacity calculation).

5. The fourth and last injection of a mixture of rotenone and antimycin A, distributes respective inhibitors of complexes I and III of the ETC. This cocktail leads to a complete blockage of electron transfers through the respiratory chain and a total suppression of oxygen consumption by the mitochondria. As the resulting OCR is linked to nonmitochondrial respiration, it is then possible to determine basal respiration, maximum respiration and oxygen consumption related to proton leak (*see* Fig. 1a and Table 1 for more details on mitochondrial function calculation).

Table 2
Calculation for the main glycolytic parameters

Glycolytic parameter	Parameter calculation
Basal glycolysis	Maximum ECAR measurement before oligomycin injection minus last ECAR measurement before glucose injection
Glycolytic capacity	Maximum ECAR measurement after oligomycin injection but prior to DNP injection minus last ECAR rate measurement before glucose injection
Glycolytic reserve	Glycolytic capacity minus basal glycolysis
Nonglycolytic acidification	Last ECAR measurement prior to glucose injection

3.6.2 Glycolytic Function Determination

1. The keys parameters of glycolytic function are determined through the successive injection of compounds which activate and potentiate glycolysis. Before the first injection, ECAR is linked to nonglycolytic acidification. This represents any enzymatic reaction responsible for a decrease of the extracellular pH (*see* Table 2 for more details on nonglycolytic acidification calculation).

2. Glucose, injected first, is the major substrate of the glycolysis pathway. It is injected in excess to allow for the measurement of the level of ECAR corresponding to basal glycolysis by subtraction of nonglycolytic acidification (*see* Table 2 for more details on the basal glycolysis calculation).

3. Injection of oligomycin in the second step provides an inhibitor of the enzyme ATP synthase. This suppresses the production of mitochondrial ATP which leads to a compensatory increase in glycolysis, dependent on cellular glycolytic capacities, in order to maintain a constant ATP level in the cells. This results in a major and maximum increase in acidification (*see* Table 2 for more details on the glycolytic capacity calculation). The ECAR obtained is the sum of glycolytic capacity and nonglycolytic acidification. From there the glycolytic reserve can be deduced, defined as the difference between glycolytic capacity and basal glycolysis. The glycolytic reserve is an important cellular feature which determines the ability of the cells to adapt to an increased glycolytic ATP demand (*see* Table 2 for more details on glycolytic reserve calculation).

4. The third and fourth injections are not involved in the calculation of glycolytic parameters.

3.6.3 Complementary Energetic Function Determination

To refine the analysis, other parameters can be calculated from the OCR and ECAR data (*see* Table 3 for more details on energetic function calculation).

Table 3
Calculation for complementary energetic parameters

Bioenergetic Health Index	[mitochondrial ATP production × spare respiratory capacity]/[proton leak × nonmitochondrial respiration]
Mitochondrial coupling efficiency	(mitochondrial ATP production/basal respiration) × 100
Respiratory control ratio	Maximal respiration/mitochondrial proton leak
OCR–ECAR ratio	Basal respiration/basal glycolysis
OCRmax–ECARmax ratio	Maximal respiration/glycolytic capacity

1. Bioenergetic Health Index (BHI): a single value that represents mitochondrial function in the cells by integrating most of the parameters of mitochondrial respiration [18].

2. Mitochondrial Coupling Efficiency (MCE): proportion of oxygen directly coupled to the production of ATP at the basal state.

3. Respiratory Control Ratio (RCR): parameter allowing a better characterization of the mitochondrial function when, at the basal state, the oxygen consumption coupled with the production of ATP is low.

4. Ratio of ATP-linked respiration to maximal respiration.

5. OCR–ECAR Ratio and OCRmax–ECARmax Ratio: to identify the metabolic preferences of the cells tested. Under certain conditions, these ratios also make it possible to highlight the Warburg effect or, conversely, an increase in mitochondrial respiration.

4   Notes

1. MSCs were obtained from total bones marrow through their ability to adhere to plastic. At the end of the first passage (Po), the cells are frozen in nitrogen in SVF 10% DMSO.

2. Seahorse XF96 Cell Culture Microplates are designed for Seahorse XFe96/XF96 Analyzers. For Seahorse XFe24 or XFp Analyzer use material-specific culture plates.

3. The use of a cell culture incubator for the hydration of the cartridge with 5% CO_2 would lead to a modification of the calibration medium pH inducing an error in ECAR probes calibration.

4. At this step, the cartridge must be hydrated for a minimum of 16 h and up to 3 days. In this case, it is important to check that

all probes are continuously submerged. If necessary, adding water preheated to 37 °C can be considered to compensate evaporation. Likewise, it is important to check that no air bubbles have formed on contact with the probe, which could alter the calibration.

5. When a coculture is performed of MSCs with hematopoietic cells, MSCs are first sorted by magnetic selection or on a cell sorter based on the expression of CD90. Then they are plated on the Seahorse culture plate and left to adhere overnight or immediately fixed using poly-D-lysine on a precoated plate before adding the hematopoietic cells.

6. Several replicates of the same sample are advised with a minimum of three technical replicates. The 96 wells can be used for this purpose to test a large number of samples in parallel. However, wells A1, A12, H1, and H12 of the XFe96 Cell Culture Microplates must always be filled only with culture growth medium (no cells) in order to determine the background noise.

7. Optimal cell concentration depends on the doubling time as well as the number of cells necessary for confluence. Preliminary tests are recommended to optimize analyzes.

8. Seed the cell suspension in the lower half of the wells and let the cells rest at room temperature for 1 h under the tissue culture hood to allow homogeneous cell distribution at the bottom of the well and reduced edge effects. Allow the cells to adhere overnight in humidified air with 5% CO_2, in a 37 °C incubator and check the adhesion of the cells (inverted microscope) before the next step (Subheading 3.3, **step 4**).

9. It is important that the Seahorse is at the stable temperature of 37 °C before starting the calibration step to ensure optimal functioning of the probes. Likewise, opening the device when loading the cartridges and the culture plate must be carried out as quickly as possible in order to avoid any thermal loss.

10. It is important to maintain the pH of the solution at 7.4 at 37 °C. The measurement of ECAR requires the absence of a strong buffer, which makes it necessary to work at a strictly controlled pH. Since the temperature can influence pH, it is important to control it at the working temperature of 37 °C.

11. Injection of the different drugs is carried out simultaneously by air pulse, making it important that all ports are loaded with the same volume for proper injection. Discard any remaining compound. When working with a new cell type or culture condition, the concentration of each inhibitor should be carefully titrated.

Acknowledgements

The authors thank Marie-Christine Bene for the editing of the manuscript.

References

1. Chen C-T, Shih Y-RV, Kuo TK et al (2008) Coordinated changes of mitochondrial biogenesis and antioxidant enzymes during osteogenic differentiation of human mesenchymal stem cells. Stem Cells 26:960–968

2. Pietilä M, Palomäki S, Lehtonen S et al (2012) Mitochondrial function and energy metabolism in umbilical cord blood- and bone marrow-derived mesenchymal stem cells. Stem Cells Dev 21:575–588

3. Pattappa G, Heywood HK, de Bruijn JD et al (2011) The metabolism of human mesenchymal stem cells during proliferation and differentiation. J Cell Physiol 226:2562–2570

4. Ren H, Cao Y, Zhao QQ et al (2006) Proliferation and differentiation of bone marrow stromal cells under hypoxic conditions. Biochem Biophys Res Commun 347:12–21

5. Kanda Y, Hinata T, Kang SW et al (2011) Reactive oxygen species mediate adipocyte differentiation in mesenchymal stem cells. Life Sci 89:250–258

6. Tormos KVV, Anso E, Hamanaka RBB et al (2011) Mitochondrial complex III ROS regulate adipocyte differentiation. Cell Metab 14:537–544

7. Hofmann AD, Beyer M, Krause-Buchholz U et al (2012) OXPHOS supercomplexes as a hallmark of the mitochondrial phenotype of adipogenic differentiated human MSCs. PLoS One 7:e35160

8. Lechpammer S, Epperly MW, Zhou S et al (2005) Adipocyte differentiation in Sod2−/− and Sod2+/+ murine bone marrow stromal cells is associated with low antioxidant pools. Exp Hematol 33:1201–1208

9. Lee H, Lee YJ, Choi H et al (2009) Reactive oxygen species facilitate adipocyte differentiation by accelerating mitotic clonal expansion. J Biol Chem 284:10601–10609

10. Pattappa G, Thorpe SD, Jegard NC et al (2013) Continuous and uninterrupted oxygen tension influences the colony formation and oxidative metabolism of human mesenchymal stem cells. Tissue Eng Part C Methods 19:68–79

11. Bonuccelli G, Avnet S, Grisendi G et al (2014) Role of mesenchymal stem cells in osteosarcoma and metabolic reprogramming of tumor cells. Oncotarget 5:7575–7588

12. Hanahan D, Weinberg RA (2011) Hallmarks of cancer: the next generation. Cell 144:646–674

13. Picou F, Debeissat C, Bourgeais J et al (2018) n-3 polyunsaturated fatty acids induce acute myeloid leukemia cell death associated with mitochondrial glycolytic switch and Nrf2 pathway activation. Pharmacol Res 136:45–55

14. Kouzi F, Zibara K, Bourgeais J et al (2020) Disruption of gap junctions attenuates acute myeloid leukemia chemoresistance induced by bone marrow mesenchymal stromal cells. Oncogene 39:1198–1212

15. Moschoi R, Imbert V, Nebout M et al (2016) Protective mitochondrial transfer from bone marrow stromal cells to acute myeloid leukemic cells during chemotherapy. Blood 128:253–264

16. Rodriguez A-M, Nakhle J, Griessinger E et al (2018) Intercellular mitochondria trafficking highlighting the dual role of mesenchymal stem cells as both sensors and rescuers of tissue injury. Cell Cycle 17:712–721

17. Desbourdes L, Javary J, Charbonnier T et al (2017) Alteration analysis of bone marrow mesenchymal stromal cells from de novo acute myeloid leukemia patients at diagnosis. Stem Cells Dev 26:709–722

18. Chacko BK, Kramer PA, Ravi S et al (2014) The bioenergetic health index: a new concept in mitochondrial translational research. Clin Sci 127:367–373

Part II

Flow Cytometry-Based Analysis of the Bone Marrow Cellular Compartments

Flow Cytometry Analysis of Mouse Hematopoietic Stem and Multipotent Progenitor Cells

Julien M. P. Grenier, Marjorie C. Delahaye, Stéphane J. C. Mancini, and Michel Aurrand-Lions

Abstract

Flow cytometry has been widely used to detect a single event by means of multiparametric fluorescence measurements. Here we describe a method to analyze subsets of hematopoietic stem and progenitor cells isolated from long bones of mice. We further show that this method allows for comparing JAM-C protein expression between subsets of hematopoietic stem and progenitor cells.

Key words Mouse, Bone marrow, Hematopoietic stem cells, Flow cytometry

1 Introduction

Hematopoiesis takes place in the bone marrow (BM) and is the process that leads to the formation of blood components throughout life. It is hierarchically organized with hematopoietic stem cells (HSCs) at the apex. HSCs are defined by their self-renewal properties and their ability to differentiate into all blood cells. Such properties have been initially addressed in mouse models using engraftment of purified cells in irradiated recipients and follow-up of spleen colony formation (CFU-S) [1]. Characterization of the cells possessing the stem cell activity has been challenging until the development of monoclonal antibodies and flow cytometry in the eighties, which lead to a quest toward the identification of specific phenotypic markers. Nowadays technologies have improved, and single-cell technologies including multiparametric flow cytometry are routinely used to characterize and purify HSCs [2].

Stéphane J. C. Mancini and Michel Aurrand-Lions are senior coauthors.

Marion Espéli and Karl Balabanian (eds.), *Bone Marrow Environment: Methods and Protocols*, Methods in Molecular Biology, vol. 2308, https://doi.org/10.1007/978-1-0716-1425-9_6, © Springer Science+Business Media, LLC, part of Springer Nature 2021

The first characterization of mouse HSCs was reported in 1986 in a study showing that CFU-S activity is enriched in cells expressing Thy1 (CD90) and lacking lineage markers of granulocytes, monocytes, T and B lymphocytes (Lin$^-$: Gr1$^-$/Mac-1$^-$/CD4$^-$/CD8$^-$/B220$^-$)[3]. Thereafter, G. Spangrude and collaborators added Sca1 to the panel in order to demonstrate that Lin$^-$ Thy^{lo} Sca1$^+$ cells reconstitute Erythroid, Myeloid, T and B lymphocyte lineages [4]. HSCs were then found to be enriched in the Lin$^-$ Thy^{lo} Sca$^+$ c-kit$^+$ compartment lacking CD34 expression [5–7]. Another key study showed that Lin$^-$ Sca1$^+$ c-kit$^+$ (LSK) cells expressing Flt3$^+$ (Flk2 or CD135) lose their self-renewal activity and are lymphoid biased, suggesting that Flt3 expression is associated with more mature progenitors [8, 9]. These findings were consistent with the higher repopulating capacities of Flt3$^-$ HSCs compared to Flt3$^+$ HSCs which were respectively called long-term and short-term HSC [10]. In 2005, the use of the SLAM family markers CD150 and CD244 allowed to subdivide LSK cells in HSCs (CD150$^+$ CD48$^-$ CD244$^-$), MPP1 (CD34$^+$ CD150$^-$ CD48$^-$ CD135$^-$), MPP2 (CD34$^+$ CD48$^-$ CD150$^+$ CD135$^-$), MPP3 (CD34$^+$ CD48$^+$ CD150$^-$ CD135$^-$), and MPP4 (CD34$^+$ CD48$^+$ CD150$^-$ CD135$^+$) which were more recently shown to be lymphoid biased [11–13]. Finally, several studies have reported specific expression of surface markers such as EPCR (Endothelial Protein C Receptor), ESAM (Endothelial cell-selective adhesion molecule), or JAM-C (Junctional Adhesion Molecule C) on subset of Hematopoietic stem and progenitor cells (HSPC) [14–16]. More specifically, we have found that JAM-C is not only a functional marker of the most quiescent long-term HSC in human or mouse but also of leukemic stem cells in acute myeloid leukemia, the pathological counterpart of HSPCs [17–20]. We describe here a method and a gating strategy to quantify JAM-C expression by subsets of hematopoietic stem and progenitor cells.

2   Materials

2.1   Reagents

1. eBioscience™ 1× Red Blood Cell (RBC) Lysis Buffer (Ref: 00-4333-57).

2. PBS: phosphate-buffered saline 1×.

3. FACS buffer composed of PBS 1×, 0.1% heat-inactivated fetal calf serum, 0.2 mM EDTA.

4. 70% ethanol.

5. Trypan blue.

6. Sytox Green (*see* **Note 1**).

7. Biotinylated Lin antibodies: anti CD4 (clone RM4-5), CD8 (clone 53-6.7), CD3 (clone 145-2C11), CD19 (clone 6D5),

CD11c (clone N418), DX5 (clone DX5), Ter119 (clone Ter119), CD11b (clone M1/70), B220 (clone RA3-6B2), Gr-1 (clone RB6-8C5) (*see* **Note 1**).

8. BV510 streptavidin (*see* **Note 1**).

9. APC/Cy7 anti-CD117 (c-kit) (clone 2B8) (*see* **Note 1**).

10. BV421 anti-Sca-1 (clone D7) (*see* **Note 1**).

11. PE/CF594 anti-CD135 (Flk2) (clone A2F10.1) (*see* **Note 1**).

12. APC anti-CD150 (clone TC15-12F12.2) (*see* **Note 1**).

13. PE/Cy7 anti-CD48 (clone HM48-1) (*see* **Note 1**).

14. Rabbit polyclonal anti-JAM-C 501 [19] (*see* **Note 1**).

15. PE anti-rabbit IgG (H + L) (*see* **Note 1**).

2.2 Equipment

1. Micro centrifuge.

2. Falcon 15 mL tubes.

3. 1.5 mL eppendorf tubes.

4. 21G × 1 ½″ needles.

5. 1.5 mL perforated eppendorf's tube (*see* **Note 2**).

6. 5 mL polystyrene round-bottom tubes.

7. 6-well plates.

8. Scalpel blades.

9. Centrifuge.

10. Flow cytometer (*see* **Note 3**).

3 Methods

3.1 BM Cell Recovery

1. Sacrifice mice by cervical dislocation, spray with 70% ethanol thoroughly and fix with ventral side up by pinning the legs through the mouse paw pads below the ankle joint. Dissect the skin with scissor starting by an incision in the abdomen down to the legs and ankle joint. Pull back the skin to expose the legs.

2. Remove muscles from tibia and femur to expose the bone. Cut the tibia at the ankle and the femur at the junction of the hips. At that stage, cutting should be easy and must concern only ligaments. Make sure femoral heads are intact. Place the legs into 3 mL of PBS in a 6-well plate.

3. Carefully clean the bones with a scalpel in order to eliminate all flesh from the bones (*see* **Note 4**) and isolate femur from tibia by cutting knee ligaments. Cut the bones in the middle using a scalpel and put them into perforated Eppendorf tubes with the cut end facing the bottom. Place the perforated tube into a

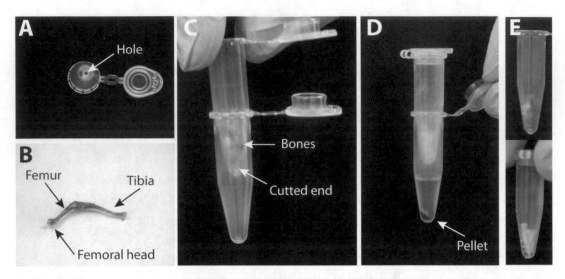

Fig. 1 BM cell extraction. (**a**) Bottom view of a perforated eppendorf tube. (**b**) Image of a posterior leg prepared for BM extraction. (**c**) Mounting of the collection tube with the perforated tube containing bones. (**d**) Appearance of the tubes shown in (**c**) after centrifugation. (**e**) Aspect of the bones before (upper image) and after centrifugation (lower image)

collection Eppendorf tube of 1.5 mL, add 300 µL of PBS in the perforated tube and centrifuge in the collection tube at $6000 \times g$ during 8 min (Fig. 1).

4. After centrifugation, trash the perforated tube containing bones which should have turned white if correctly flushed by centrifugation (Fig. 1). Retrieve cells from the collection tube and remove the supernatant as much as possible before proceeding with the lysis of the erythrocytes. To this end, resuspend cells in the remaining volume of supernatant by taping the tube and add 1 mL of $1 \times$ RBC lysis buffer. Mix by inverting (do not vortex) and incubate for 5 min at room temperature.

5. Stop the reaction by transferring the cell suspension into 9 mL of PBS already prepared in 15 mL tubes. Centrifuge at $500 \times g$ for 5 min at room temperature to pellet the cells. Discard the supernatant and resuspend cells in an appropriate volume of PBS (10 mL for two rear legs). Cells are ready to be processed.

3.2 Cell Staining

1. Count living cells using trypan blue exclusion and adjust the cellular density to at least 10×10^6 cells per mL (*see* **Note 5**).

2. Distribute 10×10^6 cells per sample into 5 mL round-bottom tubes and complete to 2 mL with FACS buffer (*see* **Note 6**). Prepare a negative control composed of a pool of unstained cells taken from each sample and complete to 2 mL with FACS buffer. Centrifuge all the tubes 3 min at $500 \times g$ at 4 °C to wash the cells. Discard the supernatant from the unstained tube and

add 200 μL of FACS buffer and place the tube at 4 °C, the negative control is ready. In the meantime discard the supernatant from other tubes and proceed to staining.

3. Add 100 μL of the mix composed of Lin antibodies cocktail and anti-JAM-C 501 prepared in FACS buffer and incubate for 30 min at 4 °C in the dark.

4. Wash cells three times with FACS buffer (centrifuge at 500 × g 3 min at 4 °C).

5. Add 100 μL of PE Anti-Rabbit IgG and incubate for 30 min at 4 °C in the dark.

6. Wash cells three times with FACS buffer and centrifuge at 500 × g for 3 min at 4 °C (*see* **Note 7**).

7. Add 100 μL of the mix composed of SYTOX Green, Streptavidin, anti-CD117, CD135, Sca-1, CD135, and CD48. Incubate for 30 min at 4 °C in the dark.

8. Wash cells three times with FACS buffer, centrifuge at 500 × g for 3 min at 4 °C and resuspend in 200 μL of FACS buffer. Keep samples in the dark until acquisition.

3.3 HSPC Characterization

1. Use a Log scale for all the parameters except for the size (Forward Scatter, FSC) and the granulosity (Side Scatter, SSC) of the cells, use a linear scale. Save the FSC-A (Area), FSC-H (Height), FSC-W (Width), and the SSC-A, SSC-H, SSC-W parameters (*see* **Note 8**). Settle the cytometer using the unstained cells. Use compensation beads before processing your samples (*see* **Note 9**).

2. Use the following gating strategy (Fig. 2).

   (a) Select the live cells on an FSC-A/SYTOX Green plot.

   (b) From the live cells, gate on a SSC-W/SSC-H dot plot to remove doublets on the granulosity parameter and then gate on an FSC-W/FSC-H dot plot to remove doublets on the size parameter.

   (c) In an FSC-A/Lin dot plot, select Lin negative cells.

   (d) From the Lin$^-$ population select the c-kit$^+$/Sca-1$^+$ corresponding to the Lin$^-$/Sca-1$^+$/c-Kit$^+$ (LSK) population.

   (e) From a CD135/CD150 dot plot on the LSK subset.

   (f) Select the CD135$^+$/CD150$^-$ cells and on a CD150/CD48 dot plot exclude the CD48$^-$ cells which correspond to the MPP4 cells.

   (g) CD135 negative cells correspond to the LT-HSC, ST-HSC, MPP2, and MPP3 (HSPC subsets).

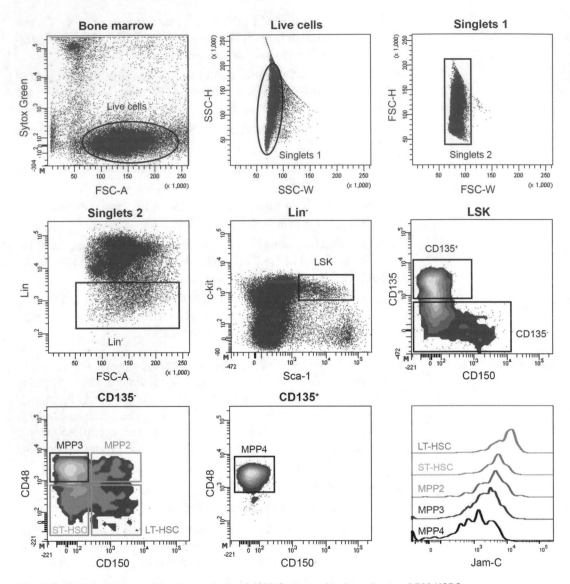

Fig. 2 Gating strategy used for the analysis of JAM-C expression by subsets of BM HSPCs

(h) Prepare a CD150/CD48 dot plot on the HSPCs subset.

(i) CD48$^-$/CD150$^+$ cells correspond to LT-HSC.

(j) CD48$^-$/CD150$^-$ cells correspond to ST-HSC.

(k) CD48$^+$/CD150$^+$ cells correspond to MPP2 cells.

(l) CD48$^+$/CD150$-$ cells correspond to MPP3 cells.

(m) On a histogram plot check JAM-C expression for long-term HSCs, short-term HSCs, and MPPs (*see* **Note 10**).

4   Notes

1. Make sure your antibodies are properly titrated before use. The concentrations given in Table 1 are those obtained in the lab following a careful titration. Be aware that the optimal antibody concentration may vary depending on the supplier, product lot, and experimental setting.

2. Perforated 1.5 mL Eppendorf's tubes are prepared using a red-hot heated needle to make a hole at the bottom of the tubes. Puncture the tube from the inside to avoid the formation of plastic asperities which would trap cells in the bottom of the upper tube.

3. Minimum requirement for the flow cytometer is four lasers, violet (405 nm), blue (488 nm), green (532 nm), and red (633 nm).

Table 1
List of antibodies used for flow cytometry and their dilutions

Antibody	Dilution	Fluorochrome	Clone
CD4	1/500	Biotin	RM4-5
CD8	1/500	Biotin	53-6.7
CD3	1/500	Biotin	145-2C11
CD19	1/500	Biotin	Clone 6D5
CD11c	1/500	Biotin	N418
DX5	1/500	Biotin	DX5
Ter119	1/500	Biotin	Ter119
CD11b	1/1000	Biotin	M1/70
B220	1/1000	Biotin	RA3-6B2
Gr1	1/2000	Biotin	RB6-8C5
Jam-C	1/500	None	501
Sca-1	1/600	BV421	D7
Strepta	1/400	BV510	
CD135	1/400	PE-CF594	A2F10.1
CD48	1/1000	PE-Cy7	HM48-1
CD150	1/500	APC	TC15-12F12.2
CD117	1/200	APC-Cy7	2B8
anti-Rabbit	1/200	PE	
SYTOX Green	1/80000	PerCP-Cy5.5	

4. Use of paper towel may help cleaning the bones properly.

5. Cell counting at that stage allows for comparing absolute number of cells in rear legs of mice between mice.

6. 1×10^6 cells is the minimal amount of cells to stain in order to reach a sufficient number of interpretable events (approximately 700 LSK, 400 MPP4, and 300 HSPCs).

7. At this stage you must perform three washes because you use a secondary antibody that can cross react with other antibodies from the mix.

8. The acquisition and the analysis can be performed by using the FSC-A and SSC-A parameters only (or FSC-H and SSC-H). However, by using all the parameters (Area, Height and Width), cellular doublets can be removed, improving the efficiency of the analysis.

9. Prepare compensation beads according to the manufacturer's recommendations. It is strongly recommended to perform the single color staining with the antibodies used in the experiment.

10. Jam-C expression decreases with differentiation [15, 18, 19].

Acknowledgments

This work was supported by the Ligue Nationale Contre le Cancer (EL2020) and Canceropole/Gefluc (2019-00110). MCD and JMPG were recipient of PhD fellowships from la Ligue Nationale Contre le Cancer (IP/SC-16060) and Fondation ARC (# 20190509010), respectively.

References

1. Till JE, Mc CE (1961) A direct measurement of the radiation sensitivity of normal mouse bone marrow cells. Radiat Res 14:213–222

2. Povinelli BJ, Rodriguez-Meira A, Mead AJ (2018) Single cell analysis of normal and leukemic hematopoiesis. Mol Asp Med 59:85–94

3. Muller-Sieburg CE, Whitlock CA, Weissman IL (1986) Isolation of two early B lymphocyte progenitors from mouse marrow: a committed pre-pre-B cell and a clonogenic Thy-1-lo hematopoietic stem cell. Cell 44:653–662

4. Spangrude GJ, Heimfeld S, Weissman IL (1988) Purification and characterization of mouse hematopoietic stem cells. Science 241:58–62

5. Morita Y, Ema H, Nakauchi H (2010) Heterogeneity and hierarchy within the most primitive hematopoietic stem cell compartment. J Exp Med 207:1173–1182

6. Morrison SJ, Weissman IL (1994) The long-term repopulating subset of hematopoietic stem cells is deterministic and isolatable by phenotype. Immunity 1:661–673

7. Osawa M, Hanada K, Hamada H et al (1996) Long-term lymphohematopoietic reconstitution by a single CD34-low/negative hematopoietic stem cell. Science 273:242–245

8. Adolfsson J, Borge OJ, Bryder D et al (2001) Upregulation of Flt3 expression within the bone marrow Lin(−)Sca1(+)c-kit(+) stem cell compartment is accompanied by loss of self-renewal capacity. Immunity 15:659–669

9. Sitnicka E, Bryder D, Theilgaard-Monch K et al (2002) Key role of flt3 ligand in regulation

of the common lymphoid progenitor but not in maintenance of the hematopoietic stem cell pool. Immunity 17:463–472

10. Christensen JL, Weissman IL (2001) Flk-2 is a marker in hematopoietic stem cell differentiation: a simple method to isolate long-term stem cells. Proc Natl Acad Sci U S A 98:14541–14546

11. Kiel MJ, Yilmaz OH, Iwashita T et al (2005) SLAM family receptors distinguish hematopoietic stem and progenitor cells and reveal endothelial niches for stem cells. Cell 121:1109–1121

12. Wilson A, Laurenti E, Oser G et al (2008) Hematopoietic stem cells reversibly switch from dormancy to self-renewal during homeostasis and repair. Cell 135:1118–1129

13. Pietras EM, Reynaud D, Kang YA et al (2015) Functionally distinct subsets of lineage-biased multipotent progenitors control blood production in Normal and regenerative conditions. Cell Stem Cell 17:35–46

14. Kent DG, Copley MR, Benz C et al (2009) Prospective isolation and molecular characterization of hematopoietic stem cells with durable self-renewal potential. Blood 113:6342–6350

15. Praetor A, McBride JM, Chiu H et al (2009) Genetic deletion of JAM-C reveals a role in myeloid progenitor generation. Blood 113:1919–1928

16. Yokota T, Oritani K, Butz S et al (2009) The endothelial antigen ESAM marks primitive hematopoietic progenitors throughout life in mice. Blood 113:2914–2923

17. Arcangeli ML, Bardin F, Frontera V et al (2014) Function of Jam-B/Jam-C interaction in homing and mobilization of human and mouse hematopoietic stem and progenitor cells. Stem Cells 32:1043–1054

18. Arcangeli ML, Frontera V, Bardin F et al (2011) JAM-B regulates maintenance of hematopoietic stem cells in the bone marrow. Blood 118:4609–4619

19. Bailly AL, Grenier JMP, Cartier-Michaud A et al (2020) GRASP55 is dispensable for Normal hematopoiesis but necessary for Myc-dependent leukemic growth. J Immunol 204:2685–2696

20. De Grandis M, Bardin F, Fauriat C et al (2017) JAM-C identifies Src family kinase-activated leukemia-initiating cells and predicts poor prognosis in acute myeloid leukemia. Cancer Res 77:6627–6640

Chapter 7

Flow Cytometry-Based Analysis of the Mouse Bone Marrow Stromal and Perivascular Compartment

Yuki Matsushita, Wanida Ono, and Noriaki Ono

Abstract

Bone marrow stromal cells (BMSCs) account for an extremely small percentage of total bone marrow cells; therefore, it is technically challenging to harvest a good quantity of BMSCs with good viability using fluorescence-activated cell sorting (FACS). Here, we describe the methods to effectively isolate BMSCs for flow cytometry analyses and subsequent FACS. Use of transgenic reporter lines facilitates FACS-based isolation of BMSCs, aiding to uncover fundamental characteristics of these diverse cell populations.

Key words Bone marrow stromal cells (BMSCs), Skeletal stem cells (SSCs), Mesenchymal stem cells (MSCs), Fluorescence-activated cell sorting (FACS), In vivo lineage-tracing experiments, Single cell RNA-seq, C-X-C motif chemokine ligand 12 (CXCL12), Transgenic reporter mouse lines

1 Introduction

Bone marrow contains a diverse array of cells, including hematopoietic cells, vascular endothelial cells, and skeletal (mesenchymal) cells [1, 2]. Among them, skeletal cells account for an extremely small percentage of total bone marrow cells [3, 4]. Skeletal cells encompass bone-making osteoblasts, bone marrow stromal cells (BMSCs) and other skeletal progenitor and precursor cells. BMSCs are mainly located in contact with blood vessels [5], a small subset of which assumingly function as skeletal stem cells [6, 7]. A large majority of BMSCs express two important genes, C-X-C motif chemokine ligand 12 (CXCL12, also known as stromal cell-derived factor 1, SDF1) [8–10] and leptin receptor (LepR), a receptor for a fat-specific hormone leptin [11]. BMSCs can be marked by their reporter strains, including *Cxcl12-GFP* [12], *Cxcl12-DsRed* [13], *Cxcl12-creER*; *R26R*^{tdTomato} [10], and *Lepr-cre*; *R26R*^{tdTomato} [11] mice; as a result, these cells are also termed as CXCL12$^+$ or CXCL12$^+$LepR$^+$ cells [14]. BMSCs can be isolated from these transgenic fluorescent reporter strains by flow cytometry and fluorescence-activated cell sorting (FACS) in a

Marion Espéli and Karl Balabanian (eds.), *Bone Marrow Environment: Methods and Protocols*, Methods in Molecular Biology, vol. 2308, https://doi.org/10.1007/978-1-0716-1425-9_7, © Springer Science+Business Media, LLC, part of Springer Nature 2021

highly reproducible manner, without involving technique-sensitive steps of antibody staining. With this approach, BMSCs can be clearly identified by strong signals emanating from products of fluorescent reporter genes, the existence of which can be further confirmed by subsequent transcriptomic analyses. Moreover, if a cell type-specific tamoxifen-inducible *creER* line is utilized, for example in *Cxcl12-creER*; *R26R*^tdTomato mice, cell fates of a specific subset of BMSCs can be interrogated through in vivo lineage-tracing experiments.

Here as an example, we describe the methods to isolate BMSCs effectively from the femurs of *Cxcl12*^GFP/+; *Cxcl12-creER*; *R26R*^tdTomato mice, which were treated with tamoxifen at postnatal day (P) 21 and dissected at P28 (Fig. 1). We further describe the methods to harvest BMSCs using FACS and subsequently apply these cells to a droplet-based single-cell RNA-sequencing analysis. We also describe the method to culture lineage-marked BMSCs ex vivo [10, 15] to circumvent the common problem associated with FACS-isolated BMSCs that do not survive well in cultured conditions [16].

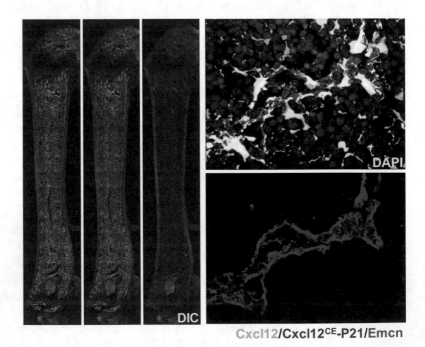

Fig. 1 Histological images of *Cxcl12*^GFP/+; *Cxcl12-creER*; *R26R*^tdTomato femurs. Cxcl12-GFP+ and/or Cxcl12-creER;tdTomato+ cells, which are located in contact with Endomucin (Emcn)+ blood vessels

2 Materials

2.1 Cell Preparation

1. Transgenic reporter mice (here, *Cxcl12*$^{GFP/+}$; *Cxcl12-creER*; *R26R*^{tdTomato} mice, as an example).

2. Tamoxifen.

3. Sunflower seed oil.

4. Absolute ethanol.

5. Isoflurane.

6. Ice.

7. Scissors (sharp and dull).

8. Forceps (1 × 2 teeth).

9. Razor (single edge blade).

10. Paper wipe.

11. 6-well plate

12. Liberase™ TM (collagenase, Sigma/Roche).

13. Molecular biology–grade water.

14. Pronase (neutral protease, Sigma/Roche).

15. Ca^{2+}, Mg^{2+}-free Hank's balanced salt solution (HBSS).

16. HBSS medium: HBSS supplemented with 1 mM ethylenediaminetetraacetic acid (EDTA) solution and 2% fetal bovine serum.

17. Mortar and pestle.

18. Luer-lock syringe 1 ml.

19. 18G Needle.

20. 70 μm cell strainer.

21. 50 ml conical tube.

22. DPBS–1% BSA: Dulbecco's phosphate-buffered saline (DPBS) supplemented with 1% bovine serum albumin (BSA).

23. DMEM medium: Low-glucose DMEM with GlutaMAX™ supplement (DMEM).
    supplemented with 10% mesenchymal stem cell–qualified FBS and 1% penicillin–streptomycin.

24. Flow cytometry staining buffer (eBioscience).

25. Allophycocyanin (APC)-conjugated CD45 antibody (30F-11).

26. 5 ml polystyrene round-bottom tube with cell-strainer cap.

27. Shaking incubator.

28. Centrifuge.

2.2 Cell Sorting and Single-Cell RNA-Seq Analysis

1. BD FACS Aria III (Ex.407/488/561/640 nm).
2. RNase zap.
3. Dulbecco's phosphate-buffered saline (DPBS).
4. Bovine serum albumin (BSA).
5. Protein LoBind 1.5 ml tube.
6. Trypan blue solution.
7. Centrifuge.
8. Automated Cell Counter.
9. Chromium Single Cell 3′ v3 microfluidics chip (10× Genomics Inc).
10. Sequencer (for example NovaSeq 6000 from Illumina).
11. R (ver 3.5) and R studio (ver 1.1).
12. Seurat R package [17].

2.3 Colony-Forming Assay, Subsequent Subcloning and In Vitro Trilineage Differentiation

1. Counting chambers (Neubauer-improved).
2. 6-well plate.
3. DMEM medium: DMEM supplemented with 10% mesenchymal stem cell-qualified FBS and 1% penicillin–streptomycin (mix #3, #4, and #5).
4. 70% ethanol.
5. 50% ethanol.
6. Methylene blue.
7. Cloning cylinders.
8. Dow high vacuum grease.
9. 0.5% trypsin–EDTA.
10. Non-tissue culture-coated V-bottom 96-well plate.
11. StemPro™ Chondrogenesis differentiation kit (Gibco).
12. Alcian Blue Staining Solution.
13. 48-well plate.
14. αMEM medium: αMEM with GlutaMAX™ supplemented with 10% FBS and 1% penicillin–streptomycin.
15. Osteogenic medium: αMEM medium with 1 μM dexamethasone, 10 mM β-glycerophosphate, and 50 μg/ml ascorbic acid.
16. 4% paraformaldehyde solution.
17. Alizarin Red S.
18. IBMX (3-isobutyl-1-methylxanthine).
19. Adipogenic medium: αMEM medium with 1 μM dexamethasone, 0.5 mM and 1 μg/ml insulin.
20. LipidTOX Green (Invitrogen).

21. Oil Red O.

22. Vortex.

23. Centrifuge.

24. Automated inverted fluorescence microscope with a structured illumination system (for example the Zeiss Axio Observer Z1 with ApoTome.2 system).

25. Zeiss Zen 2 software (blue edition).

3 Methods

3.1 Cell Preparation

1. Label CXCL12$^+$ BMSCs with green and red fluorescent proteins, by administering tamoxifen to *Cxcl12*$^{GFP/+}$; *Cxcl12-creER*; *R26R*^{tdTomato} mice: typically, a single dose of 1 mg tamoxifen is injected intraperitoneally (100 μl of 10 mg/ml tamoxifen-oil mix) for mice at 3–6 weeks of age (*see* **Note 1**).

2. Sacrifice the mice after a short-chase (typically 2–7 days after tamoxifen injection) using carbon dioxide overdose or isoflurane overdose in a drop jar followed by induction of bilateral pneumothorax.

3. Make an incision in the skin of the leg by scissors and peel it off. Dissect the femur by excising the ligament at the femur head and at the knee joint using sharp scissors. Remove soft tissues carefully from dissected femurs using scissors and paper wipes.

4. Remove distal epiphyseal growth plates using dull scissors and cutting off proximal ends (*see* **Note 2**).

5. Open up the marrow cavity and expose bone marrow. To achieve this, cut the cortex of the femurs longitudinally, roughly by a razor for 1–2 times, into several pieces. Do not cut the femur into small pieces.

6. Transfer the roughly cut bone pieces into a 6-well plate containing 2 ml HBSS.

7. Digest the bone pieces with enzymes, by adding 2 Wunsch units of Liberase™ TM with or without 1 mg of Pronase to 2 ml HBSS in the 6-well plate (*see* **Notes 3** and **4**).

8. Incubate the solution at 37 °C, 300 rpm for 60 min on a shaking incubator (*see* **Note 5**).

9. Aspirate and mix the supernatant containing dissociated cells using an 18-gauge needle with a 1 ml Luer-Lok syringe, and filter it through a 70 μm cell strainer into a 50 ml tube on ice.

10. Add 2 ml fresh HBSS to the 6-well plate containing digested bone pieces. Aspirate and mix the supernatant rigorously using the needle with the syringe used above, and filter it through the 70 μm cell strainer into the 50 ml tube on ice.

11. Transfer the digested bone pieces to a mortar containing 2 ml fresh HBSS medium, then mechanically triturate the bone pieces using the 18-gauge needle with the 1 ml Luer-Lok syringe used above, as well as using a pestle, to release any remaining cells on bone pieces. Filter dissociated cells through the 70 μm cell strainer into the 50 ml tube on ice to prepare single cell suspension. Repeat this step for five times and collect the supernatant into the same tube.

12. Centrifuge, 280 × *g*, at 4 °C for 10 min, then discard supernatant by decanting (*see* **Note 6**).

13. Resuspend the pellet in 2 ml ice-cold DPBS–1% BSA (for single-cell RNA-seq) or 10 ml DMEM medium (for cell culture, skip to Subheading 3.3 below).

14. Filter cell suspension into a 5 ml polystyrene round-bottom tube with a cell-strainer cap, then keep it on ice until loading onto a cell sorter.

3.2 Cell Sorting and Single-Cell RNA-Seq Analysis of FACS-Isolated Cells

1. Set up a four-laser BD FACS Aria III (Ex.407/488/561/640 nm) high-speed cell sorter with a 100 μm nozzle (*see* **Note 7**).

2. Rinse the nozzle with RNase zap then backflush.

3. Run no-fluorescence control, and GFP⁺ or tdTomato⁺ single-color controls to set up the gate and the compensation to sort the desired target cells. Single-color samples are separately collected from control mice carrying only single colors (*see* **Note 8**) (Fig. 2).

4. Load the sample onto a cell sorter and collect the target cells (sorting speed should be less than 10,000 events/s).

5. Sort GFP⁺ cells and/or tdTomato⁺ cells directly into 600 μl ice-cold DPBS–1% BSA in a Protein LoBind 1.5 ml tube (*see* **Note 9**).

6. Pellet cells by centrifugation (300 × *g*, at 4 °C for 10 min) and resuspend the pellet in 10–30 μl DPBS–1% BSA to achieve a concentration of 1000 cells/μl (*see* **Note 10**).

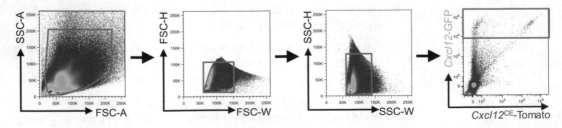

Fig. 2 Gating strategy for cell sorting of Cxcl12-GFP⁺Cxcl12-creER⁺ cells isolated from *Cxcl12*^GFP/+; *Cxcl12-creER; R26R*^tdTomato femurs. (Figures modified and adapted with permission from [10])

7. Mix 5 µl sample (or cell count control) and 5 µl trypan blue solution for cell counting.

8. Count total and live cell numbers by automated Cell Counter.

9. Load cells onto the Chromium Single Cell 3′ v3 microfluidics according to the manufacturer's protocol.

10. Sequence cDNA libraries by Illumina NovaSeq 6000.

11. Preprocess the sequencing data using the 10× Genomics software Cell Ranger.

12. Generate and use a custom genome fasta and index file by including the sequences of *eGFP* and *tdTomato-WPRE* to the mouse genome (mm10) for alignment purposes.

13. Perform further downstream analyses using the Seurat R package.

14. Filter out cells with less than 1000 genes per cell and with more than 15% mitochondrial read content.

15. Perform downstream analyses, including normalization, identification of highly variable genes across the single cells, scaling based on number of UMI, dimensionality reduction (PCA, CCA, and UMAP), unsupervised clustering, and the discovery of differentially expressed cell-type specific markers (Fig. 3).

16. Perform differential gene expression analyses to identify cell-type specific genes.

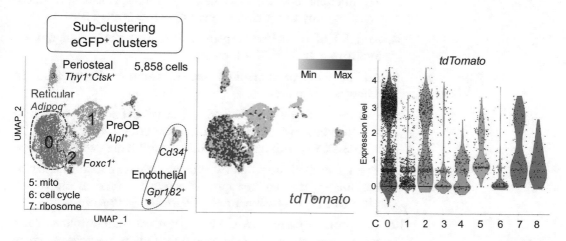

Fig. 3 Single cell RNA-seq analysis of Cxcl12-GFP⁺ (including Cxcl12-creER;tdTomato⁺) BMSCs harvested from *Cxcl12* GFP/+; *Cxcl12-creER*; *R26R*tdTomato femur bone marrow at P28 (tamoxifen injected at P21). Left panel: UMAP-based visualization of major classes of FACS-sorted cells, center panel: feature plots of *tdTomato* expression, right center panels: violin plots, *tdTomato* expression in each cluster. (Figures modified and adapted with permission from [10])

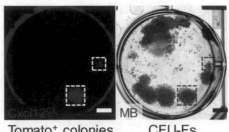

Tomato⁺ colonies   CFU-Fs

Fig. 4 tdTomato⁺ cells and total CFU-Fs

3.3 Colony-Forming Assay, Subcloning, and In Vitro Trilineage Differentiation

1. Voltex DMEM-suspended cells.

2. Count cells manually using a hemocytometer.

3. Plate cells into a 6-well plate at a density of 10^5 cells/cm^2 with DMEM medium for 10–14 days (*see* **Note 11**).

4. Locate tdTomato⁺ colonies in a 6-well plate: capture tdTomato epifluorescence from live cultured cells by tile-scanning the entire 6-well plates with an inverted fluorescence microscope equipped with a fully automated stage (Fig. 4). For subcloning, skip to **step 7**.

5. Aspirate the supernatant, then fix cells by adding 70% ethanol, for 5 min.

6. Discard the ethanol, then add 2% methylene blue–50% ethanol (methylene blue staining) (Fig. 4).

7. To subclone individual tdTomato⁺ colonies, isolate each tdTomato⁺ colony using cloning cylinders (*see* **Note 12**).

8. Add 50 μl of 0.25% trypsin-EDTA into the cylinder, then incubate it at 37 °C for 5 min.

9. Add 50 μl of a basal medium, and transfer the cell suspension to a new 6-well plate.

10. Culture a single cell-derived clone of tdTomato⁺ cells in a 6-well plate with a basal medium described above at 37 °C with 5% CO_2 with exchanges into fresh media every 3–4 days.

11. Passage each clone when the culture reaches subconfluency to confluency, transferring approximately 1/10 of an old 6-well plate to a new 6-well plate with a fresh basal medium.

12. Continue to passage clones to analyze self-renewability, or use passage 4–7 clones for trilineage differentiation assays (*see* **steps 13, 19**, or **22**).

13. To induce chondrocyte differentiation, transfer cells (approximately 1/6 of a confluent 6-well plate) into a non-tissue culture-coated V-bottom 96-well plate.

14. Centrifuge at $150 \times g$ for 5 min at room temperature to pellet the cells, then aspirate the supernatant carefully.

15. Add StemPro Chondrogenesis medium, and centrifuge the plate at $150 \times g$ for 5 min at room temperature to pellet the cells.

16. Culture the pellet in the differentiation medium with exchanges into fresh media every 3–4 days for up to 3 weeks, each time with centrifugation at $150 \times g$ for 5 min at room temperature to repellet the cells.

17. Fix cell pellets with 70% ethanol for 5 min.

18. Stain for Alcian-Blue Staining Solution for 30 min.

19. To induce osteoblast differentiation, plate cells (approximately 1/6 of a confluent 6-well plate) on a 48-well plate and cultured with αMEM medium until confluency.

20. Change to the osteogenic medium with exchanges into fresh media every 3–4 days for up to 3 weeks.

21. Fix cells with 4% paraformaldehyde for 5 min and stain for 2% Alizarin Red S for 30 min.

22. To induce adipocyte differentiation, plate cells (approximately 1/6 of a confluent 6-well plate) on a 48-well plate and cultured with αMEM medium until confluency.

23. Change to the adipogenic medium with exchanges into fresh media every 3–4 days for up to 2 weeks.

24. Stain cells with LipidTOX Green (1:200 in basal medium) for 3 h at 37 °C, or fix cells with 4% paraformaldehyde for 5 min and stain for Oil Red O for 30 min.

4 Notes

1. To make 10 mg/ml tamoxifen solution, dissolve 100 mg tamoxifen in 2.5 ml absolute ethanol in a 50 ml conical tube. Voltex persistently until crystals become completely dissolved. Add 10 ml sunflower seed oil and voltex persistently again. Open the cap and incubate at 65 °C overnight in a chemical fume hood until ethanol completely evaporates. Close the cap and keep the tamoxifen-oil mix at room temperature until use. A dose of 25–175 mg/kg body weight is typically approved by an IACUC protocol. Average body weights of male C57BL/6J are 10.6 ± 1.9 g at 3-week, 16.5 ± 2.6 g at 4-week, 21.9 ± 1.8 g at 6-week and 25.0 ± 1.4 g at 8-week, according to The Jackson Laboratory. Tamoxifen should be administered to mice under a certified biosafety cabinet, and the mice should be housed in a special containment space.

2. Epiphyses of femurs can be easily popped out using dull scissors.

3. Reconstitute and aliquot Liberase solution according to the manufacturer's protocol. Add 10 ml molecular biology-grade water to the amber bottle of Liberase™ TM (50 mg), and swirl the bottle well. Keep it on ice for 30 min, while inverting the bottle for every 5 min (do not use vortex). Aliquot 500 μl enzyme solution in a 1.5 ml tube, and keep it at -30 °C until use.

4. Adding Pronase in the digestion cocktail increases the yield of BMSCs, especially those located in the central marrow space, without compromising RNA integrity number. However, Pronase digestion degrades protease-sensitive antigens such as CD31 (encoded by *Pecam1*).

5. Length of digestion needs to be optimized for the age of mice. A 60 min digestion is sufficient for mice at 3–8 weeks of age.

6. For standard flow cytometry analysis, cells should be resuspended in a 500 μl flow cytometry staining buffer containing appropriately diluted antibodies (such as 1:500 CD45-APC), then incubate the cells at 4 °C for 30 min. Add 2 ml flow cytometry staining buffer to wash. Centrifuge, $280 \times g$, at 4 °C for 10 min, then discard supernatant by decanting. Resuspend the cells in flow cytometry staining buffer (500–1000 μl), then run on flow cytometer.

7. RNA quality of sorted cells depends on the size of the nozzle. Wider nozzles, such as 100 μm or 130 μm nozzles, give much better results than narrower nozzles, such as 70 or 85 μm nozzles. The sheath pressure is 20 psi using a 100 μm nozzle, which is close to the atmospheric pressure (15 psi). A 130 μm nozzle (10 psi) is ideal to prevent damage to the cells; however, the sorting speed becomes substantially slow. It may be difficult to harvest a sufficient number of cells within a limited amount of sorting time. A 70 μm nozzle (70 psi) or an 85 μm nozzle (45 psi) may inflict damage to the cells and compromise the RNA quality.

8. Exclude FSC^{high}/SSC^{high} events as they contain duplets or multiplets of cells. We often experience that Cxcl12-GFP^{+} BMSCs are physically coupled with hematopoietic cells in a way indistinguishable from purely single cells.

9. Sort more than 50,000 cells ideally. If less than 50,000 cells are harvested, a separate control tube should be set up solely for the purpose of cell counting, to estimate the cell number in an experimental tube. For the cell counting control tube, sort at least 10,000 GFP-low or tdTomato-low cells and dilute it at the same ratio as the experimental tube.

10. Generally, 30% of the cells counted on FACS are lost due to centrifugation, between the sorting step and the actual cell

counting step. As the ideal cell concentration for a $10\times$ microfluidics device is 1000 cells/µl, sorted cells should be diluted to approximately 1400 cells/µl.

11. Cell cultures are maintained at 37 °C in a 5% CO_2 incubator.

12. Print tile-scanned tdTomato$^+$ colony images as the actual size of a 6-well plate. For this purpose, make 1.42×1.42-in. circles in 10×7.5-in. PowerPoint slide, and print it to a letter size paper. Then insert the printed images under the 6-well plate. After aspirating the medium, attach a cloning cylinder using grease to the position of the plate corresponding to the position of tdTomato$^+$ colonies.

Acknowledgments

This research was supported by grants from National Institute of Health (R01DE026666 to N.O., R01DE029181 to W.O.), and the Japan Society for the Promotion of Science KAKENHI Grant JP19K19236 to Y.M. We thank for K. Mizuhashi for inventing innovative approaches described in this chapter.

Declaration of interests: The authors declare no competing interests.

References

1. Mendelson A, Frenette PS (2014) Hematopoietic stem cell niche maintenance during homeostasis and regeneration. Nat Med 20 (8):833–846

2. Morrison SJ, Scadden DT (2014) The bone marrow niche for haematopoietic stem cells. Nature 505(7483):327–334

3. Pittenger MF, Mackay AM, Beck SC et al (1999) Multilineage potential of adult human mesenchymal stem cells. Science 284 (5411):143–147

4. Zhu H, Guo ZK, Jiang XX et al (2010) A protocol for isolation and culture of mesenchymal stem cells from mouse compact bone. Nat Protoc 5(3):550–560

5. Sacchetti B, Funari A, Michienzi S et al (2007) Self-renewing osteoprogenitors in bone marrow sinusoids can organize a hematopoietic microenvironment. Cell 131(2):324–336

6. Bianco P, Cao X, Frenette PS et al (2013) The meaning, the sense and the significance: translating the science of mesenchymal stem cells into medicine. Nat Med 19(1):35–42

7. Matsushita Y, Ono W, Ono N (2020) Skeletal stem cells for bone development and repair: diversity matters. Curr Osteoporos Rep 18 (3):189–198

8. Greenbaum A, Hsu YM, Day RB et al (2013) CXCL12 in early mesenchymal progenitors is required for haematopoietic stem-cell maintenance. Nature 495(7440):227–230

9. Asada N, Kunisaki Y, Pierce H et al (2017) Differential cytokine contributions of perivascular haematopoietic stem cell niches. Nat Cell Biol 19(3):214–223

10. Matsushita Y, Nagata M, Kozloff KM et al (2020) A Wnt-mediated transformation of the bone marrow stromal cell identity orchestrates skeletal regeneration. Nat Commun 11(1):332

11. Zhou BO, Yue R, Murphy MM et al (2014) Leptin-receptor-expressing mesenchymal stromal cells represent the main source of bone formed by adult bone marrow. Cell Stem Cell 15(2):154–168

12. Ara T, Tokoyoda K, Sugiyama T et al (2003) Long-term hematopoietic stem cells require stromal cell-derived factor-1 for colonizing bone marrow during ontogeny. Immunity 19 (2):257–267

13. Ding L, Morrison SJ (2013) Haematopoietic stem cells and early lymphoid progenitors occupy distinct bone marrow niches. Nature 495(7440):231–235

14. Seike M, Omatsu Y, Watanabe H et al (2018) Stem cell niche-specific Ebf3 maintains the bone marrow cavity. Genes Dev 32 (5–6):359–372

15. Mizuhashi K, Ono W, Matsushita Y et al (2018) Resting zone of the growth plate houses a unique class of skeletal stem cells. Nature 563(7730):254–258

16. Omatsu Y, Sugiyama T, Kohara H et al (2010) The essential functions of adipo-osteogenic progenitors as the hematopoietic stem and progenitor cell niche. Immunity 33(3):387–399

17. Butler A, Hoffman P, Smibert P et al (2018) Integrating single-cell transcriptomic data across different conditions, technologies, and species. Nat Biotechnol 36(5):411–420

Chapter 8

Immunophenotyping of the Medullary B Cell Compartment In Mouse Models

Amélie Bonaud, Karl Balabanian, and Marion Espéli

Abstract

B cell development is a stepwise process encompassing several B cell precursor stages that can be phenotypically distinguished, and that is taking place in the bone marrow in adults. Interestingly, within the bone marrow B cell precursors coexist with the most differentiated actors of this lineage, the plasma cells. In this chapter, we describe a method to recover cells from the bone marrow and a flow cytometric approach to identify each subpopulation of the B cell lineage that resides within the bone marrow compartment. This protocol focuses on membrane markers to discriminate all the B cell subpopulations in order to preserve cell integrity during experimentation and for further analyses.

Key words B cell lineage, Prepro-B cell, Pro-B cell, Pre-B cell, Immature B cell, Mature B cell, Plasma cell, Bone marrow, Flow cytometry

1 Introduction

B cells are essential actors of the humoral immune response especially through the production of antibodies by plasma cells, the more mature stage of the B cell differentiation. In adult, this lineage is generated in the bone marrow and derives from hematopoietic stem cells (HSC), at the origin of all mature blood cells. The lymphoid lineage arises from the common lymphoid progenitor (CLP) and the expression of the master regulator gene *Pax5* is key for engagement toward the B cell lineage at the Prepro-B cell stage [1]. B cells are characterized by the expression of a key antigen-specific membrane receptor, the B cell receptor (BCR) composed of two immunoglobulin (Ig) heavy chains and two Ig light chains, associated with the coreceptor Igα/Igβ. The different steps leading to the surface expression of a complete BCR define the different subpopulations of B cell precursors. The genes encoding Ig heavy chains are rearranged first at the Pro-B cell stage. The rearranged heavy chains are tested at the large Pre-B cell stage via the surface expression of a pre-BCR composed of two Ig heavy

Marion Espéli and Karl Balabanian (eds.), *Bone Marrow Environment: Methods and Protocols*, Methods in Molecular Biology, vol. 2308, https://doi.org/10.1007/978-1-0716-1425-9_8, © Springer Science+Business Media, LLC, part of Springer Nature 2021

chains and two pseudo-light chains. After completion of this check-point, the rearrangement of the light chains happens during the small Pre-B cell stage and a complete BCR is finally expressed by immature B cells [2]. Immature B cells expressing a functional BCR are able to egress to the periphery and to reach secondary lymphoid organs where they will terminate their maturation. Mature B cells will then recirculate between lymphoid organs until they encounter the antigen(s) they are specific for or until their death. After antigen exposure, activated B cells can differentiate into one of the two last steps of this lineage, memory B cells and plasma cells [3]. In mice, memory B cells are still difficult to characterize phenotypically, and the use of specific mouse models allowing tracking with fluorescent proteins is recommended [4]. Concerning plasma cells, for many years this cell population has being restrained to two dichotomous states based on the proliferative status: short or long-lived cells. Short lived plasma cells proliferate and persist only for 3–5 days while long lived plasma cells, also referred to as memory plasma cells, are quiescent and are able to survive for many years, potentially during all the lifespan of an individual, into dedicated niches within the bone marrow [5]. Over the last few years, however, new subpopulations of plasma cells have been characterized depending on their maturation stage and the loss of expression of classical B cell markers [6].

The main method to study B cell subpopulations (Prepro-B, Pro-B, Large Pre-B, Small Pre-B, Immature B cell, Recirculating B cell, and Plasma cell) is flow cytometry, through the staining of a combination of markers well described, approved by the scientific community and covering all the differentiation steps [7] (Fig. 1). Historically, the first flow cytometry staining used for identifying medullary B cell subpopulations was designed according to the expression of three markers, B220, IgM, and Tdt (terminal deox-ynucleotidyl transferase) [8–11]. In the 1980s, the marker CD19, considered now as essential for the identification of B cells, was characterized for the first time [12, 13]. CD93 is expressed by all B cell progenitors like CD24. The latter one is expressed gradually during all the stages of B cell development [14, 15]. Distinct markers were subsequently described and allow a clear segregation of each subpopulations of B cell precursors, such as BP1, expressed during the Pro-B cell and Pre-B cell stages [16]. In the same way, CD25, a subunit of the IL2 receptor, is expressed only during Pre-B cell stages [17]. To better discriminate B cell progenitor subpopulations, CD43 was demonstrated to have a relatively stable expression until Large Pre-B stage, allowing notably a distinction between Large and Small Pre-B which do not express this marker [18]. Finally, c-kit also named CD117 was demonstrated as a marker expressed from CLP and earlier stages until the Pro-B stage [19–21]. Regarding plasma cells, they are characterized in mice by their expression of CD138 and TACI (Transmembrane Activator and CAML Interactor) [22, 23]. Some plasma cells are

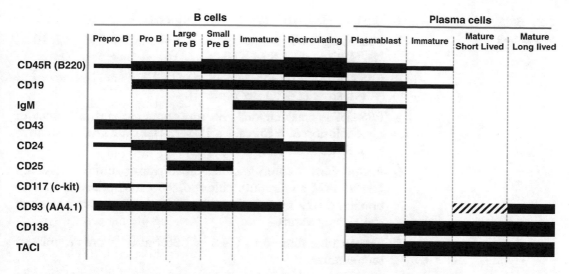

Fig. 1 Membrane expression of B cells markers among B cell lineage. Table showing expression of membrane markers used to identify each subpopulation of medullary B cells and plasma cells. Another classification in "Fraction" is also described in literature Prepro-B correspond to Fraction A, Pro-B to Fraction B, Large Pre-B to Fraction C, Small Pre-B to Fraction D, Immature B cells to Fraction E

reported to re-express CD93 that might be required for their maintenance in the bone marrow [24]. Recently, three distinct subgroups of plasma cells were identified in terms of their residual expression of the B cell markers CD19 and B220, named respectively according to the maturation stage "plasmablasts," "early plasma cells," and "mature resting plasma cells," allowing a segregation on their maturation stage and not only on the dichotomy short versus long lived [6].

In this chapter we provide a method to recover bone marrow cells and to identify by flow cytometry each subpopulation of the B cell lineage from the earlier precursors to the more mature stages.

2   Materials

2.1   Material and Tools

1. Sharp scissors.
2. Forceps.
3. Scalpel.
4. 6- or 12-well plate.
5. 0.5 mL and 2 mL tubes.
6. 18 g needle.
7. 15 mL and 50 mL conical tubes.
8. 70 μm nylon cell strainers.
9. Counting chamber.
10. 5 mL FACS tubes (*see* **Note 1**).

2.2 Buffers and Media

1. Ethanol 70%, store at ambient temperature.

2. Phosphate Buffer Saline (PBS): 137 M NaCl, 2.7 M KCl, 10 M Na_2HPO_4, and 1.8 M KH_2PO_4, pH 7.4, store at 4 °C.

3. PBS–2%FBS: PBS supplemented with 2% of fetal bovine serum (FBS), store at 4 °C (*see* **Note 2**).

4. PBS–1% Formaldehyde: PBS supplemented with a formaldehyde solution 37% to reach a 1% formaldehyde final concentration. Store at ambient temperature.

5. Ammonium chloride–potassium lysis buffer (ACK): 154.95 mM ammonium chloride, 9.99 mM potassium bicarbonate, 0.099 mM EDTA diluted in distilled water, store at room temperature.

6. Trypan blue diluted to 0.4% in PBS buffer. Store at ambient temperature.

7. Antibodies (*see* Table 1 and **Note 3**).

2.3 Equipment

1. Centrifuges for 2 mL and 15 mL/50 mL tubes.

2. Flow cytometer apparatus equipped with a minimum of three lasers, compatible with the acquisition of 11 different fluorochromes.

Table 1
Antibodies. Antibodies with validated clones used by experts of the field are indicated and recommended. Fluorochrome combinations are specified only for information and can be changed as user preference. Fc block is highly recommended to prevent aspecific binding on Fc receptors. We propose in the text (Subheading 3.4) the fixable viability dye "Live/Dead eFluor 506," but other dyes are possible

Antibody	Clone	Fluorochrome
B220 (CD45R)	RA3-6B2	Pe-Cy7
CD19	1D3	PeCF594
IgM	II/41	APC-efluorCy7
CD93	AA4.1	BV650
CD43	S7	BV786
CD24	M1/69	PE
CD25	PC61	BV421
CD117 (c-kit)	2B8	FITC
CD138	281-2	BB700
TACI	8F10-3	APC
CD16/32 (fc block)	2.4G2	Uncoupled
Viability stain		

3. FACS data must be analyzed using compatible software such as FlowJo (Treestar, Inc.) or FACSDiva (BD Bioscience) as example.

3 Methods

3.1 Mouse Experimentation

1. Kill mice by cervical dislocation or CO_2 inhalation, in accordance with local ethical guidelines and protocols.

2. Wet the mouse with EtOH 70% to clean the skin and avoid mouse hair contamination.

3. Perform a buttonhole opening on the side of each mouse. With forceps hold the skin firmly and place sharp scissors between skin and muscle to cut the skin along the leg. Finish removing the skin from the leg by hand to expose the leg. Extract the tibias, femurs and hips using a forceps and a scalpel or a pair of sharp scissors. Remove all muscles with a scalpel or with the help of absorbent paper. As much flesh as possible needs to be removed.

4. Place the bones in sufficient amount of PBS to submerge them (*see* **Note 4**).

3.2 Cell Preparation

1. When the bones are clean of all muscles, carefully remove the cartilage at one extremity of the femur and the tibia with a scalpel (*see* **Note 5**).

2. Place one hip, one femur and one tibia into a 0.5 mL tube previously pierced at its bottom with an 18 g needle, with the open extremity facing down (*see* **Note 6**).

3. Place the 0.5 mL tube containing the three bones into a 2 mL tube filled with 200 µL of PBS–2% FBS (Fig. 2).

4. Centrifuge at $15,000 \times g$ for 1 min at room temperature. All marrow is flushed down into the 2 mL tube. Empty bones are still in the 0.5 mL tube (*see* **Notes 7** and **8**).

5. Add directly 1 mL of ACK for red blood cell lysis and resuspend immediately the cells by gentle pipetting.

6. Incubate for 2 min at 4 °C (*see* **Note 9**).

7. Transfer the cell suspension into 15 mL tubes and add 10 mL of PBS–2% FBS to dilute the ACK and stop the red blood cell lysis.

8. Centrifuge at $300 \times g$ for 5 min at 4 °C.

9. Discard supernatant.

10. Resuspend the cells in 1 mL of PBS–2% FBS.

11. Filter the cell suspension trough a 70 µm cell strainer, already placed on a 50 mL tube, to remove debris and aggregates.

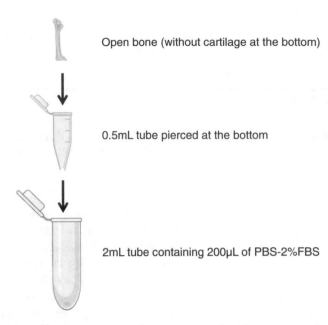

Open bone (without cartilage at the bottom)

0.5mL tube pierced at the bottom

2mL tube containing 200µL of PBS-2%FBS

Fig. 2 Schematic representation of the process to flush bone marrow from bone using centrifugation

12. Wash the cell strainer with 1 mL of PBS–2% FBS to collect the maximum of cells.

13. Count the cells with trypan blue to calculate the number of live and dead cells (*see* **Note 10**).

3.3 Flow Cytometry Staining

1. Perform staining on 1×10^6 cells. Adjust the cell concentration to have 1×10^6 cells in 50 µL of PBS–2% FBS.

2. Dispatch 50 µL of the cell suspension in an FACS tube.

3. Add 50 µL of fixable viability dye diluted as recommended in PBS–2% FBS.

4. Incubate 15 min at 4 °C.

5. Wash the cells by adding 1 mL of PBS–2% FBS to each tubes and centrifuge at $300 \times g$ for 5 min at 4 °C.

6. Discard supernatant.

7. Add 20 µL of Fc-block diluted at 0.5 ng/mL in PBS–2% FBS.

8. Incubate 10 min at 4 °C.

9. If not done ahead, prepare during this incubation the antibody mix containing all the antibodies indicated in the material part (*see* Table 1), in PBS–2% FBS. Prepare 50 µL of mix per sample (*see* **Notes 11–13**).

10. Without washing the cells, add 50 µL of the antibody mix and mix well by gently vortexing the tubes.

11. Incubate the cells for 30 min at 4 °C in the dark.

12. Wash the cells by adding 1 mL of PBS–2% FBS to each tube and centrifuge at 300 × g for 5 min at 4 °C.

13. Discard the supernatant.

14. Resuspend the cells in 500 µL of PBS–2% FBS if the tubes are acquired on the flow cytometer within 2 h. If the acquisition is planned more than 3 h later, fix the cells to conserve them by resuspending them in 500 µL of PBS–1% formaldehyde after the last wash and store at 4 °C protected from light.

15. Acquire the cells on a flow cytometer apparatus according to the manufacturer's instructions (*see* **Note 14**).

3.4 Flow Cytometry Analysis

Data analysis can be performed with several software programs. Figure 3 shows gating strategy, in detail below, step-by-step using the FlowJo software.

1. Gate on the lymphocyte population on the basis of size/scatter, to exclude dead cells, debris, and monocytes/macrophages and neutrophils (Fig. 3a).

2. Gate out cell doublets (Fig. 3b).

3. Gate live cells by excluding dead cells (positive for the staining with viability dye) (Fig. 3c). From this stage on, our analysis is separated in two parts, B cell subpopulations on the left (Fig. 3d–h) and plasma cell subpopulations (Fig. 3d′–f′) on the right.

4. For B cell subpopulations (Fig. 3d–h), gate B220 positive cells to discriminate all B cells from the others (Fig. 3d). On total B cells make a dot plot between CD19 and c-kit, to isolate Prepro-B cells that are c-kit$^+$ CD19$^-$, and continue the gating strategy on total B cells (Fig. 3e). Using B220 and IgM, immature B cells (B220$^+$ IgM$^+$), recirculating mature B cells (B220^{++} IgM$^+$) and precursor B cells (B220$^+$ IgM$^-$) can be discriminated (Fig. 3f). On precursor B cells, a dot plot with CD24 and CD25 allows to distinguish Pro-B cells (CD24$^+$ CD25$^-$) from Pre-B cells (CD24$^{+/high}$ CD25$^+$) (Fig. 3g). Finally, CD43 expression is used to discriminate large Pre-B cells (CD43$^-$) from small Pre-B cells (CD43$^+$) (Fig. 3h).

5. For the analysis of plasma cell subpopulations (Fig. 3d′–f′), live cells gated on Fig. 3c are analyzed based on CD138 and TACI expression. Double positive cells represent total plasma cells (Fig. 3d′). This population can be subdivided in 3 populations using B220 and CD19 staining and corresponding to plasmablasts (B220$^+$ CD19$^+$), early plasma cells (B220$^{low/-}$ CD19$^+$) and mature resting plasma cells (B220$^-$ CD19$^-$) respectively (Fig. 3e′).

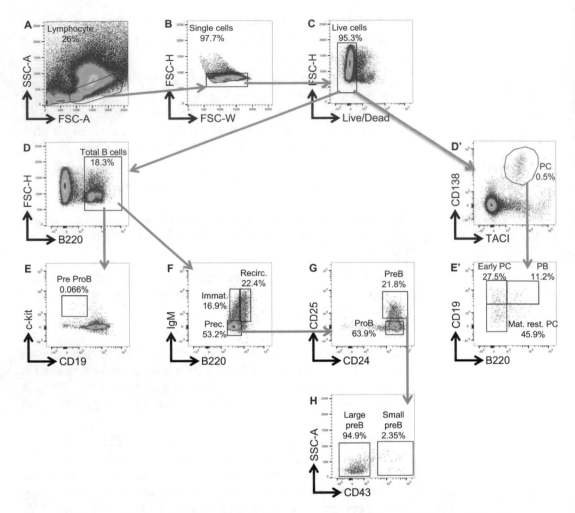

Fig. 3 Gating strategy for identification B cell and Plasma cell subpopulation identification. Data show mouse bone marrow cells. Cells were isolated and stained as described in Subheadings 3.1–3.3. Gating strategy is described in Subheading 3.4. (**a-h**) Indicated percentage correspond to the percentage in function of the parent population. Recirc. correspond to recirculating mature B cell, Immat. to immature B cell, Prec. to precursor B cell, PC to Plasma Cell, PB to plasmablast and Mat. Rest. PC to mature resting plasma cell

4   Notes

1. For flow cytometry staining, use preferentially polypropylene tubes rather than polystyrene tubes to minimize cell adhesion to the tubes. If compatible with your flow cytometer, it is also possible to perform staining in 96-well plate-V bottom. In this case adapt the volume and time of centrifugation according to your plate.

2. Before using FBS, inactivate it by heating for 30 min at 56 °C.

3. In absence of good markers for the identification of memory B cells in mouse, this cell population is not included in this

analysis. Currently, the best way to study memory B cells in mouse is the use of reporter model like AID-GFP mice, as cited in the introduction or the use of labelled antigen to detect antigen-experienced memory B cells.

4. Put bones into a 6 or 12 well plate or petri dish depending on the reagents at your disposition. At this stage, the only important point is to maintain bones in a wet environment.

5. For hip, dissection is sufficient to "open" these bones and allow cell extraction during centrifugation step.

6. Place maximum one complete leg (femur, tibia, hip bone) per 0.5 mL tube.

7. As a general rule, keep cells at 4 °C during all the experiment, in cold room, fridge or on ice but never in direct contact with ice. Ice contact induces cell death. Plasma cells are very fragile cells that die quickly once they are extracted from their environment. To study efficiently these very rare cell populations perform the experiment in the hours after sacrifice or if not possible conserve the cells in untouched bones (maximum 24 h at 4 °C immerged in PBS–2% FBS). Precursor B cells are more robust than plasma cells but do not exceed 24 h for your experiment to preserve cell integrity and viability. For cell conservation over 24 h prefer a RPMI medium supplemented with 10% FBS.

8. If some marrow ("red" bones) is still detectable into the bones after centrifugation, the cartilage may not have been removed correctly. Check that at least one extremity of the bone is properly open and repeat the centrifugation step.

9. Time required for efficient red blood cell lysis with the ACK method is comprised between 1 and 5 min. Do not exceed 5 min as longer incubations can affect cell viability.

10. Dilute 10 µL of your cell suspension to 1/20 in PBS and to 1/2 (volume/volume) in trypan blue previously diluted at 0.4%, to have a total dilution of your cell suspension of 1/40 to have an optimal concentration for the cell count. Note whether your total cell count is for 1 or 2 completes legs. Trypan blue is one of the most used staining to determine cell viability but other alternative techniques to count cells are possible.

11. Using the adequate antibody concentration is mandatory for reliable and reproductive flow cytometry experiments. All antibodies need to be properly titrated to determine the optimal concentration in your personal settings as it can vary depending on the flow cytometer used, on the fluorochrome and on the supplier.

12. Antibody mix can be prepared before the experiment (max 24 h), in PBS–2% FBS, ready to use and store at 4 °C protected from light.

13. When antibody mix is performed for many samples, always prepare an extra volume of 10% to anticipate some pipetting loss.

14. In all multiparametric flow cytometry experiments compensations need to be made to correct the spill-over from one fluorochrome to another. In parallel to your experiment, perform single fluorochrome staining either on cells or on specific beads. Both techniques can be used depending on the preference of the user or the availability of the cells. Do not forget unlabeled cells and the fixable viability dye control. To increase the proportion of dead cells in your control viability tube, you can incubate your cells at 56 °C for 5 min before the staining. During sample acquisition, unlabeled cells and all single fluorochrome staining need to be acquired first to calculate the compensations before the acquisition of the experimental tubes. Compensations can be corrected after sample acquisition during the analysis step if required.

Acknowledgments

The study was supported by an ANR @RAction grant (ANR-14-ACHN-0008), an ANR JCJC grant (ANR-19-CE15-001901), a Fondation ARC grant (PJA20181208173), a grant from IdEx Université de Paris (ANR-18-IDEX-0001) to ME, and an ANR PRC grant (ANR-17-CE14-0019), an INCa grant (PRT-K 2017) and the Association Saint Louis pour la Recherche sur les Leucémies to KB.

References

1. Fuxa M, Busslinger M (2007) Reporter gene insertions reveal a strictly B lymphoid-specific expression pattern of Pax5 in support of its B cell identity function. J Immunol 178:8222–8228. https://doi.org/10.4049/jimmunol.178.12.8221-a

2. Winkler TH, Mårtensson I-L (2018) The role of the pre-B cell receptor in B cell development, repertoire selection, and tolerance. Front Immunol 9:2423. https://doi.org/10.3389/fimmu.2018.02423

3. Corcoran LM, Tarlinton DM (2016) Regulation of germinal center responses, memory B cells and plasma cell formation-an update. Curr Opin Immunol 39:59–67. https://doi.org/10.1016/j.coi.2015.12.008

4. Dogan I, Bertocci B, Vilmont V et al (2009) Multiple layers of B cell memory with different effector functions. Nat Immunol 10:1292–1299. https://doi.org/10.1038/ni.1814

5. Lindquist RL, Niesner RA, Hauser AE (2019) In the right place, at the right time: spatiotemporal conditions determining plasma cell survival and function. Front Immunol 10:788. https://doi.org/10.3389/fimmu.2019.00788

6. Pracht K, Meinzinger J, Daum P et al (2017) A new staining protocol for detection of murine antibody-secreting plasma cell subsets by flow cytometry. Eur J Immunol 47:1389–1392. https://doi.org/10.1002/eji.201747019

7. Hardy RR, Kincade PW, Dorshkind K (2007) The protean nature of cells in the B lymphocyte lineage. Immunity 26:703–714. https://doi.org/10.1016/j.immuni.2007.05.013

8. Raff MC, Megson M, Owen JJ, Cooper MD (1976) Early production of intracellular IgM by B-lymphocyte precursors in mouse. Nature 259:224–226. https://doi.org/10.1038/259224a0

9. Coffman RL, Weissman IL (1981) A monoclonal antibody that recognizes B cells and B cell precursors in mice. J Exp Med 153:269–279. https://doi.org/10.1084/jem.153.2.269

10. Park YH, Osmond DG (1987) Phenotype and proliferation of early B lymphocyte precursor cells in mouse bone marrow. J Exp Med 165:444–458. https://doi.org/10.1084/jem.165.2.444

11. Park YH, Osmond DG (1989) Post-irradiation regeneration of early B-lymphocyte precursor cells in mouse bone marrow. Immunology 66:343–347

12. Nadler LM, Anderson KC, Marti G et al (1983) B4, a human B lymphocyte-associated antigen expressed on normal, mitogen-activated, and malignant B lymphocytes. J Immunol 131:244–250

13. Uckun FM, Jaszcz W, Ambrus JL et al (1988) Detailed studies on expression and function of CD19 surface determinant by using B43 monoclonal antibody and the clinical potential of anti-CD19 immunotoxins. Blood 71:13–29

14. McKearn JP, Baum C, Davie JM (1984) Cell surface antigens expressed by subsets of pre-B cells and B cells. J Immunol 132:332–339

15. Wenger RH, Ayane M, Bose R et al (1991) The genes for a mouse hematopoietic differentiation marker called the heat-stable antigen. Eur J Immunol 21:1039–1046. https://doi.org/10.1002/eji.1830210427

16. Cooper MD, Mulvaney D, Coutinho A, Cazenave PA (1986) A novel cell surface molecule on early B-lineage cells. Nature 321:616–618. https://doi.org/10.1038/321616a0

17. Rolink A, Grawunder U, Winkler TH et al (1994) IL-2 receptor alpha chain (CD25, TAC) expression defines a crucial stage in pre-B cell development. Int Immunol 6:1257–1264. https://doi.org/10.1093/intimm/6.8.1257

18. Gulley ML, Ogata LC, Thorson JA et al (1988) Identification of a murine pan-T cell antigen which is also expressed during the terminal phases of B cell differentiation. J Immunol 140:3751–3757

19. Ogawa M, Nishikawa S, Nishikawa S (1991) The role of c-kit on intra-marrow hemopoiesis. Tanpakushitsu Kakusan Koso 36:696–703

20. Rolink A, Streb M, Nishikawa S, Melchers F (1991) The c-kit-encoded tyrosine kinase regulates the proliferation of early pre-B cells. Eur J Immunol 21:2609–2612. https://doi.org/10.1002/eji.1830211044

21. Hardy RR, Carmack CE, Shinton SA et al (1991) Resolution and characterization of pro-B and pre-pro-B cell stages in normal mouse bone marrow. J Exp Med 173:1213–1225. https://doi.org/10.1084/jem.173.5.1213

22. Sanderson RD, Lalor P, Bernfield M (1989) B lymphocytes express and lose syndecan at specific stages of differentiation. Cell Regul 1:27–35. https://doi.org/10.1091/mbc.1.1.27

23. Moreaux J, Hose D, Jourdan M et al (2007) TACI expression is associated with a mature bone marrow plasma cell signature and C-MAF overexpression in human myeloma cell lines. Haematologica 92:803–811. https://doi.org/10.3324/haematol.10574

24. Chevrier S, Genton C, Kallies A et al (2009) CD93 is required for maintenance of antibody secretion and persistence of plasma cells in the bone marrow niche. Proc Natl Acad Sci U S A 106:3895–3900. https://doi.org/10.1073/pnas.0809736106

Chapter 9

Metabolic Analysis of Mouse Hematopoietic Stem and Progenitor Cells

Liang He, Amina Aouida, Amira Mehtar, Houda Haouas, and Fawzia Louache

Abstract

The intrinsic properties of self-renewal and differentiation of hematopoietic stem and progenitor cells (HSPCs) play a critical role in the regeneration of mature hematopoietic cells at steady state and in stress conditions including bleeding, inflammation and aging. Common techniques such as flow cytometry and genetic methods have answered many questions about their intrinsic and extrinsic regulation. Using these approaches, it was demonstrated that HSPCs in the bone marrow demonstrate low rates of proliferation and apoptosis. This dormant phenotype is associated with a low production of reactive oxygen species and low mitochondrial activity.

Here, we describe the methodology to characterize the physiologic state of HSPCs isolated from their native hematopoietic organ using flow cytometry-based assays. These protocols allow evaluation of their ROS levels and activated signaling pathways under various conditions.

Key words Bone marrow processing, HSPCs, Flow cytometry, Cell metabolism, ROS levels, ROS signaling

1 Introduction

Specialized bone marrow (BM) microenvironments or niche contribute to the regulation of the intrinsic long-term potential of HSPCs all along the life and understanding how intrinsic and extrinsic factors are connected have important implications in HSPC biology and diseases. There is actually emerging evidence that HSPCs are located in specific niches characterized by local limited oxygen contents. With this low-oxygen microenvironment, HSPCs developed a specific metabolic phenotype corresponding to an anaerobic glycolysis characterized by a low rate of oxygen consumption, low ATP level, and a higher production of lactate [1]. This metabolic profile helps HSPCs to avoid excessive reactive oxygen species (ROS) production as excessive endogenous ROS are

Marion Espéli and Karl Balabanian (eds.), *Bone Marrow Environment: Methods and Protocols*, Methods in Molecular Biology, vol. 2308, https://doi.org/10.1007/978-1-0716-1425-9_9, © Springer Science+Business Media, LLC, part of Springer Nature 2021

the major source of DNA damage contributing to chromosome instability and accumulation of mutation and deletion [2]. In the HSPC compartment, endogenous ROS production regulates HSPC-renewal and their elevated levels have been shown to limit lifespan of HSPC via activation of p38 MAPK pathway [3].

In the BM niches, the CXCR4/CXCL12 signaling helps in regulating ROS signaling contributing to their long term maintenance [4]. Beyond their deleterious action, these same ROS when produced at physiological levels regulate various biological responses such as HSPC mobilization and migration, cell proliferation, cell-cycle progression, survival, differentiation, and gene expression [5, 6].

There are many technical difficulties when studying the physiologic and pathologic redox state of HSPCs in their native environment [7, 8]. The first limitation is related to their rarity as they represent only about 30 out of one million BM mononuclear cells [7, 8]. In addition, these cells are only identifiable by the use of a combination of multiple cell surface markers [9]. The second limitation derives from the fact that cell handling itself produces an increase in ROS levels which may exceed normal physiologic production [10]. The powerful technique of flow cytometry is particularly well suited to the field of HSPCs insofar as it allows rapid and simultaneous analysis of multiple parameters at a single-cell level in suspended nonhomogeneous populations by measuring fluorescence emissions of cellular markers associated with proliferation, apoptosis or oxidative stress. The combination of phenotyping markers with markers specific for proliferation status, apoptotic features, and functional probes for ROS level assessment provides important insights on the various states of HSPCs and understanding the impact of intrinsic and extrinsic factors on these states.

In this chapter, we will describe methods to estimate the physiologic state of HSPCs. They include protocols to recover hematopoietic cells from the bones and spleen and to perform cell surface staining for HSPC identification. We will also provide protocols to assess their oxidative state and associated signaling pathways.

2 Materials

2.1 Cell Preparation

1. Isoflurane chambers.

2. Prepare 0.8% sodium citrate solution in distilled water.

3. Cell processing solution: prepare PBS- fetal bovine serum (FBS) (PBS $1\times$ supplemented with 5% FBS) (*see* **Note 1**).

4. Red blood cell lysing solution containing 150 mM NH_4Cl, 10 mM $KHCO_3$, and 0.1 mM EDTA diluted in distilled water, pH 7.4.

5. Dissection tools.

6. 0.4% trypan blue solution.

7. Bright-line hemacytometer.

8. Light microscope.

9. 5 mL Round Bottom Polystyrene Test Tube (FACS tubes).

10. 70 μm nylon cell strainer.

11. 21 gauge-needle with a 1 mL syringe.

12. Swinging bucket centrifuge.

2.2 Cell Surface Staining

1. 500 μg/mL Fc blocker (anti-mouse CD16/32).

2. Labeled monoclonal antibodies against lineage-specific antigens myeloid Gr-1, erythroid lineage cells, B220, CD3e, and CD11b (*see* Table 1 and **Notes 2** and **3**).

3. Labeled monoclonal antibodies for identification of HSPCs, that is, anti-c-Kit, anti-Sca-1, anti-CD34, anti-CD150, anti-CD48, and anti-CD41 (*see* Table 1 and **Notes 3** and **4**).

4. 10 mg/mL stock solution of Propidium Iodide (PI). This solution is stable at 4 °C in the dark (*see* **Note 5**).

5. 5 mg/mL stock solution of DAPI (*see* **Note 6**).

6. 2% paraformaldehyde (PFA).

7. Data of stained cells are collected using flow cytometry acquisition instrument (e.g., BD FACS LSR or Canto, BD Biosciences/Cytek) and analyzed with flow cytometry analysis software such as BD FACSDiva or FlowJo.

Table 1
list of antibodies used

Antibody	Dilution	Clone
Gr1	1:100	RB-8C5
Erythroid	1:100	TER119
B220	1:100	RA3-6B2
CD3e	1:100	145-2C11
CD11b	1:100	M1/70
CD117	1:100	2B8
Sca-1	1:100	E13-167.7
CD34	1:50	HM34
CD150	1:100	TC15-12F12.2
CD48	1:100	HM48-1
CD41	1:100	MW Reg 30

2.3 Assessment of ROS Levels

1. HBSS Hank's Balanced Salt Solution (HBSS) with Calcium and Magnesium and without Red Phenol (HBSS/Ca/Mg).

2. 5-(and 6-)chloromethyl-2′,7′-dichlorodihydrofluorescein diacetate, acetyl ester (CM-H2DCFDA).

3. MitoSOX™ red mitochondrial superoxide indicator.

4. 10 mg/mL PI solution.

5. 5 mg/mL DAPI solution.

2.4 Intracellular Staining for Activated Stress Signaling Pathways

1. Rabbit monoclonal antibody to phosphorylated p38 MAPK (Thr180/Tyr182).

2. Anti-rabbit specific IgG, conjugated with Alexa Fluor 488 as a secondary antibody.

3. Cytofix/Cytoperm Kit.

3 Methods

3.1 Cell Preparation

Cells from murine hematopoietic tissues (e.g., blood, BM, and spleen) are obtained from adult (8–12-week-old) C57BL/6 mice that should be handled according to procedures approved by the housing institution.

1. For each mouse, prepare petri dishes of 60×15 mm containing 5 mL of PBS-FBS.

2. After euthanizing the mouse, if needed remove all blood via cardiac puncture in 0.8% sodium citrate (9/1) as anticoagulant.

3. Dissect tissues to expose bones and spleen.

4. Remove the bones (2 tibias, 2 femurs and two pelvic bones per mouse) and remove carefully any surrounding tissue. Put organs in the petri dishes. Cut the edges of the bones to expose the interior of the marrow and collect the cells into 1–2 mL PBS-FBS by repeated flushing using 21-gauge needle attached to a 1 mL syringe.

5. Remove spleen and place it in the petri dish. Using the plunger of a 10-mL syringe, grind spleen against a cell strainer in a circular motion until mostly fibrous debris remains.

6. Filter the cell suspensions by passage through a sterile 70-μm nylon mesh. Dilute cell suspension in Eppendorf tube by adding 50 μL of cell suspensions to 450 μL PBS-FBS.

7. Count the number of BM and spleen cells using trypan blue to exclude nonviable cells.

8. Spin, pour off supernatant, add PBS-FBS to adjust to 10^8 cells/mL and gently resuspend. Keep at 4 °C.

9. To recover nucleated blood cells, proceed to erythrocyte lysis by incubating each 100 μL blood with 1 mL of cold Red blood cell lysing solution in 5 mL tube. Mix by turning the tubes over. Incubate for 3 min at room temperature (RT) and stop the lysing reaction by filling the tube with cold PBS-FBS. Use the equivalent of 200 μL blood for each test tube.

10. Centrifuge immediately at 400 × g for 5 min at 4 °C, discard supernatant and resuspend pellet in 4 mL of cold PBS-FBS. Centrifuge at 400 × g for 5 min at 4 °C and discard supernatant. Resuspend pellets in PBS-FBS and keep at 4 °C.

3.2 Cell Surface Staining

Detection and quantification of HSPCs by flow cytometry are based on the following phenotype: Lin⁻ (Lineage negative) Sca-1⁺ c-Kit⁺ CD34-CD150⁺ CD48⁻ CD41⁻ and Flk-2⁻. At the minimum, staining for cell surface expression of LSK and CD34 should be performed.

1. Mix carefully target cell suspension containing 10^8 cells/mL and dispense 100 μL in FACS tube.

2. Complete the volume to 173 μL using PBS-FBS. To inhibit nonspecific binding of antibodies, add 1 μL of Fc blocking buffer (1:200), mix by gently tapping tubes and incubate 10 min at RT.

3. Prepare an antibody cocktail containing 2 μL of each antibody for each test tube with the exception of anti-CD34 for which 2 μL should be used (*see* Table 1 and **Notes 7** and **8**).

4. A 45 min incubation time together with higher antibody concentration are needed when using anti-CD34 antibody to allow optimal saturation binding equilibrium to be reached.

5. Dispense 26 μL of antibody cocktail in each tube, mix by briefly vortexing and incubate in the dark at 4 °C for 40 min (*see* **Note 7**).

6. Prepare negative controls (*see* **Notes 9–11**) and single stained controls (*see* **Note 12**) as well as fluorescent minus one (*see* **Note 12**).

7. Wash by adding 3 mL PBS-FBS, centrifuge at 400 ×g for 6 min and discard the supernatant.

8. For immediate flow cytometric analysis, resuspend pellets in 300 μL PBS-FBS containing 2 μg/mL PI or DAPI and keep tubes at 4 °C. For storage until delayed analysis, gently tap tubes to resuspend pellets and add 200 μL of 2% PFA. Store at 4 °C overnight in dark. If performing intracellular staining, do not fix cells and immediately use CytoFix/Perm kit according to the manufacturer's protocol (e.g., BD Biosciences).

3.3 Assessment of Intracellular and Mitochondrial ROS

Intracellular ROS production is detected thanks to the redox-sensitive dye 5-(and-6)-chloromethyl-2′,7′dichlorodihydrofluorescin diacetate, acetyl ester (CM-H2DCF-DA) (*see* **Note 13**).

1. Place BM cells to immunofluorescent staining of cell surface antigens as described above to allow later gating on mature and immature cell populations.

2. Resuspend labeled cells in 500 μL of prewarmed HBSS/Ca/Mg containing 1 μM CM-H2DCF-DA and incubate for 30 min at 37 °C in the dark.

3. Prepare positive control containing 1 μM CM-H2DCF-DA and 1 μM H_2O_2 in 500 μL of prewarmed HBSS/Ca/Mg.

4. Prepare negative control tube without CM-H2DCF-DA to set the appropriate negative staining.

5. Wash labeled cells two times with 2 mL of prewarmed HBSS/Ca/Mg to remove the excess of the probes.

6. To test the viability of the cells, after the second wash resuspend the cells in 500 μL of PBS containing 2 μg/mL of DAPI.

7. Keep the tubes on ice and in the dark and analyze by flow cytometry immediately.

8. For detection of mitochondrial superoxide anion production, load labeled cells with 2 μM Mitosox 30 min at 37 °C.

9. Wash the cells three times with 1 mL of prewarmed HBSS/Ca/Mg to remove the excess of the probes.

10. To exclude dying cells, resuspend labeled cells in PBS containing DAPI as a viability dye as above.

11. Proceed for flow cytometry immediately (*see* **Note 14**).

12. An example of ROS evaluation is depicted in Fig. 1.

3.4 Assessment of Activated Stress Signaling Pathways

1. Perform cell surface staining for HSPC identification.

2. Fix and permeabilize HSPCs to allow staining for intracellular markers of ROS activated signaling pathway such as p38 MAPK.

3. Add to each tube 250 μL of Cytofix/Cytoperm solution, mix thoroughly by pipetting up and down, and vortex.

4. Incubate for 20 min on ice (vortex halfway through and at the end).

5. Add 2 mL of Perm/Wash solution (Cytofix/Cytoperm Kit), vortex thoroughly, and incubate about 10 min on ice.

6. Centrifuge at 300 × g for 5 min at 4 °C and pour off supernatant.

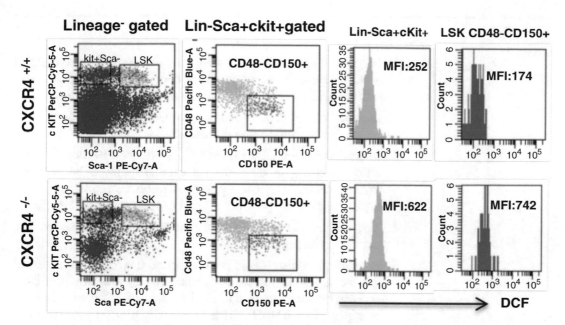

Fig. 1 Analysis of ROS levels in hematopoietic progenitor cells (Lin⁻Sca⁺cKit⁺) and HSPCs gated in lineage negative BM cells from CXCR4 wild type and knockout mice. CM-H2DCF-DA staining was performed in BM cells from CXCR4+/+ and CXCR4−/− after cell surface staining. The mean fluorescence intensity (MFI) of CM-H2DCF-DA was measured after gating in Lin⁻Sca1⁺Kit⁺ (hematopoietic progenitor cells) and the more immature population Lin⁻Sca1⁺Kit⁺ CD48-CD150⁺

7. Vortex briefly, add 2 mL of Perm/Wash solution, vortex thoroughly, centrifuge at $300 \times g$ for 5 min at 4 °C, and pour off supernatant.

8. Add to each tube 50 μL of antibody of interest (e.g., 5 μL/mL anti-p38 MAPK in Perm/Wash buffer) with P-200, mix by pipetting up and down, and vortex thoroughly.

9. Incubate for 1 h at RT in dark (vortex halfway through and at the end).

10. Wash by adding 1 mL Perm/Wash solution. Centrifuge at $400 \times g$ for 5 min and discard the supernatant.

11. Add to each tube 50 μL of labeled anti rabbit antibody in Perm/Wash buffer with P-200, mix by pipetting up and down, and vortex thoroughly.

12. Incubate for 30 min at RT in dark (vortex halfway through and at the end).

13. Wash by adding 3 mL PBS. Centrifuge at $400 \times g$ for 5 min, discard the supernatant and suspend pellet in PBS. keep at 4 °C until FACS analysis.

4 Notes

1. Prepare all reagents under sterile conditions and store them at 4 °C to avoid bacterial or fungi contamination. Do not add sodium azide to reagents as it can change cell behavior.

2. Monoclonal antibodies against lineage-specific antigens must be labeled with the same fluorochrome.

3. All antibodies are either from BD Biosciences, BioLegend, and eBiosciences. The dilutions for each Ab should be carefully titrated. Note that the optimal Ab concentration may vary depending on the product lot and experimental settings.

4. The combination of fluorochromes chosen depends on the nature of the experiment.

5. Propidium iodide (PI), a fluorescent dye that binds to all double-stranded nucleic acids is excited by 488 nm laser light and can be detected with in the PE/Texas Red® channel with a bandpass filter 610/10. It is commonly used in evaluation of cell viability or DNA content in cell cycle analysis by flow cytometry. PI is stable in solution at 4 °C in the dark.

6. Alternatively DAPI (4′,6-diamidino-2-phenylindole, dilactate), a blue fluorescent dye excited by a UV laser which also binds dsDNA and is used to evaluate cell viability or DNA content. Use DAPI at the final concentration of 2 μg/mL. DAPI is stable when stored refrigerated and protected from light. For long-term storage, aliquot and store at −20 °C.

7. In most cases, we use labeled antibodies at the final concentration of 5 μg/mL (1:100) with the exception of anti-CD34 used at 10 μg/mL (1:50) dilution. However, the quantity of each antibody and its dilution depend on the cell number needed and on the suppliers. Thus, optimization of each antibody concentration should be performed before use.

8. Preparing a master mix of the antibodies creates less variability between multiple samples in the same FACS run.

9. If possible, prepare a mix of all test cell suspensions that will be used for control tubes. This mixed cell suspension allows for optimization of the flow cytometry parameters and compensations.

10. Prepare unstained control tubes to set the appropriate forward (FSC) and side (SSC) scatter.

11. Prepare single stained sample control tubes containing only one fluorochrome-conjugated antibody used in the antibody cocktail. These single stained control tubes help to set gain values for the fluorescent channels and will discriminate

between negative and positive populations for each fluorophore used. They also allow to set appropriate compensation values.

12. Perform control tubes containing all the appropriate fluorophore-conjugated antibodies except one (fluorescent minus one or FMO). For each control, replace the missing antibody by a fluorochrome-labeled antibody of the same isotype at the same concentration and in same conditions.

13. This probe is a nonfluorescent molecule that is readily converted to a green-fluorescent form when its acetate groups are removed by oxidation under the activity of ROS.

14. During flow cytometry acquisition, place tubes on ice in the dark to minimize artifactual fluctuations of ROS detection.

Acknowledgments

This study was supported by the Institut National de la recherche médicale, Agence Nationale de la Recherche, the Foundation de France, and Canceropole, IDF.

References

1. Ito K, Bonora M, Ito K (2019) Metabolism as master of hematopoietic stem cell fate. Int J Hematol 109:18–27

2. Walter D, Lier A, Geiselhart A et al (2015) Exit from dormancy provokes DNA-damage-induced attrition in haematopoietic stem cells. Nature 520:549–552

3. Ito K, Hirao A, Arai F et al (2006) Reactive oxygen species act through p38 MAPK to limit the lifespan of hematopoietic stem cells. Nat Med 12:446–451

4. Zhang Y, Dépond M, He L et al (2016) CXCR4/CXCL12 axis counteracts hematopoietic stem cell exhaustion through selective protection against oxidative stress. Sci Rep 6:37827

5. Samimi A, Khodayar MJ, Alidadi H et al (2020) The dual role of ROS in hematological malignancies: stem cell protection and cancer cell metastasis. Stem Cell Rev Rep 16:262–275

6. Herault O, Hope KJ, Deneault E et al (2012) A role for GPx3 in activity of normal and leukemia stem cells. J Exp Med 209:895–901

7. Goodell BMA, Brose K, Paradis G et al (1996) Isolation and functional properties of murine hematopoietic stem cells that are replicating in vivo. J Exp Med 183:1797–1806

8. Bryder D, Rossi DJ, Weissman IL (2006) Hematopoietic stem cells: the paradigmatic tissue-specific stem cell. Am J Pathol 169:338–346

9. Morita Y, Ema H, Nakauchi H (2010) Heterogeneity and hierarchy within the most primitive hematopoietic stem cell compartment. J Exp Med 207:1173–1182

10. Simsek T, Kocabas F, Zheng J et al (2010) The distinct metabolic profile of hematopoietic stem cells reflects their location in a hypoxic niche. Cell Stem Cell 7:380–390

Part III

Imaging of the Bone Marrow

Chapter 10

Multicolor Immunofluorescence Staining of Paraffin-Embedded Human Bone Marrow Sections

Francesca M. Panvini, Simone Pacini, Stefano Mazzoni, and Simón Méndez-Ferrer

Abstract

Immunofluorescence is an indispensable method for the identification, localization and study of the expression of target antigens in formalin-fixed, paraffin-embedded (FFPE) tissue sections of human bone marrow. However, the procedure shows technical limitations because of the chemical and physical treatments required for sample processing before imaging. Here we describe a revisited protocol to obtain high-resolution images of human bone marrow trephine biopsies, improving the antigen-antibody recognition and preserving the morphology and the architecture of the bone marrow microenvironment.

Key words Immunofluorescence, FFPE, Antigen retrieval, Microwave heating, Fixative

1 Introduction

The hematopoietic stem cell (HSC) niche is an anatomical and functional microenvironment where HSCs are maintained in an undifferentiated and self-renewing state and receive stimuli that determine their fate [1]. Several cellular components of the HSC niche such as mesenchymal cells, endothelial cells, perivascular stromal cells, nerves belonging to the sympathetic nervous system (SNS), nonmyelinating Schwann cells, adipocytes, osteolineage cells, structural components of the extracellular matrix, ions (calcium, magnesium, and phosphate), vesicles, cytokines, chemokines, hormones and growth factors influence HSC self-renewal, proliferation, differentiation, or migration [2–4]. Given this complexity, studying the HSC niche requires choosing optimal strategies for sample preparation and imaging to evaluate the expression and the localization of proteins [5–9]. Immunofluorescence on tissue sections represents one of the most amenable techniques which has been very powerful to study mouse HSC niches using cryopreserved samples. However, cryoblocks are rarely available for

Marion Espéli and Karl Balabanian (eds.), *Bone Marrow Environment: Methods and Protocols*, Methods in Molecular Biology, vol. 2308, https://doi.org/10.1007/978-1-0716-1425-9_10, © Springer Science+Business Media, LLC, part of Springer Nature 2021

human samples since these samples are routinely fixed with formalin and embedded in paraffin for histological examination and diagnostic procedures. Unfortunately, this processing limits the capacity to detect antigens in paraffinized human samples [5, 10]. In fact, tissue treated with aldehyde fixatives and included in hydrocarbon compounds show less antigenicity and generate autofluorescence and/or background fluorescence which decrease the quality of the final result. Antigenic heat-mediated exposure or unmasking of antigens can improve the quality of the antigen-antibody reaction, but tends to damage thin bone marrow [11]. Microwave heating methods alternating heating intervals can sometimes yield better results. The antibody washing steps are particularly relevant to reduce the background noise, especially in the case of samples pretreated with signal enhancers.

One key advantage of paraffin inclusion of the specimen is its long storage, which allows retrospective clinical examination. The availability of these samples is becoming increasingly on demand to understand the pathological alterations of the human bone marrow microenvironment. Accomplishing this goal requires in-depth knowledge of the physiopathological human bone marrow architecture, but our knowledge in this regard is very limited compared with the information generated in the mouse system. Postdiagnostic paraffinized human bone marrow (trephine) biopsies could represent an invaluable source to increase our knowledge of the human bone marrow. For this reason, we describe here a protocol that allows multicolor staining of thin human bone marrow sections for high resolution fluorescence imaging. Briefly, after dewaxing samples are maintained always hydrated by using washing buffers (D-PBS and 0.5% Triton X-100). Antigen retrieval is carried out by heating in a bain-marie three times ($7' + 4' + 4'$) with steps of thermal shocks. Nonspecific fluorescence (background), which is a commonly issue when applying fluorescent conjugates to paraffinized sections, is largely eliminated by using glycine, washing sections three times (after antigen–antibody reactions) and separating the incubations of secondary antibodies. These optimized critical steps and examples of the resulting images (Fig. 1) are described below in detail.

2 Materials

1. Xylene (or other dewaxing/clearing reagents), 200 ml.

2. Absolute ethanol, 200 ml.

3. 96% Ethanol: Add 192 ml of absolute ethanol to a 200 ml graduate cylinder and add 8 ml of distilled water. Mix well and transfer the solution in a glass Coplin jar.

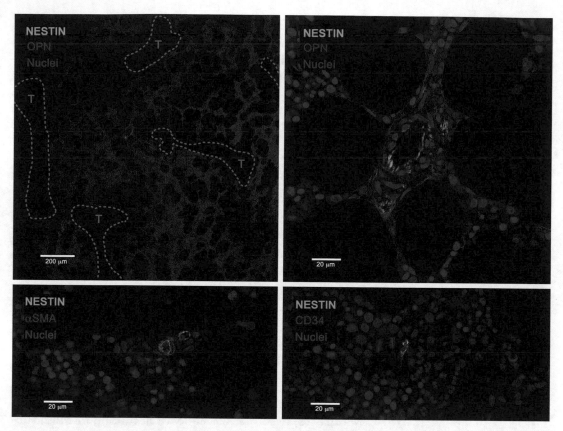

Fig. 1 Immunofluorescence of blood vessels and perivascular cells in the human bone marrow. Confocal images (at 20× and 63× magnification) of formalin-fixed, paraffin-embedded sections (3μm) of human bone marrow (trephine) biopsies processed as described here. The immunofluorescence of antibodies against human NESTIN (green), osteopontin (OPN), a-smooth muscle actin (αSMA), or CD34 (red) is shown in normal human bone marrow samples. Nuclei were counterstained with To-PRO-3 (blue). T, bone trabeculae

4. 70% Ethanol: Add 140 ml of absolute ethanol to a 200 ml graduate cylinder and add 60 ml of distilled water. Mix well and transfer the solution in a glass Coplin jar.

5. 50% Ethanol: Add 100 ml of absolute ethanol to a 200 ml graduate cylinder and add 100 ml of distilled water. Mix well and transfer the solution in a glass Coplin jar.

6. Citrate buffer 0.1 M, pH 6.0: add 100 ml of distilled water to a 100 ml graduate cylinder and transfer into a glass beaker. Weigh 0.210 g of citric acid and transfer to the glass beaker. Use a magnetic stirrer to dissolve citric acid. Add 500 ml of distilled water to a 1 l graduate cylinder and transfer into a glass beaker. Weigh 1.47 g of sodium citrate dehydrate and transfer to the glass beaker. Use a magnetic stirrer to dissolve dehydrated sodium citrate. Discard 80 ml of sodium citrate dehydrated and adjust pH with citric acid solution (pH 6).

7. Two microwavable containers and a glass tray.

8. Dulbecco's phosphate buffered saline solution (D-PBS).

9. 10% (v/v) Permeabilization stock solution (PRM): Add 90 ml of D-PBS to a 100 ml graduated cylinder. Transfer the D-PBS into a glass bottle (100 ml) and pipette 10 ml of Triton X-100 slowly and dissolve completely. At the time of staining, dilute the stock solution with D-PBS to obtain 0.5% (v/v) of working solution. Store stock solution at RT temperature (2 months) or at 4 °C.

10. Glycine (mw: 75.07 g/mol): dissolve glycine 0.3 M to D-PBS. Mix with a magnetic stirrer, store at 4 °C (1 month) and filter the solution before the use.

11. Blocking solution: mix 10% (v/v) normal serum and 0.5% (w/v) BSA (Bovine Serum Albumin). Dissolve BSA in sterile D-PBS at 4 °C and add normal serum. Mix with a magnetic stirrer. Store solution at 4 °C.

12. Primary antibody solution: mix 1% (v/v) normal serum and 0.5% (w/v) BSA in 0.5% Triton X-100. Dissolve BSA in sterile D-PBS at 4 °C.

13. Secondary fluorochrome-conjugated antibody solution: mix 1% (v/v) normal serum and 0.5% (w/v) BSA in 0.5% Triton X-100.

14. Nuclear dye.

15. Cover glasses, tweezers, and aqueous mounting medium supplemented with antifade reagent.

3   Methods

1. Cut tissue sections using a low-profile disposable blade to obtain sections (3–15μm).

2. Transfer sections to microscope slides at room temperature (RT) and dry at RT for at least 48 h before starting the immunostaining (*see* **Note 1**).

3. Immerse slides in Xylene (or other dewaxing/clearing reagents, 200 ml) for 10 min (repeat twice). Remove slides from the solution.

4. Hydrate (immersion) by graded alcohol series (200 ml): absolute ethanol for 10 min (repeat twice), 96% ethanol for 10 min, 70% ethanol for 10 min, 50% ethanol for 10 min.

5. Immerse the slides for 10 min in distilled water (*see* **Note 2**).

6. Perform the antigen retrieval. Prepare two microwavable containers containing 80–100 ml of 0.1 M citrate buffer pH 6.0 (4 °C). Place slides in one of them. Prepare a glass tray and fill it in half with running water. Immerse the container with slides

and Microwave at 750–800 W in a bain-marie for 7 min to initiate boiling of antigen retrieval solution. Monitor this process carefully to prevent drying of the sample. Take the slides out of the microwave with a heating protection and place them in the microwavable container with fresh citrate buffer at RT for 1 min (thermal shock). Immerse slides into the hot microwavable container and heat them for 4 min. Repeat the thermal shock. Heat again for 4 min (*see* **Note 3**). Cool down sections for 30 min at RT.

7. Wash slides by adding D-PBS, leave for 2 min and pour. Repeat twice.

8. Prepare humidified incubation chamber to be used later during the incubation with antibodies. Place a foil of lab paper into the chamber and add some drops of distilled water (avoiding excess).

9. Discard D-PBS using a micropipette. Leave a small amount to prevent drying of tissue sections. Wipe the slide carefully around the sections creating margins (*see* **Note 4**).

10. Dilute 10% (v/v) PRM stock solution with D-PBS to obtain 0.05% (v/v) of working solution. Add 100–200µl of PRM solution and incubate the tissue sections at RT for 15 min (*see* **Note 5**).

11. Discard the PRM solution and incubate slides in D-PBS containing 0.3 M glycine for 20 min (*see* **Note 6**).

12. Wash slides three times for 2 min each in D-PBS.

13. Discard D-PBS using a micropipette and add 100–200µl of freshly prepared blocking solution. Incubate at RT for 30 min (*see* **Note 7**).

14. Remove the blocking solution and wipe carefully around the sections. Immediately add 100–200µl of primary antibody solution to each section and incubate at 4 °C overnight (*see* **Note 8**) in a humidified chamber. For multiple staining, primary antibodies of different host species can be used together.

15. Use isotypic control antibodies to evaluate not specific fluorescence signals.

16. Wash slides three times for 2 min each in D-PBS.

17. Add prespinned conjugated secondary antibody (to avoid precipitates) diluted at the desired concentration in secondary antibody solution and incubate at 4 °C for 1 h protected from light exposure (*see* **Note 9**).

18. Wash slides three times for 2 min each in D-PBS.

19. Repeat **step 17** for each secondary antibody.

20. Wash slides three times for 2 min each in D-PBS.

21. Prepare nuclear staining solution. Add 100–200 μl to each section and incubate at RT (*see* **Note 10**).

22. Remove the staining solution and wash sections three times for 2 min each in D-PBS.

23. Discard D-PBS and wipe carefully around the sections. Drop 26 μl of aqueous mounting medium supplemented with anti-fade reagent. Avoid bubbles and excessive mounting medium. Place a clean coverslip on the section and apply pressure gently using tweezers.

24. Seal off the slides using nail polish. From that moment the slides are ready for imaging (*see* **Note 11**).

4   Notes

1. Alternatively, dry tissue sections at 37 °C for 48 h.

2. This is a pause point. Samples can be stored in distillated water at RT for about 3 h.

3. Alternatively, heat samples at 750–800 W for 3 min, reduce microwave power (400 W) and continue to heat samples for 12 min. Monitor carefully and add more buffer if it evaporates. Antigen retrieval breaks methylene bridge cross-links formed between proteins during formaldehyde-mediated fixation. Salt concentration and pH of retrieval buffer solutions should be evaluated to guarantee cross-bridge removal. The use of tissue-adhesive coated (poly-L-lysine coated) slides is strongly recommended because the antigenic unmasking is a critical step that can nevertheless cause detachment and damage of the tissue sections. Avoid boiling of the solution to prevent this.

4. Alternatively, PAP pen can be used to draw margins, as far as enough space is allowed between margins and the section. From this step on do not let the sections dehydrate.

5. Drop solutions slowly to the tissue sections in order to prevent damage to the sections. Liquids should be pipetted adjacent to the tissue sections forming drops. It is recommended to create drops of solution on the surface of the sample and avoid overspill.

6. Glycine reduces background fluorescence by binding to free aldehyde groups, possibly reducing the fluorescent signal of secondary antibodies. Commercially available signal enhancer solutions can be directly applied on the tissue section (drop 26 μl) to increase the fluorescent intensity. If used, avoid mixing with other solutions (blocking solutions) and wash well in D-PBS.

7. Normal serum helps to prevent cross-reaction between the secondary antibody and endogenous immunoglobulins present in the tissue. Normal serum should be from the same species as the secondary antibody. For instance, for a goat anti-mouse secondary antibody, normal goat serum should be used as blocking reagent. BSA has the same purpose to reduce nonspecific binding by blocking hydrophobic binding sites in the tissue.

8. The primary antibody needs to derive from a species different from the sample. Antibody dilution and incubation time should be optimized for each antibody. Generally, overnight incubation (4 °C) is preferred to ensure that all specific antigens have reached their saturation point. Protect the slides from light when using fluorochrome-conjugated antibodies.

9. The fluorochromes used should match the excitation and detection spectra of the microscope system used. The power of light should be adjusted to avoid photobleaching the sample. The excitation and the emission of secondary antibodies should be spanned to avoid spectrum overlapping. For multicolor immunofluorescence, we suggest to check the match with the primary antibody animal species. Avoid using more than one secondary antibody raised in the same animal species to avoid signal overlapping and false-positive signal colocalization. The coincubation of secondary antibodies is possible but not encouraged. Spin down (4 °C) the diluted antibody before use to reduce molecules aggregates that could disturb the signal.

10. Nuclear staining can be performed by using DNA intercalating dyes 4′, 6-diamidino-2-phenylindole (DAPI) or TO-PRO-3. Check excitation and emission of DNA intercalating dyes to create a good match with the secondary antibodies. Incubation time and dilution factor are specified on the product's datasheet. The use of nuclear staining is strongly encouraged to help identify cells.

11. Leave to dry the slides before the imaging session. Store slides at 4 °C and protect from light to avoid photobleaching.

Acknowledgments

This work was supported by core support grants from the Wellcome Trust (203151/Z/16/Z) and the MRC to the Cambridge Stem Cell Institute, National Health Service Blood and Transplant (UK), European Union's Horizon 2020 research (ERC-2014-CoG-648765), and a Programme Foundation Award (C61367/A26670) from Cancer Research UK to S.M.-F.

References

1. Pinho S, Frenette PS (2019) Haematopoietic stem cell activity and interactions with the niche. Nat Rev Mol Cell Biol 20(5):303–320. https://doi.org/10.1038/s41580-019-0103-9

2. Tamma R, Ribatti D (2017) Bone niches, hematopoietic stem cells, and vessel formation. Int J Mol Sci 18(1):151. https://doi.org/10.3390/ijms18010151

3. Beerman I, Luis TC, Singbrant S, Lo Celso C, Mendez-Ferrer S (2017) The evolving view of the hematopoietic stem cell niche. Exp Hematol 50:22–26. https://doi.org/10.1016/j.exphem.2017.01.008

4. Mendez-Ferrer S, Bonnet D, Steensma DP, Hasserjian RP, Ghobrial IM, Gribben JG, Andreeff M, Krause DS (2020) Bone marrow niches in haematological malignancies. Nat Rev Cancer 20(5):285–298. https://doi.org/10.1038/s41568-020-0245-2

5. Tjin G, Flores-Figueroa E, Duarte D, Straszkowski L, Scott M, Khorshed RA, Purton LE, Lo Celso C (2019) Imaging methods used to study mouse and human HSC niches: current and emerging technologies. Bone 119:19–35. https://doi.org/10.1016/j.bone.2018.04.022

6. Battifora H, Kopinski M (1986) The influence of protease digestion and duration of fixation on the immunostaining of keratins. A comparison of formalin and ethanol fixation. J Histochem Cytochem 34(8):1095–1100. https://doi.org/10.1177/34.8.2426335

7. Takaku T, Malide D, Chen J, Calado RT, Kajigaya S, Young NS (2010) Hematopoiesis in 3 dimensions: human and murine bone marrow architecture visualized by confocal microscopy. Blood 116(15):e41–e55. https://doi.org/10.1182/blood-2010-02-268466

8. Kristensen HB, Andersen TL, Marcussen N, Rolighed L, Delaisse JM (2013) Increased presence of capillaries next to remodeling sites in adult human cancellous bone. J Bone Miner Res 28(3):574–585. https://doi.org/10.1002/jbmr.1760

9. Kusumbe AP, Ramasamy SK, Starsichova A, Adams RH (2015) Sample preparation for high-resolution 3D confocal imaging of mouse skeletal tissue. Nat Protoc 10 (12):1904–1914. https://doi.org/10.1038/nprot.2015.125

10. Robertson D, Savage K, Reis-Filho JS, Isacke CM (2008) Multiple immunofluorescence labelling of formalin-fixed paraffin-embedded (FFPE) tissue. BMC Cell Biol 9:13. https://doi.org/10.1186/1471-2121-9-13

11. Gu L, Cong J, Zhang J, Tian YY, Zhai XY (2016) A microwave antigen retrieval method using two heating steps for enhanced immunostaining on aldehyde-fixed paraffin-embedded tissue sections. Histochem Cell Biol 145 (6):675–680. https://doi.org/10.1007/s00418-016-1426-7

Chapter 11

3D Microscopy of Murine Bone Marrow Hematopoietic Tissues

YeVin Mun and César Nombela-Arrieta

Abstract

The soft marrow tissues, which are found disseminated throughout bone cavities, are prime sites for hematopoietic cell production, development, and control of immune responses, and regulation of skeletal metabolism. These essential functions are executed through the concerted and finely tuned interaction of a large variety of cell types of hematopoietic and nonhematopoietic origin, through yet largely unknown sophisticated molecular mechanisms. A fundamental insight of the biological underpinnings of organ function can be gained from the microscopic study of the bone marrow (BM), its complex structural organization and the existence of cell-specific spatial associations. Albeit the application of advanced imaging techniques to the analysis of BM has historically proved challenging, recent technological developments now enable the interrogation of organ-wide regions of marrow tissues in three dimensions at high resolution. Here, we provide a detailed experimental protocol for the generation of thick slices of BM from murine femoral cavities, the immunostaining of cellular and structural components within these samples, and their optical clearing, which enhances the depth at which optical sectioning can be performed with standard confocal microscopes. Collectively, the experimental pipeline here described allows for the rendering of single-cell resolution, multidimensional reconstructions of vast volumes of the complex BM microenvironment.

Key words BM microenvironment, 3D imaging, Optical clearing, Confocal microscopy

1 Introduction

Bone marrow (BM) tissues encompass remarkable cellular diversity and a yet unchartered functional complexity [1]. Indeed, the BM is home to self-renewing hematopoietic stem cells, and hosts the vast majority of processes leading to their differentiation into almost every mature blood cell type released to peripheral circulation [2]. This primary hematopoietic activity is intertwined with a highly diverse immune cellular landscape, which confers the BM the ability to act as a secondary lymphoid organ and sustain the development of efficient immune responses [3, 4]. Beyond this, the BM is an important tissue reservoir where antigen-specific pools of cells

Marion Espéli and Karl Balabanian (eds.), *Bone Marrow Environment: Methods and Protocols*, Methods in Molecular Biology, vol. 2308, https://doi.org/10.1007/978-1-0716-1425-9_11, © Springer Science+Business Media, LLC, part of Springer Nature 2021

involved in immunological memory, including T cells and B cells, as well as plasma cells, find shelter for prolonged periods of time [5]. All these cellular components of hematopoietic origin interact with numerous nonhematopoietic stromal cell types, which form intricate networks, and act as tissue organizers, as well as orchestrators of hematopoietic and immune processes [6]. As for all multicellular organs, the higher organ functions of the BM depend on the sophisticated and finely tuned communication, established between all individual cellular components, through direct contact or secretion of soluble mediators, such as chemokines, cytokines, or growth factors.

The dissection of the fundamental principles that underlie the integration of such cellular activities, ultimately requires a thorough study of the global structural configuration that different cell types adopt in homeostasis, and the perturbations that these arrangements undergo in disease [7]. Beyond a comprehensive view of tissue structure, key information on fundamental pathways of molecular communication is gathered from the analysis of cell specific interactions within the so-called microniches [8]. The most widely studied example is that of hematopoietic stem cells, which are known to rely close association with different BM mesenchymal and endothelial cell types in perivascular locations. Thus, imaging technologies, and specifically, fluorescence-based microscopy have been integral to advance our knowledge on BM function [8].

Recent breakthroughs in the design of new and faster optical setups, the development of excitation and detection systems of higher efficiency and sensitivity, novel protocols for tissue processing, fluorescent probes with improved spectral properties and optimal computational tools for rapid, quantitative imaging analysis, are shaping a profound revolution in the way in which histological studies can be performed [9–11]. Major headway has been made possible by the advent of numerous optical clearing methods, which minimize the scattering of light as it travels through tissues, and exponentially increase in the depth at which optical sections can be generated into thick and opaque specimens [10]. With these advances at hand, large 3D images of entire organs or large, thick tissue slices can be rendered, with single cell and even subcellular resolution. Although many of these developments have been fostered by a strong desire to visualize long-distance neuronal connections in brain tissues, 3D imaging technologies are currently widely applied to the study of lymphoid and hematopoietic tissues, including the BM. When combined with computational tools that enable image-based analysis, these methods have the added advantage of providing a realistic quantitative assessment of stromal cell types, that are otherwise not extracted nor detected by conventional flow cytometry methods [12].

The hard and opaque nature of bones has hampered the straightforward use of imaging techniques to analyze BM, and for the most part precludes the generation of whole organ images [12–16]. However, a number of groups have attempted to devise strategies to either optically clear entire bones and image through osseous structures, or more frequently, generate slices of defined thickness from whole BM of different bone locations. The available protocols strongly vary on the fashion in which BM tissues are processed, whether decalcification methods are employed, and mostly on the clearing methods of choice [8]. Here we outline a nuanced description of the methodology developed in our laboratory to generate, immunostain, clear, and image nondecalcified sections of murine femoral cavities [12]. Of note, subtle variations of the described methodology can be applied to generate transversal sections or slices of BM locations, other than those inside murine femurs. Furthermore, the immunostaining and clearing protocols presented here have also been adapted to successfully image thick slices of different soft tissues generated with a vibratome.

2 Materials

All reagents mentioned can be prepared at room temperature (RT) and should be stored at 4 °C unless stated otherwise. Follow all specific waste disposal regulations when disposing reagents.

2.1 Fixation Buffer

1. Prediluted 16% paraformaldehyde (PFA) (*see* **Note 1**).

2.2 Cryopreservation and Sectioning

1. 30% Sucrose–PBS: Weigh 30 g sucrose and transfer to a glass container. Fill up with sterile PBS to 100 ml and stir until sucrose is dissolved. Store at 4 °C.

2. Benchtop liquid nitrogen dewar.

3. Forceps.

4. Disposable plastic cryomolds (24 mm × 24 mm × 5 mm, or similar).

5. Cryopreserving medium optimal cutting temperature (OCT).

6. Cryostat.

7. Steel blade, *profile D** (*see* **Note 2**).

2.3 Immunohistology

1. Blocking buffer: 10% donkey serum in PBS/0.2% Tween 20. For long/medium range storage keep at 4 °C.

2. Washing buffer: PBS/0.2% Tween 20. Store at 4 °C.

3. Flat bottom 6-well tissue culture plastic plates.

4. Portable Rotor.

5. Glass slides.

6. 24 mm × 32 mm glass coverslips.

7. 4′,6-Diamidin-2-phenylindol (DAPI).

8. One-Step Tissue Clearing reagent/kit (e.g., RapiClear 1.52™, SUNJin Lab or Focus Clear™ Celexplorer Labs).

2.4 Microscopy and Image Analysis

1. Confocal microscope equipped with several lasers and detectors enabling spectral separation of 4–5 fluorophores.

3   Methods

3.1 BM Harvesting and Tissue Processing

1. Prepare PFA-based fixation buffer, precool to 4 °C and keep on ice (*see* **Note 1**).

2. Euthanize mice by CO_2 asphyxiation and proceed to isolate bones immediately.

3. Collect femoral bones from the kneecap to the hip and clean thoroughly using tissue Kimwipes (*see* **Note 3**).

4. Directly place the clean femoral bones in fixation buffer. Incubate for 8–12 h at 4 °C to ensure complete fixation of BM tissues.

5. Thoroughly wash bones by prolonged and repeated incubations in PBS. We recommend at least three washes of 1 h under gentle agitation at 4 °C.

6. After repeated washing, incubate fixed femoral bones in PBS–30% sucrose rehydrating solution and incubate for 48–72 h at 4 °C (*see* **Note 4**).

7. Take bones out of the rehydrating solution, and gently dry excess liquid by placing in precision wipes or absorbent paper. It is important to optimally position the femoral bones in cryomolds by leveling the flattest, most regular side of the femur facing down and longitudinally along the longest axis of the mold (*see* **Note 5**). Add OCT slowly to avoid the formation of air bubbles until the whole bone is covered in freezing medium. If bubbles, nevertheless, form between the bone and the surface of the cryomold, gently lift and move the bones to allow displacement of smallest bubbles toward the edge of the cryomold, from where they can be directly removed with forceps.

8. Once the bones are positioned and completely covered by OCT, use long metal forceps to grab the cryomold and freeze sample by placing it such that only the lower surface is in contact and partially immersed in liquid nitrogen (*see* **Note 6**).

9. Frozen samples can be temporally stored in dry ice but long-term preservation should be done in a −80 °C freezer until further processing.

3.2 Generation of Thick Slices of BM Tissues

1. Set temperature of the cryostat and sample holder to −25 °C. Insert the steel blade in the cryostat and place it in its holder, at least 30 min prior to sectioning, to allow cooling of the material to appropriate temperatures before use (Fig. 1a–c).

2. Add one drop of OCT on the metal adaptor and immediately place sample block on top. The rapid freezing of the medium will cement the sample to the adaptor. Once OCT is solidified place metal adaptor containing the adhered OCT block in the corresponding sample holder of the cryostat (Fig. 1d, e).

3. Start facing block in 30 μm steps until reaching the bone. Make sure from the very beginning to start sectioning the block as homogeneously as possible throughout its entire surface. If the bone has been optimally placed flat in the mold, this will allow the sectioning of the entire femur throughout its longitudinal axis in an even fashion, which is needed for homogeneous exposure of the BM tissues (Fig. 1e, zoomed in image). Once you approach the bone and it becomes visible below the thinned OCT layer, reduce the step size (to 5 μm) and carefully start cutting into the bone. While doing this, discard all the thin tissue section that result from the process. Repeat this until the red BM cavity is evenly visible throughout the all regions, including both metaphyses as well as the diaphysis (*see* **Notes 7** and **8**) (Fig. 1f).

4. Once a homogeneous surface of whole BM cavity is visible, remove the sample from the holder using a razor blade to cut through the OCT layers on the inferior side of the sample while holding the metal piece as shown in Fig. 1g. Clean the sample holder of any residual OCT fragments.

5. As described in **step 2**, add a large drop of fresh liquid OCT to the surface of the holder and wait for several seconds in order for OCT to slightly freeze so that exposed BM would not directly touch the holder and get damaged. Before OCT freezes completely, turn the sample block containing the sectioned bone upside down and adhere the sectioned side to the metal block by pressing against this thick layer of OCT (Fig. 1h, i).

6. Once the OCT completely freezes, place the block on the holder and repeat the exact same process, iteratively shaving the sample block to finally uncover the other face of the femoral BM cavity. The end result will be a bilaterally exposed think slice of BM, embedded in a thin layer of frozen OCT (Fig. 1j, k, and *see* **Note 9**).

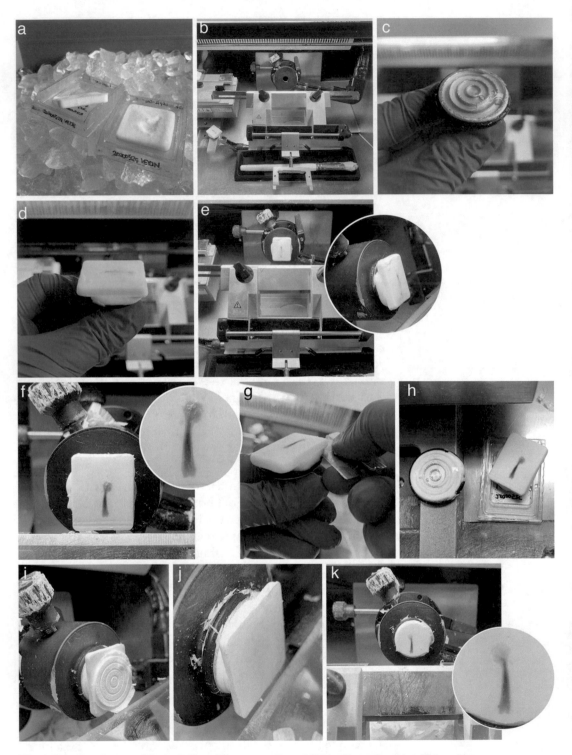

Fig. 1 Step-by-step graphic depiction of the generation of thick slices of murine BM cavities

7. Keep the thick BM slice inside the refrigerated chamber of the cryostat, or alternatively in dry ice until you are ready to proceed with the thawing and start the immunostaining process.

8. Once ready to continue with tissue processing, take the block out of the cryostat and before thawing of the OCT completes at RT, carefully remove as much excess OCT surrounding the thick sample as possible using razor blade (Fig. 2a–c) until only a small region of surrounding OCT is left around the sample. To completely clean the remaining OCT, take the sample with forceps and dip it in one well of a 6-well plate filled with room temperature PBS (Fig. 2c, d). After the OCT layer appears completely thawed in the PBS solution, use forceps to very gently encircle the bone slice, which will help to separate remaining OCT as much as possible from the proximity of the bone slice, as indicated in Fig. 2d. This should avoid that once the bone has been picked up with forceps, the high viscosity of the excess OCT will cause separation of BM from the bone frame, while dripping. To manipulate the slice, always carefully hold it with forceps from the tip of one of the femoral metaphysis, or pieces of by the remaining threads of connective tissue that may be left in the bone ends (Fig. 2e).

9. To wash the bone slice, thoroughly sequentially dip it three times into new, clean wells of the plate filled with fresh clean PBS and incubate in wells with fresh PBS at least twice (Fig. 2f).

3.3 Immunostaining and Tissue Clearing

1. Incubate clean BM slices in 1 ml of fresh fixation solution inside a 1.5 ml eppendorf tube, for 4–8 h at 4 °C. This additional fixation step preserves the integrity of the issue and improves antigenicity (Fig. 2g).

2. Wash BM slices in fresh PBS–0.2% Tween inside the Eppendorf tube. At least three washing steps of 1 h under gentle agitation at 4 °C are recommended.

3. Block sample overnight in 500 µl of blocking buffer at 4 °C under gentle rotation (Fig. 2h).

4. Remove blocking solution by slightly tilting eppendorf tube to displace the sample toward the bottom side of the tube and aspirating the solution from the top.

5. Add primary antibody mix in a total volume of 500 µl of blocking buffer (*see* **Note 10**). Incubate at 4 °C under gentle agitation for 24–72 h (*see* **Note 11**).

6. Remove primary antibody solution by aspiration. Wash sample in fresh washing solution for 2 h at 4 °C under agitation. Repeat procedure by changing the washing solution at least three times.

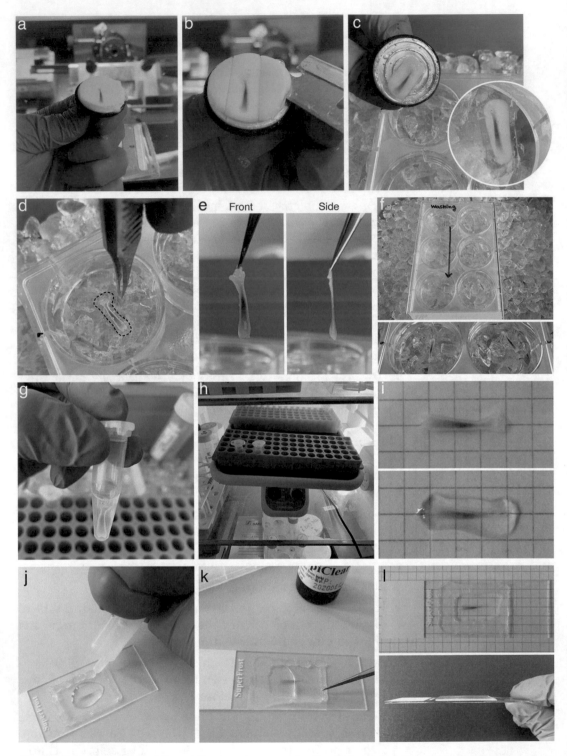

Fig. 2 Detailed graphic depiction of the sample processing, immunostaining, and optical clearing procedure

7. Repeat **step 5** with the solution containing secondary antibodies and incubate for 24–72 h. From this step on, if secondary antibodies are conjugated from fluorophores, minimize exposure from light.

8. Wash as described in **step 6**.

9. If an additional step using tertiary antibodies or streptavidin conjugated antibodies is needed, repeat the procedure (**steps 7** and **8**).

10. After the last washing step, remove washing solution from the tube containing the sample and add 1 ml of 0.5–1.0 μM DAPI. Leave for 1–2 h at 4 °C.

11. Perform washing step as described in **step 6**.

12. Proceed with optical clearing process. Remove excess washing buffer and add clearing solution RapiClear (*see* **Note 12**) to completely cover the sample. Incubate overnight at 4 °C in an environment protected from light. After clearing, the tissue slice becomes at least partially transparent (Fig. 2i).

3.4 Deep Tissue Imaging

1. Mount the optically cleared sample for imaging. Take BM slice with forceps and place in a clean glass slide. Position the sample to ensure that it lies along the flattest sectioned surface.

2. Apply vacuum grease around the sample to seal a large region around the BM slice, within which the clearing solution will be added until it superficially covers the sample (Fig. 2j).

3. Place a cover slip on top of the sample and apply gentle pressure to the sides of slide along the boundaries of the vacuum grease layer, until the bottom side of the slip is in touch with the BM slice. Excess clearing solution may leak out and should be removed with a Kimwipes (Fig. 2k, l).

4. Imaging should be performed with an advanced confocal upright or inverted microscope with specific objectives adapted for deep tissue imaging, as well as the laser and detector sets for the excitation and detection of the used fluorophores (*see* **Notes 12–14**).

4 Notes

1. The choice of PFA may significantly influence the quality of the immunostaining process. We recommend prediluted PFA at 16% available from a number of commercial vendors including Electron Microscopy Sciences. Alternatively, freshly prepared PFA solutions should be used.

2. The steel blade should be maintained and sharpened on a regular basis. The maintenance of the blade is critical as the presence of small nicks, dents, or scratches will affect the quality

of the sections. Uneven sectioning or the presence of irregular cuts will typically result increased autofluorescence and unspecific deposition of antibodies and will make the mounting of thick slices challenging.

3. If imaging of all relevant regions of the BM is to be performed, it is important not to damage the bones during harvesting, as well as to thoroughly remove surrounding muscle and connective tissue. With the use of a scalpel and/or precision wipes, carefully detach these adhered tissues, which are a typical source of autofluorescence and artifacts during sample preparation and imaging.

4. Prolonged incubation of at least 6 days in 30% sucrose at 4 °C has been shown not to significantly affect the preservation of the tissue and immunostaining process. This period provides a perfect time window within which bones maybe shipped or transported in ice.

5. The epiphyses and metaphyses of the femoral bones are voluminous and stick out compared to the diaphysis making it challenging to find a flat surface along which to place the femur. However, finding the optimal position to position the bone is key for the subsequent sectioning steps as it will facilitate exposing the entire BM cavity uniformly, and the generation of get complete tissue sections spanning all regions of the bone.

6. Complete dipping of the cryomold may result in the rapid formation of large air bubbles in the OCT mold, fractures, and/or uneven freezing. It is crucial to avoid this by ensuring that molds are only partial dipped in nitrogen, which will result in gradual and homogeneous solidification of the OCT medium.

7. The integrity of the tissue slice obtained is a key factor that will absolutely determine the quality of the immunostaining procedure, the appearance of potential artifacts (notches, unspecific staining patterns, antibody precipitates), the signal to noise ratio, the depth of penetration at which images may be generated, and thus the end result of the imaging. It is therefore of utmost importance to pay specific attention to this step and try to cut slices that are homogeneous in thickness and in which marrow cavities are exposed as cleanly and evenly as possible.

8. It is invariably necessary to constantly and gradually adapt the sectioning angle throughout the process, in order to align the sectioning plane with the main longitudinal plane of the femoral bone.

9. The described process does not allow to standardized or predetermine the final thickness of the slice. We estimate that, when successfully generated, bilaterally exposed sections have

an even thickness, which varies between 300 and 700 μm. The aim is to obtain the thinnest even slice possible in the entire breadth of the bone is preserved.

10. The optimal concentration should be standardized for each antibody. Keep in mind that thick specimens, as opposed to thin histological sections are being stained, may require antibody concentrations similar to those employed for whole mount immunostainings of large organs.

11. On average we incubate antibody mixes for 48 h to ensure signals at imaging depths of 150–200 μm. However, the duration of these steps maybe adapted, depending on the type and depth of acquisition of the experiment. We have observed efficient antibody penetration after 24 h of incubation.

12. When performing imaging of thick specimens, special attention should be paid to the objectives employed. Long working distance should be selected and, to avoid spherical aberrations, special attention should be paid to the immersion media, whose refractive index should be matched as much as possible to that of the clearing solution. In the case of the protocol here described the RI of RapiClear ranges from 1.47 to 1.52, therefore matching the RI of typically used immersion oils. For a detailed discussion on clearing methods and objective selection please *see* [10].

13. As in any fluorescence-based technique, signal intensity will depend on multiple factors related not only to the preparation of the sample, fixative employed, affinity of the antibody and intensity of the fluorochrome, but also to the sensitivity of the microscopy set-up and detection system employed. Background, unspecific fluorescence may be substantial, especially within certain ranges of the electromagnetic spectrum. We thus recommend the use of high-sensitivity last generation detectors found in advanced confocal microscopy setups, which should maximize the signal to noise ratio.

14. Beyond femurs we have employed a similar protocol to image other bones, and by modifying the sectioning procedure and using a vibratome, this protocol maybe adapted to generate images of other organs: spleens, thymi, and livers.

Acknowledgments

This work was supported by grants from the Swiss National Science Foundation (310030_185171), the Swiss Cancer Research Foundation (KFS-3986-08-2016) and the Clinical Research Priority Program of the University of Zurich (to C.N-A).

References

1. Nombela-Arrieta C, Manz MG (2017) Quantification and three-dimensional microanatomical organization of the bone marrow. Blood Adv 1:407–416. https://doi.org/10.1182/bloodadvances.2016003194

2. Morrison SJ, Scadden DT (2014) The bone marrow niche for haematopoietic stem cells. Nature 505:327–334. https://doi.org/10.1038/nature12984

3. Feuerer M, Beckhove P, Garbi N et al (2003) Bone marrow as a priming site for T-cell responses to blood-borne antigen. Nat Med 9:1151–1157. https://doi.org/10.1038/nm914

4. Mercier FE, Ragu C, Scadden DT (2012) The bone marrow at the crossroads of blood and immunity. Nat Rev Immunol 12:49–60. https://doi.org/10.1038/nri3132

5. Chang H-D, Tokoyoda K, Radbruch A (2018) Immunological memories of the bone marrow. Immunol Rev 283:86–98. https://doi.org/10.1111/imr.12656

6. Pinho S, Frenette PS (2019) Haematopoietic stem cell activity and interactions with the niche. Nat Rev Mol Cell Biol 20:1–18. https://doi.org/10.1038/s41580-019-0103-9

7. Ludewig B, Stein JV, Sharpe J et al (2012) A global "imaging" view on systems approaches in immunology. Eur J Immunol 42:3116–3125. https://doi.org/10.1002/eji.201242508

8. Gomariz Á, Isringhausen S, Helbling PM et al (2019) Imaging and spatial analysis of hematopoietic stem cell niches. Ann N Y Acad Sci 94:284–212. https://doi.org/10.1111/nyas.14184

9. Ntziachristos V (2010) Going deeper than microscopy: the optical imaging frontier in biology. Nat Methods 7:603–614. https://doi.org/10.1038/nmeth.1483

10. Richardson DS, Lichtman JW (2015) Clarifying tissue clearing. Cell 162:246–257. https://doi.org/10.1016/j.cell.2015.06.067

11. Pittet MJ, Garris CS, Arlauckas SP et al (2018) Recording the wild lives of immune cells. Sci Immunol 3:eaaq0491. https://doi.org/10.1126/sciimmunol.aaq0491

12. Gomariz Á, Helbling PM, Isringhausen S et al (2018) Quantitative spatial analysis of haematopoiesis-regulating stromal cells in the bone marrow microenvironment by 3D microscopy. Nat Commun 9:407–415. https://doi.org/10.1038/s41467-018-04770-z

13. Coutu DL, Kokkaliaris KD, Kunz L et al (2017) Multicolor quantitative confocal imaging cytometry. Nat Methods 15:39–46. https://doi.org/10.1038/nmeth.4503

14. Grüneboom A, Hawwari I, Weidner D et al (2019) A network of trans-cortical capillaries as mainstay for blood circulation in long bones. Nat Metab:1–19. https://doi.org/10.1038/s42255-018-0016-5

15. Greenbaum A, Chan KY, Dobreva T et al (2017) Bone CLARITY: clearing, imaging, and computational analysis of osteoprogenitors within intact bone marrow. Sci Transl Med 9:eaah6518. https://doi.org/10.1126/scitranslmed.aah6518

16. Acar M, Kocherlakota KS, Murphy MM et al (2015) Deep imaging of bone marrow shows non-dividing stem cells are mainly perisinusoidal. Nature 526:126–130. https://doi.org/10.1038/nature15250

Chapter 12

Ex Vivo Whole-Mount Imaging of Leukocyte Migration to the Bone Marrow

Stephan Holtkamp and Christoph Scheiermann

Abstract

The bone marrow is the major hematopoietic organ, consisting of distinct microenvironmental niches for the production of hematopoietic cells. Advanced visualizing methods are required to define and better understand the interactions between stromal and hematopoietic cells. In this chapter, we describe an ex vivo whole-mount imaging technique of the bone marrow, which allows for a fast, high-quality, and three-dimensional visualization of different bone marrow components. We provide a guide for conducting adoptive transfer experiments of fluorescently labeled leukocytes and visualizing their location in the bone marrow with respect to the bone marrow vasculature. This method presents a quick, easy, and inexpensive approach to image the bone marrow in three dimensions.

Key words Bone marrow, Adoptive transfer, Whole-mount three-dimensional imaging, Vasculature staining

1 Introduction

The bone marrow provides the environment for the generation of hematopoietic stem cells (HSCs), progenitor cells as well as most mature leukocyte subsets. Different microenvironmental domains or niches such as the vasculature have been implicated in this process [1, 2]. Via the vasculature, leukocytes are mobilized from the bone marrow (BM) but also home back to this tissue. The bone marrow vasculature is composed of one central vein and several arteries, aligned in parallel to the vein. From the center of the bone (medulla) toward the cortex (calcified), numerous sinusoids branch off (Figs. 1f and 2b). These vessels then perpendicularly traverse the cortical bone and connect to the periosteal circulation [3].

Our current understanding of the BM complexity is largely based on different microscopy technologies. Two-dimensional (2D) immunohistology involving classical cryosectioning of bones has generated first important imaging insights into HSC physiology [4–7]. However, generation of intact BM sections is challenging

Marion Espéli and Karl Balabanian (eds.), *Bone Marrow Environment: Methods and Protocols*, Methods in Molecular Biology, vol. 2308, https://doi.org/10.1007/978-1-0716-1425-9_12, © Springer Science+Business Media, LLC, part of Springer Nature 2021

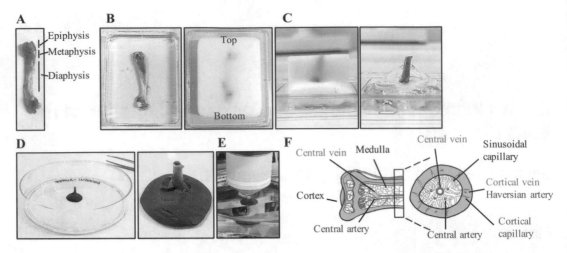

Fig. 1 Preparation and mounting of bones for whole-mount imaging. (**a**) The femur is harvested and cleaned. (**b**) The femur is embedded in a cryomold filled with OCT compound and the small, bottom end is glued to the cryostat sample holder. (**c**) The femur is vertically cut from the top to the center of the diaphysis. It is left to thaw and carefully extracted from the melting OCT. The OCT is immediately washed off in PBS, making sure OCT has not come in contact with the surface area. (**d**) The bone is mounted in modelling clay and positioned upright, trying to perfectly level the imaging surface. (**e**) Using water dipping objectives, the bone is imaged in a petri dish filled with PBS, making sure no bubbles having been introduced to the top of the imaging area. (**f**) Structure of blood vessels in the bone, (adapted from [2])

and requires special equipment or long decalcification processes alongside with loss of structural and functional information [8, 9]. Furthermore, it does not allow for a three-dimensional (3D) assessment of structures.

During the past decade, microscopy-based investigations of the BM have shifted from 2D to technologies that allow acquiring 3D representations. One of these advances is the use of tissue clearance protocols, during which the tissue becomes translucent providing drastically increased imaging depth and preserved resolution [10–12]. Nonetheless, clearing reagents often modify integrity and structure of proteins as well as antigenicity of epitopes [13]. Moreover, most clearing methods described require a significant amount of time.

Overcoming this, 3D volumetric whole-mount imaging of the BM is a fast, efficient, and cheap method allowing for the exploration of different microenvironments implicated in HSC formation and leukocyte function [6, 14–17]. Here, we provide an in-depth step-by-step guide to the 3D whole-mount imaging of the BM of long bones. This includes adoptive intravenous (i.v.) transfer of isolated fluorescently labeled leukocytes, staining of the BM vasculature as well as whole mounting of bones and the imaging of transferred cells within the BM microenvironment. During these adoptive transfers, intravenously injected leukocytes home to the BM and emigrate into the interstitial space. Despite its limitation in

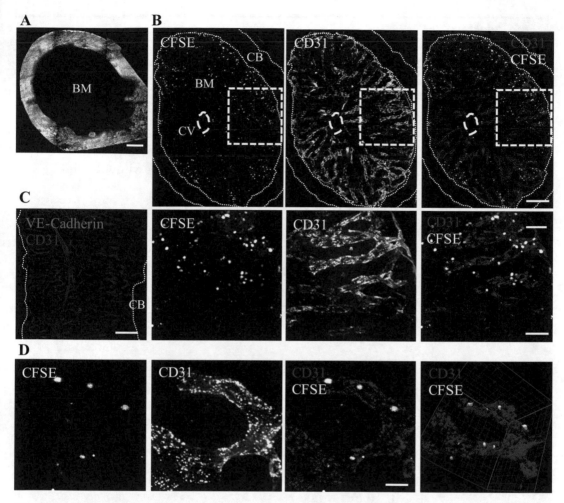

Fig. 2 Whole-mount imaging of bone marrow to define the location of CFSE$^+$ adoptively transferred cells after perfusion. (**a**) Autofluorescence of the bone. The calcified bone appears bright, the central marrow dark. BM = bone marrow. Scale bar = 150 μm. (**b**) Low (10×)- and high (20×)-magnification overview of the transverse imaged BM vasculature. Scale bar = 150 μm and 50 μm, respectively. CB = calcified bone, CV = central vein, BM = bone marrow. (**c**) Sagittally imaged BM vasculature. Scale bar = 150 μm. (**d**) 63× (3D) magnification of the BM vasculature. Scale bar = 20 μm, grid line = 10 μm

imaging depth and the need for sectioning, no further chemicals are required, allowing for the use of reporter mice and preservation of epitope antigenicity. This technique thus offers a quick and inexpensive way to achieve high spatial resolution of the migratory behavior of leukocytes and the influence of the BM microenvironment in situ.

2   Materials

2.1   Harvest and Staining of Cells

1. Femoral and tibial bones from a 6–10-week-old mouse.

2. A pair of scissors (small and straight).

3. A pair of blunt and sharp forceps (straight).

4. 1.5 mL microfuge tubes.

5. 1 mL/200 μL pipette tips.

6. 1× phosphate-buffered saline (PBS) without $MgCl_2$ and $CaCl_2$.

7. Tissue paper.

8. 1 mL syringe.

9. 21G needle.

10. Polypropylene conical centrifuge tube (50 mL).

11. 40 μm/70 μm strainer.

12. 10× RBC lysis buffer: mix 82.6 g NH_4Cl, 10 g $KHCO_3$, and 1 mL 0.5 M EDTA with a pH of 7.25 with ultrapure water up to 1 L.

13. Cell incubation buffer: mix 4 mL 0.5 M EDTA (final concentration: 0.02 M) and 20 mL of FCS (final: 2%) with ultrapure water up to 1 L.

14. Carboxyfluorescein succinimidyl ester (CFSE).

2.2   Adoptive Transfer, Blood Vasculature Staining, and Bone Harvest

1. 6–10-week-old recipient mice.

2. 0.5 mL insulin syringe (with 30G×1/2 needle).

3. 10 mL syringe.

4. 22G needle.

5. 1.5 mL microfuge tubes.

6. Rat anti-mouse CD31/PECAM-1 antibody (Ab) at 0.2 mg/mL coupled to APC fluorophore (clone 390 (nonblocking)) is recommended based on our experience.

7. 1 mL and 200 μL pipette tips.

8. Small and straight microsurgical scissors.

9. OCT compound.

10. Cryomold.

11. Dry ice.

2.3   Preparation of Bone Marrow

1. S35 microtome blade.

2. Any cryostat will be suitable (we use a Leica CM3050 S cryostat).

3. Blunt forceps.

4. 1.5 mL microfuge tubes.

5. 4% PFA in PBS.

6. Modeling clay.

7. Petri dishes (sterile, 100 × 15 mm).

8. Ultrapure water.

9. For 3D fluorescence imaging, any standard laser scanning or spinning disk confocal microscope with a movable stage, a 20× and 10× water dipping objective, as well as 485 nm and 650 nm laser is suitable.

3   Methods

The whole-mount imaging of ex vivo BM containing adoptively transferred, fluorescently labeled cells requires several preparation steps: leukocytes are harvested from donor mice and subsequently stained (about 2–2.5 h), cells and antibodies labeling the vasculature are injected into recipients prior to organ harvest (at least 1 h) and bones are prepared for imaging (preparation takes 1 h, imaging 1 h/bone).

3.1   Extraction and Staining of Cells for Injection

1. The cells for the adoptive transfer are isolated from spleen and BM (*see* **Note 1**). One donor mouse yields enough cells for injection of approximately 3–4 mice.

2. Euthanize the mouse according to local regulations and animal protocols. Harvest the spleen first. Carefully open the upper abdomen using a small, straight pair of scissors. Release the spleen from the fat and ligaments using blunt and sharp forceps and store the spleen in 1 mL PBS in a microfuge tube.

3. Harvest two femora and tibia. Release the legs from the pelvic bones by cutting the skin using a straight pair of scissors and start separating the femur and tibia by cutting major ligaments and muscle strands. By using either reverse rotations of the bones or scissors to cut the joint you can easily separate the femur from the tibia. Clean the bones as much as possible with tissue paper. Use a scalpel to cut open the end of the bones closest to the knee joint. A single cell suspension is obtained by flushing the BM gently into a 1.5 mL microfuge tube/bone using a 21G needle attached to a small 1 mL syringe filled with PBS. Make sure not to suck any air and not to create any bubbles as this reduces the viability of cells. The flushed bone can be discarded and the four tubes are then mixed together in one 50 mL falcon tube (*see* **Note 2**).

4. To obtain the splenic single cell solution, gently smash the spleen through a 40 μm or 70 μm cell strainer (depending on the cell types of interest) into a 50 mL falcon tube (from this step onward, the cells are always maintained in a 50 mL falcon tube).

5. Both single cell solutions are washed by spinning down at $300 \times g$ at RT for 5 min. The supernatant is discarded and the cell pellets are resuspended in 5 mL $1\times$ RBC lysis buffer for 5 min (the longer the incubation, the more cells are damaged). Neutralize the reaction by adding 5 mL PBS (RT) and spin down the cell solution at $300 \times g$ at RT for 5 min.

6. Resuspend the cells in 1 mL cell incubation buffer. Create a 1/10 dilution in PBS for counting in a 1.5 mL microfuge tube. The cells are counted on an automated cell counter with a reaction volume of 50 µL (*see* **Note 3** if no cell counter is available). Usually around 40×10^6 live cells are yielded from one spleen; around 50×10^6 live cells are extracted from four bones. Use 10^7 BM cells and 10^7 splenic cells from the single cell solutions and mix for a 1:1 mixture in a 50 mL falcon.

7. The donor cells are labeled with 1.5 µM CFSE for 20 min at 37 °C (*see* **Note 4**) in darkness. Gently shake the tube every 5 min. Afterward, wash cells three times with 5 mL incubation buffer with 5 min $300 \times g$ spins at RT each. Resuspend in 700 µL PBS. Take 100 µL for another count and 600 µL for injection (use sterile PBS as this is for injection). The measured concentration should be around 25–50 million cells/mL. Consequently, around 5–10 million cells/mL are injected into one mouse in a volume of 200 µL (*see* **Note 5**).

8. Store the cells at RT in the dark until the adoptive transfer.

3.2 Adoptive Transfer, Blood Vasculature Staining, and Bone Harvest

1. For the injection, prepare a 500 µL syringe ($30G \times 1/2$ needle) per mouse with 200 µL stained cell solution.

2. Inject the cells i.v. into the recipient mouse (i.v. injection method is based on local animal protocols). Let the cells recirculate for 1 h. In case of multiple injected mice, it is essential to mark the animals in a cage, which one has been injected first so that the 1 h incubation time does not differ amongst mice injected.

3. 55 min after injection, antibodies labeling the vasculature are injected i.v. For thorough staining of the BM vasculature we recommend injecting an APC-labeled αCD31/PECAM-1 antibody (clone 390) in a final concentration of 0.04 mg/mL in 200 µL PBS (8 µg total; *see* **Note 6**). The injection of an anti-CD31/PECAM-1 antibody can also be combined with injection of an anti-VE-Cadherin antibody (clone BV13) with a final concentration of 0.04 mg/mL in 200 µL PBS (8 µg total each), which together improve the staining quality (Fig. 2c) [18]. Incubate for 5 min. Prepare 10 mL PBS in a 10 mL syringe attached to a 22G needle for perfusion.

4. Euthanize the recipient mouse according to local regulations but be attentive not to damage the circulation by breaking the neck. Carefully but quickly open abdomen (to expose the liver)

and thorax of the mouse with straight scissors to fully expose the heart. Use small surgical scissors to make a small incision into the right atrium. Take blunt forceps and hold the heart by grabbing the lower right ventricle. Simultaneously, use the syringe to slowly perfuse the circulation by injecting 10 mL PBS into the left ventricle. During the perfusion, the liver and paws should turn pale (which acts as a control of the perfusion; *see* **Note** 7) and the perfused blood should leave through the small incision in the right atrium.

5. Harvest of bones (at least one femur). Release the legs from the pelvic bones and separate the femur and tibia by cutting major ligaments and muscle strands. By using either reverse rotations of the bones or scissors to cut the joint you can easily separate the femur from the tibia. Clean the bones as much as possible with tissue paper (Fig. 1a).

6. After cleaning, the bones are embedded in OCT compound with each bone being embedded in a single cryomold in parallel to the long edges (Fig. 1b). This is important for the correct orientation of the bone during the cutting procedure. The bones are then shock frozen on dry ice and stored at −80 °C until further processing (Fig. 1c).

3.3 Preparation of Bones for Whole-Mount Imaging and Analysis

1. Take the bones from −80 °C and store them at −20 °C for 45 min to adjust to the cutting temperature in the cryostat. Glue the cryomold with OCT with the bottom (Fig. 1b) to the cryostat sample holder (*see* **Note 8**). Cut the bones vertically from top to bottom (=transverse cutting) with a 50 μm step size using a cryostat (Fig. 1c). Start counting the cuts once you have reached the bone and trimmed the OCT. This can be done very fast. To reach the center of the bone cut around 1 cm deep (i.e., 20 cuts). For the final cuts use a cutting depth of 10 μm and cut more slowly so as not to cause any damage to the bone marrow as this is the surface to be imaged (*see* **Note 9**).

2. Let the OCT thaw at RT until it softens and the bone can be easily released by carefully taking it out using blunt forceps (Fig. 1c). Make sure that the imaging surface does not come in contact with OCT at any point. Wash the bone in 1 mL PBS in a 1.5 mL microfuge tube for 10 min and place it in 1 mL 4% PFA for 30 min in another 1.5 mL microfuge tube. Make sure not to shake the bones too much as cells could detach from the surface of the BM.

3. Place a 0.5–1 cm layer of modeling clay on a 10 mL petri dish and place the bone in the modelling clay with the imaging surface facing upward (Fig. 1d). Make sure the bone is quickly and completely immersed in PBS and placed exactly vertical so

that the surface is completely even. This will reduce the number of Z-stacks that will need to be acquired during the imaging step of the whole surface.

4. Once the bone is immersed in liquid, imaging of the whole mount can be started. Imaging is possible with any fluorescence microscope capable of 3D imaging and a moving stage with water dipping objectives (in our case $10\times$ and $20\times$ objectives) (Fig. 1e and Subheading 2). To be able to visualize the CFSE$^+$ labeled cells and the labeled CD31/PECAM-1 on the endothelial cell surface 485 nm and 650 nm lasers, respectively, are required.

5. First, use the objective with the lower magnification to localize the calcified bone and BM (*see* **Note 10** and Fig. 2a). For analysis, use the objective with the best resolution available (ideally a magnification of around $20\times$ as this ensures capture of multiple vessels and cells per image, Fig. 2b). Imaging can be done up to a depth of around 80 μm, depending on the microscope components (*see* **Note 11**).

6. For analysis, obtain 3D images from different areas of the bone with an interval between individual sections of around 1 μm in the z-stack (Fig. 2b, c). Eight image stacks (obtained with the $20\times$ objective) are usually enough to cover the whole BM of the femoral bone (Fig. 1f and 2b). The CD31 staining appears clustered but the vascular wall can be well defined (Fig. 1f and 2b).

7. For the final analysis, software such as Slidebook (version 6, Intelligent Imaging Innovations, 3i) or ImageJ (version 1.51n) can be used. In principle, any 3D viewer is fine. Open the z-stack file and create a 3D model and simply count the cells by rotating the BM (Fig. 2d). Due to the vasculature staining and the subsequent perfusion, transferred cells can easily be defined to be located either inside (adherent) or outside (extravasated) the vasculature. Afterward, calculate the total volume of the image. Once the volume has been calculated, the cell numbers can be normalized (if necessary). Furthermore, an inside versus outside ratio can be generated based on the normalized cell numbers.

4   Notes

1. Depending on the experiment and research question, cells can also be harvested from the lymph node and spleen if the aim is to investigate mostly lymphocytes. BM (myeloid cells) and spleen will yield a mix of most leukocyte subsets.

2. Using different tubes and thus different volumes of PBS, the reflushing of cells is minimized, which leads to higher cell viability.

3. If there is no ProCyte or cell counting machine available, it is recommended to count the cells by hand using a trypan blue exclusion test of cell viability. If the cells are counted without a live–dead stain, the living fraction of injected cells might be too low.

4. The CFSE concentration described is optimized for 2×10^7 cells. If more or less cells are injected, CFSE concentrations are to be adjusted. If cells from two different experimental settings are injected into the same mouse, use two different labels for 10^7 mixed donor cells, for example CellTracker Deep Red dye (0.1 μM; Thermo Fisher) and CellTracker Violet dye (concentrations need to be titrated; Thermo Fisher).

5. Do not inject more than 200 μL and more than 20 million cells/mouse at a time as this could lead to mouse death due to cells getting stuck in the lungs.

6. Due to i.v. injection, only the blood vasculature is stained and the imaging is thus very crisp. Do not use other clones of anti-CD31/PECAM-1 antibodies such as MEC13.3 as they can block the transmigration of leukocytes and thus interfere with the experiment. However, since the labeling of the vasculature is performed at the very end of the experiment, the potential effect is minimal. The clone 390 has been shown to not block transmigration [19]. Other directly labeled antibodies, directed against isolectin B4 [20], endomucin [21], or the nuclear dye rhodamine 6G [22] can also be used to visualize the vasculature. However, in this setting, anti-CD31/PECAM-1 or the combination of CD31/PECAM-1 and VE-Cadherin have worked best for us.

7. PBS perfusion washes out unbound anti-CD31/PECAM-1 Abs and erythrocytes, as well as rolling and circulating leukocytes, which minimizes the autofluorescence of unspecific signal and allows for determining the step in the adhesion cascade of transferred leukocytes. If a thoroughly perfused circulation is crucial, the extent of the perfusion can be investigated using a TER-119 staining to visualize the remaining erythrocytes (we can recommend the PE anti-mouse TER-119/Erythoid Cells antibody from BioLegend) within the vasculature.

   Depending on the steps after harvest, you can also perfusion-fix with 4 mL of 4% PFA after PBS injection if further immune-stains after bone harvest are desired. However, it is recommended to test subsequent stains with and without PFA perfusion. If no further stain after bone harvest is performed, PFA perfusion is not necessary.

8. The cryomolds will be glued with OCT to the cryostat with the small bottom end. This is unusual, as the mold is now in a vertical position, allowing the transverse sectioning. Make sure you use enough OCT so that the mold does not break off. Also, you will now need more room between the cryostat sample holder and the blade.

9. The individual sections cannot be collected unless the bone is either decalcified beforehand or a specific knife is used that is able to cut calcified bones. Therefore, we discard the cuts. The direction the bone is cut depends on your experiment. This imaging and analysis aims for a bone that has been cut vertically from top to bottom. You can also cut the bone horizontally from side to side (=sagittal imaging, Fig. 2c).

10. The calcified part appears bright due to its strong autofluorescence. Make sure the objective is free of any air bubbles; they easily get trapped between sample and objective if the objective is completely dipped in water. You can gently flush them out with a syringe.

11. The microscope we used is an upright Zeiss Axio Examiner.Z1 confocal spinning disk microscope, operated with Slidebook software (3i, Intelligent Imaging Innovations). It is equipped with a 10x ocular, 387 nm, 485 nm, 560 nm, 650 nm lasers and the following objectives: $10\times$ objective: w-plan—apochromat, NA = 0.3; $20\times$ objective: w-plan—apochromat, NA = 1.0; $63\times$ objective: w-plan—apochromat, NA = 1.0. Together, this equipment allows for low and high magnification of migrated cells in combination with a multi color staining panel.

Acknowledgments

We thank Jasmin Weber for generating the artwork in Fig. 1f. C.S. was supported by the European Research Council (ERC StG 635872), and the Swiss National Science Foundation (SNF, 182417).

References

1. Morrison SJ, Scadden DT (2014) The bone marrow niche for haematopoietic stem cells. Nature 505(7483):327–334. https://doi.org/10.1038/nature12984

2. Ramasamy SK (2017) Structure and functions of blood vessels and vascular niches in bone. Stem Cells Int 2017:5046953. https://doi.org/10.1155/2017/5046953

3. Gruneboom A, Hawwari I, Weidner D, Culemann S, Muller S, Henneberg S, Brenzel A, Merz S, Bornemann L, Zec K, Wuelling M, Kling L, Hasenberg M, Voortmann S, Lang S, Baum W, Ohs A, Kraff O, Quick HH, Jager M, Landgraeber S, Dudda M, Danuser R, Stein JV, Rohde M, Gelse K, Garbe AI, Adamczyk A, Westendorf

AM, Hoffmann D, Christiansen S, Engel DR, Vortkamp A, Kronke G, Herrmann M, Kamradt T, Schett G, Hasenberg A, Gunzer M (2019) A network of trans-cortical capillaries as mainstay for blood circulation in long bones. Nat Metab 2:236–250. https://doi.org/10.1038/s42255-018-0016-5

4. Kiel MJ, Yilmaz OH, Iwashita T, Yilmaz OH, Terhorst C, Morrison SJ (2005) SLAM family receptors distinguish hematopoietic stem and progenitor cells and reveal endothelial niches for stem cells. Cell 121(7):1109–1121. https://doi.org/10.1016/j.cell.2005.05.026

5. Zhao M, Perry JM, Marshall H, Venkatraman A, Qian P, He XC, Ahamed J, Li L (2014) Megakaryocytes maintain homeostatic quiescence and promote post-injury regeneration of hematopoietic stem cells. Nat Med 20(11):1321–1326. https://doi.org/10.1038/nm.3706

6. Nombela-Arrieta C, Pivarnik G, Winkel B, Canty KJ, Harley B, Mahoney JE, Park SY, Lu J, Protopopov A, Silberstein LE (2013) Quantitative imaging of haematopoietic stem and progenitor cell localization and hypoxic status in the bone marrow microenvironment. Nat Cell Biol 15(5):533–543. https://doi.org/10.1038/ncb2730

7. Mokhtari Z, Mech F, Zehentmeier S, Hauser AE, Figge MT (2015) Quantitative image analysis of cell colocalization in murine bone marrow. Cytometry A 87(6):503–512. https://doi.org/10.1002/cyto.a.22641

8. Prasad P, Donoghue M (2013) A comparative study of various decalcification techniques. Indian J Dent Res 24(3):302–308. https://doi.org/10.4103/0970-9290.117991

9. Callis G, Sterchi D (2013) Decalcification of bone: literature review and practical study of various decalcifying agents. Methods, and their effects on bone histology. J Histotechnol 21(1):49–58. https://doi.org/10.1179/his.1998.21.1.49

10. Richardson DS, Lichtman JW (2015) Clarifying tissue clearing. Cell 162(2):246–257. https://doi.org/10.1016/j.cell.2015.06.067

11. Gomariz A, Helbling PM, Isringhausen S, Suessbier U, Becker A, Boss A, Nagasawa T, Paul G, Goksel O, Szekely G, Stoma S, Norrelykke SF, Manz MG, Nombela-Arrieta C (2018) Quantitative spatial analysis of haematopoiesis-regulating stromal cells in the bone marrow microenvironment by 3D microscopy. Nat Commun 9(1):2532. https://doi.org/10.1038/s41467-018-04770-z

12. Coutu DL, Kokkaliaris KD, Kunz L, Schroeder T (2018) Multicolor quantitative confocal imaging cytometry. Nat Methods 15(1):39–46. https://doi.org/10.1038/nmeth.4503

13. Susaki EA, Ueda HR (2016) Whole-body and whole-organ clearing and imaging techniques with single-cell resolution: toward organism-level systems biology in mammals. Cell Chem Biol 23(1):137–157. https://doi.org/10.1016/j.chembiol.2015.11.009

14. Kunisaki Y, Bruns I, Scheiermann C, Ahmed J, Pinho S, Zhang D, Mizoguchi T, Wei Q, Lucas D, Ito K, Mar JC, Bergman A, Frenette PS (2013) Arteriolar niches maintain haematopoietic stem cell quiescence. Nature 502(7473):637–643. https://doi.org/10.1038/nature12612

15. Pinho S, Marchand T, Yang E, Wei Q, Nerlov C, Frenette PS (2018) Lineage-biased hematopoietic stem cells are regulated by distinct niches. Dev Cell 44(5):634–641. e634. https://doi.org/10.1016/j.devcel.2018.01.016

16. Bruns I, Lucas D, Pinho S, Ahmed J, Lambert MP, Kunisaki Y, Scheiermann C, Schiff L, Poncz M, Bergman A, Frenette PS (2014) Megakaryocytes regulate hematopoietic stem cell quiescence through CXCL4 secretion. Nat Med 20(11):1315–1320. https://doi.org/10.1038/nm.3707

17. He W, Holtkamp S, Hergenhan SM, Kraus K, de Juan A, Weber J, Bradfield P, Grenier JMP, Pelletier J, Druzd D, Chen CS, Ince LM, Bierschenk S, Pick R, Sperandio M, Aurrand-Lions M, Scheiermann C (2018) Circadian expression of migratory factors establishes lineage-specific signatures that guide the homing of leukocyte subsets to tissues. Immunity 49(6):1175–1190. e1177. https://doi.org/10.1016/j.immuni.2018.10.007

18. Lucas D, Scheiermann C, Chow A, Kunisaki Y, Bruns I, Barrick C, Tessarollo L, Frenette PS (2013) Chemotherapy-induced bone marrow nerve injury impairs hematopoietic regeneration. Nat Med 19(6):695–703. https://doi.org/10.1038/nm.3155

19. Muller WA (1995) The role of PECAM-1 (CD31) in leukocyte emigration: studies in vitro and in vivo. J Leukoc Biol 57(4):523–528. https://doi.org/10.1002/jlb.57.4.523

20. Nolan DJ, Ginsberg M, Israely E, Palikuqi B, Poulos MG, James D, Ding BS, Schachterlc W, Liu Y, Rosenwaks Z, Butler JM, Xiang J, Rafii A, Shido K, Rabbany SY, Elemento O,

Rafii S (2013) Molecular signatures of tissue-specific microvascular endothelial cell heterogeneity in organ maintenance and regeneration. Dev Cell 26(2):204–219. https://doi.org/10.1016/j.devcel.2013.06.017

21. Vandoorne K, Rohde D, Kim HY, Courties G, Wojtkiewicz G, Honold L, Hoyer FF, Frodermann V, Nayar R, Herisson F, Jung Y, Desogere PA, Vinegoni C, Caravan P, Weissleder R, Sosnovik DE, Lin CP, Swirski FK, Nahrendorf M (2018) Imaging the vascular bone marrow niche during inflammatory stress. Circ Res 123(4):415–427. https://doi.org/10.1161/CIRCRESAHA.118.313302

22. Scheiermann C, Kunisaki Y, Lucas D, Chow A, Jang JE, Zhang D, Hashimoto D, Merad M, Frenette PS (2012) Adrenergic nerves govern circadian leukocyte recruitment to tissues. Immunity 37(2):290–301. https://doi.org/10.1016/j.immuni.2012.05.021

Chapter 13

Intrafemoral Delivery of Hematopoietic Progenitors

Maximilien Evrard, Immanuel Kwok, and Lai Guan Ng

Abstract

Hematopoiesis is a central process and is essential for the replenishment of short-lived leukocytes such as neutrophils. However, the molecular events underlining the developmental transition of quiescent hematopoietic stem cells into downstream progenitors and mature blood cells are not completely understood. Here, we describe the intrafemoral delivery of hematopoietic progenitors as a method to trace their development and differentiation lineage patterns within the bone marrow (BM) niche. Unlike other approaches, the direct adoptive transfer of progenitors into the BM cavity does not require prior irradiation preconditioning of recipient mice, and enables the delivery of lower cell numbers into the marrow space in a minimally perturbed environment. As a demonstrative example, we provide a protocol for the isolation of granulocyte–monocyte progenitors (GMP) by cell sorting, the delivery of these cells into recipient animals by intrafemoral transfer, and finally, the analysis of GMP-derived progenies by flow cytometry.

Key words Hematopoiesis, Progenitors, Neutrophils, Bone marrow, Intrafemoral, Adoptive transfer, Cell sorting

1 Introduction

Hematopoiesis involves the formation of all blood cell lineages, including erythrocytes, platelets, and leukocytes, and is a process that mainly occurs in the bone marrow (BM) of human and mice adults. Hematopoiesis is a hierarchical and step-wise process with defined molecular controls to instruct the various lineage choices each cell takes in order to fully mature into a functioning immune cell [1]. Hematopoietic stem cells (HSCs) are typically found at the top of the hierarchy and give rise to all downstream progenitors. Unlike HSCs which rarely divide, hematopoietic progenitors such as the granulocyte–monocyte progenitor (GMP) actively proliferate to meet the demands for mature blood lineages [2].

Neutrophils represent the most abundant leukocyte population in the human blood, and are essential for rapid elimination of microbial threats. However, neutrophils are short-lived cells and

Marion Espéli and Karl Balabanian (eds.), *Bone Marrow Environment: Methods and Protocols*, Methods in Molecular Biology, vol. 2308, https://doi.org/10.1007/978-1-0716-1425-9_13, © Springer Science+Business Media, LLC, part of Springer Nature 2021

require constant replenishment from BM-derived progenitors. The developmental pathway toward functional blood neutrophils has been studied in detail, with early descriptions from over a century ago. Historically, neutrophil developmental stages have been characterized by using both density-based separation, and morphological criteria such as the condensation of the nucleus and the production of diverse types of granules [3–5]. However, this identification method does take into account the differences at the molecular and cellular level during neutrophil precursor development and maturation. Recently, we identified three stages of neutrophil differentiation downstream of GMP: (1) preneutrophils (preNeu) which are committed neutrophil precursors that proliferate and have the capacity to expand during inflammatory stress, (2) immature neutrophils which exit cell cycle and are absent from the blood under basal conditions, and (3) mature neutrophils which express all the granules required for their function and can enter the peripheral circulation [6, 7].

Adoptive transfer of hematopoietic progenitors into a lethally irradiated host is often used to demonstrate the potential of transferred cells in giving rise to specific cell lineages [8–11]. However, a drawback of this approach is that irradiation could cause alterations in the BM stromal niche, and may produce inflammatory cytokines that do not reflect physiological hematopoiesis [12, 13]. Some studies have employed the adoptive transfer of progenitors into nonirradiated hosts via intravenous delivery [14], although cells that have been transferred through this manner may occasionally not be able to home back to the BM. To circumvent these issues, we and others have used intrafemoral delivery to adoptively transfer hematopoietic progenitors [6, 7, 15–17]. This technique does not require the irradiation of recipient animals, and thus allows for the study of differentiation of transferred cells in a minimally perturbed environment. In addition, the direct delivery of cells into the BM cavity circumvents the issues pertaining to the adequate trafficking of transferred progenitors back to BM, and thus reduces the number of cells required for such adoptive transfer experiments.

In this chapter, we provide a step-by-step description on how to purify hematopoietic progenitors from the donor BM, and how to transfer these enriched progenitors via intrafemoral delivery into recipient mice. We further provide an example of the analysis of differentiation of transferred cells by flow cytometry. Although the current example focuses on the transfer of GMP into wild-type recipients, we anticipate that this technique will be highly adaptable for the study of other hematopoietic lineages, including the delivery of human progenitors into humanized mice [18].

2 Materials

1. Mice that are approved by institutional ethics committee. In this chapter, we use C57BL/6-Tg(Ubc-GFP)30Scha/J (also known as Ubi-GFP) as donors and C57BL/6 as recipients (*see* **Note 1** for alternatives).

2. FACS buffer containing PBS (with Ca^{2+} and Mg^{2+}) at pH = 7.4 with 2 mM EDTA and 2% fetal calf serum (FCS).

3. Complete RPMI containing RPMI 1640 and 10% FCS.

4. DAPI (4′,6-diamino-2-phenylindole,dilactate) 0.5 μM diluted in FACS buffer.

5. Ammonium-Chloride-Potassium Red Blood Cell (RBC) Lysis Buffer.

6. EasySep Release Mouse PE Positive Selection kit (Stem Cell Technologies) (*see* **Note 2**).

7. Ketamine–xylazine or Avertin.

8. 70% Ethanol

9. 50 mL and 15 mL conical centrifuge tubes.

10. 5 mL flow cytometry tubes.

11. Electric clipper/shaver.

12. Forceps, surgical scissors, scalpel holder, scalpel blade.

13. 3 mL syringes and 21-gauge needles.

14. 29-gauge insulin syringes.

15. 70 μm cell strainers and 70 μm nylon meshes.

16. P1000, P200, P20 pipettes and appropriate tips.

17. Fluorescently conjugated antibodies (*see* Table 1 for cell-sorting antibody panel, and Table 2 for flow cytometry analysis panel).

18. Centrifuge.

19. Flow cytometry cell sorters and analyzers, and analysis software (e.g., FlowJo, FSC express).

3 Methods

3.1 Cell Sorting

1. Euthanize mice via CO_2 inhalation or cervical dislocation in agreement to relevant ethics procedures.

2. Pin down mice onto a Styrofoam board, spray with 70% ethanol and remove skin on both legs.

3. Separate the legs from the hip by sectioning the femoral ligament using a scalpel or a pair of surgical scissors (*see* **Note 3**). Remove the muscles from the femurs and tibias, remove bone epiphysis on both ends to create openings.

Table 1
Cell-sorting antibody panel

Excitation laser	Antibody	Clone	Fluorochrome	Dilution (per bone)	Incubation (min)	Step
UV						
			DAPI			
Violet						
	CD34	RAM	eF450	1/100	90	Staining
	CD11b	M1/70	BV650	1/400	45	Staining
Blue						
			GFP			
Yellow-Green						
	B220	RA3-6B2	PE	1/200	25	Depletion
	CD90.2	53-2.1	PE	1/400	25	Depletion
	NK1.1	PK136	PE	1/200	25	Depletion
	Ter119	Ter-119	PE	1/400	25	Staining
	Ly6G	1A8	PE	1/400	45	Staining
	Sca-1	D7	PE	1/400	45	Staining
	cKit	2B8	PE/cF594	1/400	45	Staining
Red						
	CD16/32	2.4G2	APC/Cy7	1/400	45	Staining

4. Equip a 3 mL syringe with a 21-gauge needle. Fill the syringe with FACS buffer and flush the marrow out via the opening into a 15 mL tube. Flush the marrow as many times as necessary, the bone should appear white once the marrow has been thoroughly flushed.

5. Place the BM suspension at 4 °C until all the samples are ready to proceed for the next step (*see* **Note 4**).

6. Filter BM cell suspension with a 70 μm cell strainer, centrifuge cells at $400 \times g$ for 5 min at 4 °C.

7. Resuspend cells in 500 μL FACS buffer and count cells with a hemocytometer (*see* **Note 5**).

8. Stain BM cell suspension with PE-conjugated anti-B220 (clone RA3-6B2; dilution 1:200), anti-CD90.2 (clone 53-2.1; dilution 1:400), and anti-NK1.1 (clone PK136; dilution 1:200) antibodies for 20 min at 4 °C (*see* **Note 6**).

Table 2
Cell-sorting antibody panel

Excitation laser	Antibody	Clone	Fluorochrome	Dilution (per bone)	Incubation (min)
UV					
	CXCR4	2B11	Biotin	1/200	45
		Streptavidin	BUV395	1/200	15
			DAPI		
Violet					
	Siglec-F	E50-2440	BV421	1/200	45
	CD115	AFS98	BV421	1/200	45
	Ly6C	HK1.4	BV605	1/100	45
	CD11b	M1/70	BV650	1/200	45
Blue					
			GFP		
	Gr1	RB6-8C5	PerCP/Cy5.5	1/400	45
Yellow-Green					
	B220	RA3-6B2	PE	1/200	45
	CD90.2	53-2.1	PE	1/400	45
	NK1.1	PK136	PE	1/200	45
	Sca-1	D7	PE	1/200	45
	Ter119	Ter-119	PE	1/200	45
	cKit	2B8	PE/cF594	1/200	45
	Ly6G	1A8	PE/Cy7	1/100	45
Red					
	CXCR2	SA044G4	APC	1/200	45
	CD16/32	2.4G2	APC/Cy7	1/200	45

9. Proceed to depletion of PE labeled cells using EasySep Release Mouse PE Positive Selection Kit (Stem Cell Technologies) according to the manufacturer's instructions, and place the negative fraction into a 5 mL tube.

10. Centrifuge cells at $400 \times g$ for 5 min at 4 °C.

11. Stain BM cells with fluorescently conjugated antibodies for 45–90 min at 4 °C (*see* Table 1 and **Note** 7).

12. Centrifuge cells at $400 \times g$ for 5 min at 4 °C.

Bone Marrow *(Singlets/DAPI^neg)*

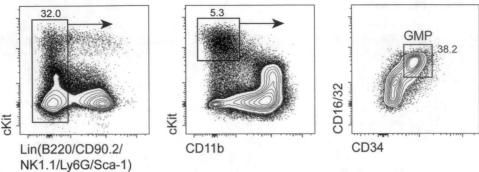

Fig. 1 Purification of GMPs. BM cells were depleted of lymphocytes (*see* Subheading 3.1, **step 7–9**) and stained with fluorescently labeled antibodies (*see* Table 1). Cells were pregated according to their morphology (FSC-A/SSC-A) and doublets were excluded using FSC-W/FSC-H. Dead cells were removed using DAPI. Figure indicates the gating strategy used for GMP cell-sorting. GMPs are identified as the Ly6G and CD11b double negative population, which are markers expressed in neutrophil precursors. Indicated numbers represent percentages for each gated cell population

13. Resuspend cells in 1 mL of 1× RBC lysis buffer and incubate at room temperature for 5 min (*see* **Note 8**).

14. Centrifuge cells at $400 \times g$ for 5 min at 4 °C.

15. Resuspend cells in FACS buffer containing 0.5 µM DAPI to exclude dead cells and filter through 70 µm nylon mesh to remove aggregates.

16. Proceed to cell-sorting. In this example, GMPs were sorted using a five-laser BD FACS Aria II (see gating strategy in Fig. 1).

17. Sorting parameters: use a 85 µm nozzle and sort cells at a pressure of 45 psi (*see* **Note 9**).

18. Sort cells into 1.5 mL tubes containing complete RPMI.

19. Top up with 1× PBS and centrifuge cells at $400 \times g$ for 5 min at 4 °C.

20. Resuspend cells in 1× PBS at a concentration of ~0.5–1.0 × 10^5 per 10 µL and keep on ice until transfer. We usually transfer 5×10^4 GMPs per mouse.

3.2 Intrafemoral Transfer

1. Anesthetize recipient mice with ketamine–xylazine or Avertin in agreement with ethics procedures.

2. Shave leg with an electric clipper to expose the patellar tendon (Fig. 2a).

3. Clean the depilated area with 70% ethanol.

4. Gently homogenize the cell suspension and load cells into a 29-gauge insulin syringe.

A   B

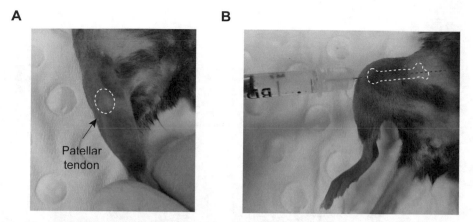

Fig. 2 Intrafemoral injection of GMPs. (**a**) Mouse leg was shaved with an electric clipper. Dotted area highlights the patellar tendon which is used as a landmark for the injection. (**b**) Knee is flexed to a ~90° angle between the femur (white dotted area) and tibia. Insulin syringe needle is inserted through the patellar tendon to reach the bone marrow cavity and a maximum volume of 10 μL can be injected into the femur

5. Flex the knee to a 90° angle and secure the leg in position (Fig. 2b).

6. Insert the needle into the femur marrow space through the patellar ligament (*see* **Note 10** and Fig. 2b).

7. Inject a maximum of 10 μL of cell suspension, hold the needle in place for 5 s before gently removing it (*see* **Note 11**).

8. Place the mouse on a 37 °C heating pad to allow it to recover from anesthesia.

3.3 Flow Cytometry Analysis

1. Sacrifice recipient mice that have previously received BM progenitors and harvest the femurs. Prepare BM cell suspension as described above (Subheading 3.1, **steps 4–6**) (*see* **Note 12**).

2. Optional step: lymphocyte depletion (Subheading 3.1, **step 7–10**) (*see* **Note 13**).

3. Stain BM cells with fluorescently conjugated antibodies for 45 min at 4 °C (*see* Table 2).

4. Centrifuge cells at 400 × *g* for 5 min at 4 °C.

5. Stain BM cells with fluorescently conjugated streptavidin for 15 min at 4 °C.

6. Centrifuge cells at 400 × *g* for 5 min at 4 °C.

7. Resuspend cells in 1 mL of 1× RBC lysis buffer and incubate at room temperature for 5 min.

8. Centrifuge cells at 400 × *g* for 5 min at 4 °C.

9. Resuspend cells in FACS buffer containing 0.5 μM DAPI to exclude dead cells and filter through 70 μm nylon mesh to remove aggregates.

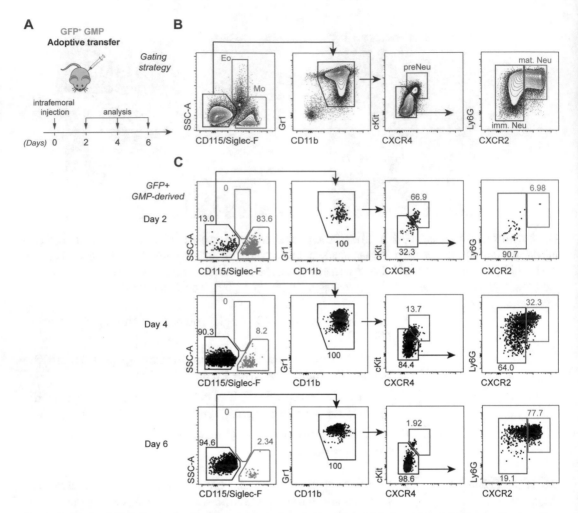

Fig. 3 Analysis of GMP-derived cells following intrafemoral transfer. (**a**) Schematic of GMP adoptive transfer. GFP⁺ GMPs were sorted and transferred intrafemorally into wildtype recipients. Femurs of recipient mice were analyzed by flow cytometry at indicated timepoints (*see* Table 2 for antibody panel). (**b**) Gating strategy of distinct myeloid cell populations. Cells are pregated according to their size (FSC-A/SSC-A). Doublets, dead cells, and lineage positive cells (B220, CD90.2, NK1.1, Sca-1, Ter119) were excluded from analysis. Eosinophils (Eo) are identified as SSC^{high}Siglec-F⁺, while Monocytes (Mo) are SSC^{low}CD115⁺. In contrast, SSC^{int}Gr1⁺CD11b⁺ Neutrophils do not express CD115 or Siglec-F, and can be further subdivided into preneutrophils (preNeu; cKit⁺CXCR4⁺), immature Neutrophils (cKit⁻CXCR4⁻Ly6G^{int}CXCR2⁻) and mature Neutrophils (cKit⁻CXCR4⁻Ly6G⁺CXCR2⁺). (**c**) Analysis of GFP⁺ cells deriving from GMPs at day 2, 4, and 6 following adoptive transfer. At day 2 posttransfer, the wide majority of GMPs give rise to monocytes, with Neutrophils being mainly represented by preNeus. From day 4 posttransfer, Neutrophils become the predominant lineage derived from GMPs and immature neutrophils actively differentiate into mature neutrophils from day 4 to day 6

10. Proceed to sample acquisition on a flow cytometer. In this example, we used a 5-laser BD LSRII (*see* **Note 14**).

11. Analyze acquired FCS files with appropriate software (e.g., FlowJo). A representative analyze obtained using this procedure is shown in Fig. 3. with cells deriving from GFP⁺ GMPs at

2, 4, and 6 days posttransfer (Fig. 3a). Figure 3b illustrates the gating strategy used to identify cells downstream of GMPs, and Fig. 3c shows the phenotype of GFP⁺ transferred cells overtime.

4 Notes

1. Alternatives for adoptive transfer include the use of other fluorescent mice (e.g., Rosa-TdTomato, Actin-CFP), and CD45.1/2 congenic system (e.g., transfer of CD45.2 cells into CD45.1 recipients), although early progenitors may typically display lower CD45 expression.

2. Enrichment is optional but decreases sorting time. This protocol has been optimized using Easysep selection kits, which typically results in fast enrichments with high depletion efficiency while minimally affecting cell viability. Other magnetic enrichment methods could nonetheless be used. Positive selection of cKit⁺ cells is a possible alternative, although the impact of this approach on cell differentiation following adoptive transfer has not been determined.

3. We typically use four bones per mouse for adoptive transfer (2 femurs + 2 tibias). Other bones could be collected (e.g., humerus and spine). Humerus can be processed in the same way as femurs and tibias. Spines require crushing using a mortar and pestle to release BM cells.

4. Certain surface markers may be downregulated at room temperature (e.g., CD115).

5. Use a solution containing 0.2% trypan blue and 3% acetic acid to exclude dead cells and Red Blood Cells respectively. About 50×10^6 BM cells are typically recovered from 2 femurs and 2 tibias of a 6–10-week-old mouse.

6. Indicated dilutions are for one bone and have been carefully titrated in our lab. Optimal antibody concentration may vary depending on the product lot and experimental setting. Other markers can be added to deplete additional cell subsets if required. Markers should, however, be carefully screened for coexpression with the cell population of interest. For instance, while the Neutrophil marker Ly6G is not expressed by GMP and could be added to the antibody depletion cocktail, we found that the monocytic marker CD115 cannot be used for depletion since it is also expressed by a significant fraction of GMP [7].

7. Here, we have used anti-PE magnetic based separation to deplete unwanted BM cells. Additional PE-labeled antibodies could be added to the staining cocktail to create a lineage

exclusion channel. In this example, we added anti-Ly6G to exclude Neutrophil committed precursors and mature Neutrophils, and Sca-1 to exclude HSCs. Of note, we observed that certain antibodies require incubations longer than 45 min for optimal staining such as anti-CD34 (clone RAM34) and anti-Flt3 (clone A2F10) antibodies.

8. In our experience, certain markers (CD115, Flt3) produce better staining profiles when performed prior to RBC lysis as these receptors can become internalized upon prolonged exposure at room temperature.

9. In our experience, hematopoietic progenitors can also be sorted using a 70 μm nozzle and 70 psi pressure for quicker sorting time, but this has a negative impact on cell viability.

10. Evans blue can be used during intrafemoral injection training to visualize the transfer of the dye. A proper injection should label the marrow but not the muscle. In addition, a 27-gauge needle can be used to make a cavity in the femur prior injection.

11. Rapid removal of the needle may result in the leakage of transferred contents out of the BM cavity.

12. The time of harvest depends on user's question. Transition from progenitors to mature cells can vary depending on cell lineage. Transitional premonocytes (TpMo) take 24 h to differentiate into mature $Ly6C^{hi}$ monocytes [17], while preNeu take 48 h to differentiate into mature neutrophils [6].

13. Depletion of nondesired cells will reduce the acquisition time on the flow cytometer.

14. It is essential to acquire the whole femur sample to recover as many transferred cells as possible. Certain flow cytometers may have restrictions on the maximum number of cells that can be acquired in a given FCS file (e.g., 5×10^6 events). If this is the case, acquire multiple files and concatenate events to a single file.

Acknowledgments

We thank the SIgN (Singapore Immunology Network) flow cytometry team for their technical assistance and support. This work was funded by the SIgN core funding, A*STAR (Agency for Science, Technology and Research), Singapore. M.E. is currently supported by the University of Melbourne McKenzie Postdoctoral Fellowship and grant DP200102753 from the Australian Research Council (ARC). The authors declare no conflict of interests. L.G.N. is supported by SIgN core funding. SIgN Flow Cytometry facility is supported by National Research Foundation (NRF) Singapore under Shared Infrastructure Support (SIS) (NRF2017_SISFP09).

References

1. Rieger MA, Schroeder T (2012) Hematopoiesis. Cold Spring Harb Perspect Biol 4(12): a008250. https://doi.org/10.1101/cshperspect.a008250

2. Boettcher S, Manz MG (2017) Regulation of inflammation- and infection-driven hematopoiesis. Trends Immunol 38:345–357. https://doi.org/10.1016/j.it.2017.01.004

3. Bjerregaard MD, Jurlander J, Klausen P et al (2003) The in vivo profile of transcription factors during neutrophil differentiation in human bone marrow. Blood 101:4322–4332. https://doi.org/10.1182/blood-2002-03-0835

4. Borregaard N (2010) Neutrophils, from marrow to microbes. Immunity 33:657–670. https://doi.org/10.1016/j.immuni.2010.11.011

5. Ng LG, Ostuni R, Hidalgo A (2019) Heterogeneity of neutrophils. Nat Rev Immunol 19:255–265. https://doi.org/10.1038/s41577-019-0141-8

6. Evrard M, Kwok IWH, Chong SZ et al (2018) Developmental analysis of bone marrow neutrophils reveals populations specialized in expansion, trafficking, and effector functions. Immunity 48:364–379 e368. https://doi.org/10.1016/j.immuni.2018.02.002

7. Kwok I, Becht E, Xia Y et al (2020) Combinatorial single-cell analyses of granulocyte-monocyte progenitor heterogeneity reveals an early Uni-potent neutrophil progenitor. Immunity 53:303–318.e5. https://doi.org/10.1016/j.immuni.2020.06.005

8. Naik SH, Sathe P, Park HY et al (2007) Development of plasmacytoid and conventional dendritic cell subtypes from single precursor cells derived in vitro and in vivo. Nat Immunol 8:1217–1226. https://doi.org/10.1038/ni1522

9. Auffray C, Fogg DK, Narni-Mancinelli E et al (2009) CX3CR1+ CD115+ CD135+ common macrophage/DC precursors and the role of CX3CR1 in their response to inflammation. J Exp Med 206:595–606. https://doi.org/10.1084/jem.20081385

10. Fogg DK, Sibon C, Miled C et al (2006) A clonogenic bone marrow progenitor specific for macrophages and dendritic cells. Science 311:83–87. https://doi.org/10.1126/science.1117729

11. Zhu YP, Padgett L, Dinh HQ et al (2018) Identification of an early Unipotent neutrophil progenitor with pro-tumoral activity in mouse and human bone marrow. Cell Rep 24:2329–2341 e2328. https://doi.org/10.1016/j.celrep.2018.07.097

12. Banfi A, Bianchi G, Galotto M et al (2001) Bone marrow stromal damage after chemo/radiotherapy: occurrence, consequences and possibilities of treatment. Leuk Lymphoma 42:863–870. https://doi.org/10.3109/10428190109097705

13. Green DE, Rubin CT (2014) Consequences of irradiation on bone and marrow phenotypes, and its relation to disruption of hematopoietic precursors. Bone 63:87–94. https://doi.org/10.1016/j.bone.2014.02.018

14. Hettinger J, Richards DM, Hansson J et al (2013) Origin of monocytes and macrophages in a committed progenitor. Nat Immunol 14:821–830. https://doi.org/10.1038/ni.2638

15. Varol C, Landsman L, Fogg DK et al (2007) Monocytes give rise to mucosal, but not splenic, conventional dendritic cells. J Exp Med 204:171–180. https://doi.org/10.1084/jem.20061011

16. Schlitzer A, Sivakamasundari V, Chen J et al (2015) Identification of cDC1- and cDC2-committed DC progenitors reveals early lineage priming at the common DC progenitor stage in the bone marrow. Nat Immunol 16:718–728. https://doi.org/10.1038/ni.3200

17. Chong SZ, Evrard M, Devi S et al (2016) CXCR4 identifies transitional bone marrow premonocytes that replenish the mature monocyte pool for peripheral responses. J Exp Med 213:2293–2314. https://doi.org/10.1084/jem.20160800

18. Rongvaux A, Willinger T, Martinek J et al (2014) Development and function of human innate immune cells in a humanized mouse model. Nat Biotechnol 32:364–372. https://doi.org/10.1038/nbt.2858

Chapter 14

Imaging of Bone Marrow Plasma Cells and of Their Niches

Carolin Ulbricht, Raluca A. Niesner, and Anja E. Hauser

Abstract

Decade-long survival of plasma cells in the bone marrow has long been a puzzling matter. To understand how plasma cells are maintained and supported by survival-niches to account for lifelong antibody production demands new intravital imaging techniques that are able to follow up a single cell and their interaction with other cell types in situ. We achieved to successfully establish longitudinal imaging of the bone marrow (LIMB) that is based on an implantable endoscopic device. In this chapter, basic approaches on how to investigate plasma cell–stroma interaction and surgical implantation procedures are introduced.

Key words Plasma cells, Bone marrow, Survival niches, Intravital imaging, Fluorescent reporter mice, Two-photon microscopy

1 Introduction

Long-lived plasma cells are known to preferentially accumulate in the bone marrow [1–3]. The finding that longevity of these cells depends strongly on their microenvironment has led to the concept of survival niches, which consist of stromal cells and other hematopoietic cell types, and which are thought to provide plasma cells with vital factors. However, standard immunological practices for the analysis of bone marrow plasma cells normally need to destroy the natural environment of plasma cells in the bone marrow, and thus the integrity of the survival niches. With flow cytometric approaches or classical histology only few markers of plasma cells can be analyzed at the same time and valuable information about location within the tissue as well as interaction with other cells is lost during the process, plus, stromal cells might be hard to isolate or stain in any case. Multiplexed histology has helped to overcome the limits in the number of analyzable markers, however, only gives a snap shot of plasma cell life within the bone marrow [4]. Moreover, although plasma cells stably dock on stromal cells, the niches are subject to a certain dynamic. This is mainly determined by other hematopoietic cell types but also by the location of the plasma cells

Marion Espéli and Karl Balabanian (eds.), *Bone Marrow Environment: Methods and Protocols*, Methods in Molecular Biology, vol. 2308, https://doi.org/10.1007/978-1-0716-1425-9_14, © Springer Science+Business Media, LLC, part of Springer Nature 2021

in relation to blood vessels [5, 6]. The latter constantly keep changing their position, and this likely impacts on the provision of plasma cell niches with nutrients which they need to upkeep their high levels of antibody production. The best way to study the dynamics of plasma cell niches is therefore intravital imaging. In order to investigate individual plasma cell properties in the bone marrow, functional and dynamic, that is, time-resolved, microscopic data must be repeatedly acquired over longer time periods, that is, in a longitudinal study manner. We recently developed longitudinal imaging of the bone marrow (LIMB) that allows the study of one and the same plasma cell within its specific microenvironment and its interaction with other cells and tissue structures, such as stromal cells and blood vessels in situ, over the time course of weeks. LIMB employs an implantation device introducing a permanent gateway into the femoral bone marrow cavity. The technique follows the principle of 3R (replace, reduce, refine) in that it collects information about cellular behavior without interindividual variations and minimizes the number of animals needed for long-term observations of immune cell turnover.

To identify cells of interest in the bulk cell mass of the bone marrow, special steps in preparation of LIMB might be needed. Especially if you want to image the interaction of plasma cells with the stromal compartment—it is recommended to first generate bone marrow chimeric mice, where hematopoietic and non-hematopoietic cells can be distinguished due to different origin and genotype. Other methods would use adoptive transfer and immunization, including optional staining of static components of the niches. In this case, cells taken from a (fluorescent reporter) donor mouse would be transferred into a host mouse, where their bone marrow homing is stimulated by immunization and thus their usually rather small number increases, as does the likelihood of their intravital observation. Another alternative is to use aged mice with an appropriate fluorescent genotype, where antigen exposure throughout the entire lifespan probably led to accumulations of several clones of long-lived plasma cells within their survival niches.

2   Materials

2.1   Animals

1. Any experiment including living animals has to be approved and licensed by the local authorities. Experiments must not be performed without preceding proper planning of all experimental steps, the administration of all substances used or an assessment of potential suffering of the species and individual animals involved.

2. From the three transcription factors essential in PC differentiation, BLIMP1, IRF4, and XBP1, the foremost BLIMP1, a master regulatory factor of plasma cell differentiation, is

suitable for lineage tracing in vivo, also because maintenance in the bone marrow is dependent on it [7–10]. There are transgenic mice available that stably express GFP variants in dependency of BLIMP1, for example, BLIMP1-GFP or BLIMP1-YFP [11–13]. Cy1cre mice offer the possibility of targeting plasma cells via conditional recombination with a (ROSA26 located) fluorescent transgene, as done for YFP and RFP [14, 15]. For IRF4 no reporter has been developed so far, although there is a IRF4-GFP conditional knockout vector available [10]. The tracing of IRF4 alone might be problematic, as its expression is not restricted to terminally differentiated PC and its expression is in part transient during development. Another alternative possibility is to transduce hematopoietic stem cells (HSC) with reporter transgenes, as done for XBP1-GFP, and transfer them to irradiated mice [16]. Also, fluorescent stromal cell reporter mice are available such as CXCL12-GFP mice and PRX1-tdRFP and -eYFP mice [6, 15, 17, 18].

3. The study of plasma cells in their niches will possibly require the generation of bone marrow chimera. In chimeric mice, all hematopoietic cells have been replaced with those from another individual. It is thus possible to investigate the interplay between the hematopoietic and nonhematopoietic compartment within the same animal, since cells of the same ontogeny will have distinct (fluorescent or surface marker) labels. For example, creating a bone marrow chimeric mouse by transferring donor bone marrow from a transgenic reporter of "green" plasma cells to a host with ubiquitous expression of "red," where all host hematopoietic cells have been deleted by irradiation, will generate mice where the interaction of green plasma cells and red stromal cells can be specifically observed. Alternatively, mixed bone marrow chimeras (transfer of cells with different genotypes) offer the possibility of simultaneously studying intrinsic or extrinsic effects of gene expression to function within the same animal. More detailed insights into experimental approaches using bone marrow chimera can be found in the detailed review of Ferreira and colleagues, 2019 [19].

4. To identify and study single plasma cells within the bone marrow, and especially if it is not planned to work with bone marrow chimera, it is advisable to investigate antigen-specific monoclonal plasma cells adoptively transferred and enriched in the bone marrow by immunization. Suitable donor mice, for example, can be gained by breeding above mentioned reporter mouse strains together with mice of the B1-8 strain, whose B cells carry BCRs specific to nitrophenyl (NP) [20]. Offspring of such breeding will thus be equipped with fluorescent plasma

cells that secrete immunoglobulins with same specificity that can be targeted specifically, and whose presence in the bone marrow can be stimulated by immunization. Systemic immunization into the peritoneal cavity with the respective antigen will lead to enhanced GC reactions producing monoclonal plasma blasts in the spleen, of which a certain percentage will migrate to bone marrow niches and become long-lived plasma cells. Several subsequent immunizations at intervals of several weeks are recommended to induce a persistent long-lived memory immune response.

Alternatively, the observation of fluorescent plasma cells within aged (1–2 years) individuals of reporter mice offers a good possibility to conduct experiments without invasive preparative treatment, if permitted by experimental schedules and available animal colonies.

5. Overall, the genotype of mice employed in the investigation of plasma cells has to be determined according to the parameters of cells and interaction partners to be observed. Also be aware of the fact that you might need to use additional fluorophores when imaging, for example for visualizing vessel structures. Quantum dots (life technologies) are a feasible suggestion for this purpose and available in a variety of colors. To avoid spectral overlap, carefully compare fluorescence properties of donor mice, host mice and additionally used, injectable labels (all matched with the properties of your microscopic system).

2.2 Buffers

1. Ethanol 70%.

2. $1 \times$ PBS.

3. $1 \times$ PBS with 0.5% BSA.

4. Tris-buffered ammonium chloride solution for red blood cell (RBC) lysis (ACT): First prepare 0.16 M NH_4Cl, then 0.17 M Tris, pH 7.65. Dilute Tris 1:10 in NH_4Cl, adjust pH 7.2. Alternatively, you can also use commercially available RBC lysis buffer and prepare it according to manufacturer's protocol.

5. 0.9% saline.

6. Transfer buffer: 100 ml sterile PBS, 1 ml 1 M HEPES, 50 U/ml penicillin–streptomycin (0.5 ml of 10,000 U/ml).

7. Acid-Citrate-Dextrose formula A (ACD-a)).

2.3 Single-Cell Suspensions (in Case of Cell Transfer/BM Chimera)

1. Ice bath.

2. Spleen and/or Bone marrow of donor animals, from both femurs and tibias.

3. Cell strainer 70 μm.

4. 50-ml tubes.

5. Plunger of a 3 ml syringe.

6. Cooling centrifuge.

7. Cell counting device (e.g., Neubauer chamber, quantitative flow cytometer).

2.4 B cell isolation (in Case of Adoptive Transfer with Subsequent Immunization)

1. Mouse B cell isolation kit (e.g., Miltenyi, StemCell).

2. Columns for magnetic separation system, suitable for expected cell numbers to be retained within.

3. Magnets suitable for column.

4. If needed: magnetic stand.

2.5 Cell Transfer and Immunization

1. Isolated B cells in transfer buffer w/ 2.5% ACD-a.

2. 1-ml syringes.

3. 27G cannulas.

4. Restrainer for injections.

5. Warming lamp or water bath (max 40 °C).

6. Protein (e.g. chicken gamma globulin) conjugated to hapten, that is, in case of B1-8 donor mice NP, 20–29 ratio recommended.

7. Alum as carrier and adjuvant for intraperitoneal (i.p.) injections.

2.6 Drugs

Drugs for anesthesia during surgery as well as for short-term sedation during microscopic measurements are listed below (Table 1). Also included are substances for analgesia and antibiosis (to apply before/during surgery and afterward). The application of all of these has to be licensed by the authority supervising your animal experiments.

2.7 Devices and Tools

1. Implantable device (e.g., RISystems) as published [6, 21].

2. Drill for microscrews, screws manufactured for the device, stand for exact positioning of drill.

3. A set of gradient index rod (GRIN) lenses (e.g., GRINTech) adequate for two-photon microscopy, as previously published [6, 21] (*see* **Note 1**).

4. Microscopic table adapter that connects to the reference plate of the implant [6].

5. Heating unit, ideally soft and flexible (e.g., heating pad) and self-regulating.

2.8 Imaging Equipment

1. Intravital microscope stage (e.g., LaVision BioTec).

2. State-of-the-art two-photon laser-scanning microscope, typically based on Titanium:Sapphire laser excitation (e.g., TriMScope II, LaVision Biotec).

Table 1
Drugs, Usage, and Application. Trade marks listed here comprise examples used with good experience in our lab, but may be exchanged for generic pharmaceuticals containing the indicated substances, or substances with comparable effects in anesthesia, pain management and inflammation prevention. Be aware that dosages may vary depending on the mouse strain. It is recommended to work out a proper medication plan with your local veterinarian

	Substance	Trade name	Dosage	Usage	Application
1.	Dexpanthenol	Bepanthen® (Bayer)	As much as needed	Hydration of cornea, postsurgical wound treatment	Direct application onto eye bulbs/ incision sites
2.	Ketamine/ Xylazine	Ketamin (Inresa) Rompun® (Bayer)	100 mg/kg/ 10 mg/kg body weight (rsp.)	Induction of anesthesia	Intraperitoneal injection
3.	Buprenorphine	Temgesic® (Indivior)	0.03 mg/kg body weight	Analgesia	Subcutaneous (s.c.) injection
4.	Enrofloxacin	Baytril® (Bayer)	10 mg/kg body weight	Antibiotic, prevents inflammation	s.c.
5.	Isoflurane (+O_2)	Forene® (AbbVie)	1.25–2%	Anesthesia	Via face mask
6.	Tramadol	Tramal® (Grünenthal)	0.05 mg/ml	Post-surgical management	Via drinking water

3. 20× objective lens with long working distance (e.g., lens with IR-coating, NA 0.45, Olympus).

4. Photomultiplier tubes for imaging of desired emission wavelengths (e.g., Hamamatsu).

2.9 Software

1. Acquisition software of the two-photon microscope (e.g., Imspector Pro, 208 or later).

2. Image analysis software, for example, Fiji Image J (open source).

3. Image analysis software with 3D-tracking module, for example Imaris 9.× (9.5.1. or higher recommended, Bitplane AG), several package licenses available (*see* **Note 2**).

2.10 Additional Material

The design of your specific experiment may require additional material. In the case of extra fluorescent probes (e.g., for vessels staining, *see* **Note 3**), check for spectral overlap.

3 Methods

3.1 Bone Marrow Chimera

1. For the generation of BM chimera, the hematopoietic system of the host individual needs to be removed by irradiation (*see* **Note 4**).

2. Deplete donor bone marrow from T cells (follow instructions for magnetic cell separation with T-cell depletion kit, as described exemplary for B cells (*see* Subheadings 2.3, 2.4, and 3.2, **step 3**).

3. After irradiation, adoptively transfer mice with approximately three million cells of whole T cell-depleted bone marrow from the donor mouse strain of choice that will reconstitute the hematopoietic system of the host. Reconstitution is done 24 h after irradiation and cells need to be sex-matched to the host. Note that if the contribution of nonhematopoietic compartment to plasma cell function is of lesser interest, reporter B cells would be transferred without prior irradiation. As a consequence, mice will have to be immunized after transfer, in order to achieve donor plasma cell homing to niches.

3.2 Cell Isolation and Adoptive Transfer

1. Sacrifice donor mice and collect spleen and/or femurs+tibias (take care not to damage joints). Work under laminar flow hood, on ice.

2. Shortly flush the bones with ethanol.

3. Open the bone marrow cavities by cutting at knee and joints.

4. Gently insert a PBS or RPMI filled syringe with cannula attached into one opening of the bone cavity push the marrow out into a petri dish.

5. Crush spleen through a cell strainer into a 50 ml tube using the plunger of a syringe.

6. Filter collected marrow through a cell strainer into a 50 ml tube.

7. Wash both types of cells by centrifugation at $300 \times g$, 8 min, 4 °C.

8. Discard supernatant, resuspend cell pellets in ACT buffer (or similar RBC lysis buffer) and incubate for 5 min at room temperature. Add $1\times$ PBS–BSA to stop reaction.

9. wash by centrifugation ($300 \times g$, 8 min, 4 °C).

10. If desired, separate cells according to manufacturer's protocol separation method of choice (*see* **Note 5**).

11. Resuspend cells in transfer buffer, add 2.5% anticoagulant ACD-a freshly.

12. Inject cells into recipient mice via intravenous application into the tail vain (*see* **Note 6**).

3.3 Immunization (in Case of Cell Transfer)

1. Emulsify one part of hapten-protein conjugate in PBS (according to your experimental concentration tested) in one part of Alum (i.e. 1:1).

2. Let emulsion stir for at least 30 min at RT.

3. Inject emulsion at a volume no larger than 200 µl into the peritoneal cavity (*see* **Note 7**).

3.4 Surgical Procedure

1. Induce anesthesia via intraperitoneal injection of ketamine–xylazine.

2. Make sure to keep the mouse warm (e.g., by a red lamp, heating device) and check reflexes on a regular basis (*see* **Note 8**).

3. Trim hair on right (or left) thigh to a short length, then, use depilating cream to remove residual hair, make sure skin is clean, hair- and dust-free.

4. Switch from injected to inhaled anesthesia (*see* **Note 9**).

5. Place mouse sideward on a cork plate and fix lower leg with a piece of silk tape. Fold swap or similar piece of cotton and place it in between the legs.

6. Stretch upper leg to approximately 180° and fix with a piece of tape. Also tightly stretch the tail and fix with an additional piece of tape; this will hold the hips in place.

7. Perform an incision horizontal to the line of the femoral bone.

8. Widen the incision for good accessibility of the femoral bone. You can use silk tape attached to the skin for it. Make sure to always keep the wound moisturized by regular application of sterile NaCl solution.

9. Cut through the connective tissue with fine scissors.

10. Separate the musculus rectus femoris from musculus vastus lateralis by cutting through the connecting fascia.

11. Carefully loosen the muscle tissue from the bone surface.

12. Bend the leg from 180° to 90°, approximately.

13. Fix the femoral bone with a fixing forceps. Use this technique to adjust the position suitable to you.

14. Insert a 0.45 mm drill hole with a fine drilling machine into the middle of the femoral bone. It is best to work with a fixed stand for the drilling machine. That is making sure there is no tilt introduced into the drill hole axis.

15. Place the implantable device (Fig. 1a) upon the femoral bone, first facing the hip, then the knee. Be careful not to injure the vessel that is located beneath the knee. This later on is placed in a way bridging the implant.

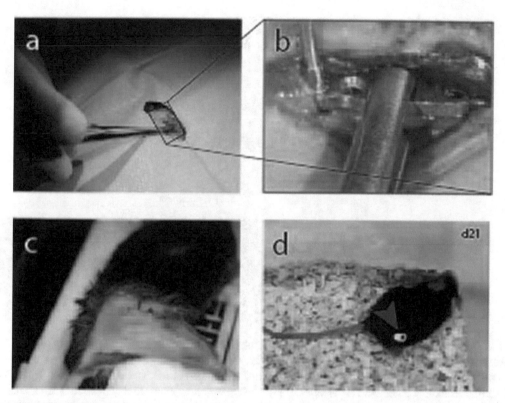

Fig. 1 (**a**) All surgical procedures have to be performed under sterile conditions. Iodine treatment of skin before incision, surgical cover, and sterile gloves are recommended. (**b**) Implant inserted between muscle and bone, fixed with forceps to held in position for hole-drilling. (**c**) After the skin is closed, only tubing and positioner are visible, this later becomes attached to the reference plate. (**d**) Fully healed mouse carrying implant 21 days after surgery

16. Use a fixing forceps to hold the implant in position. Place it in the middle of the implant and tighten it.

17. Use the 0.31 mm drill for drilling the positioning holes, guided by the holes in the implant.

18. Tighten the screws into their respective holes.

19. Remove the fixing forceps.

20. Reconnect the musculus rectus femoris to the musculus vastus lateralis. Using catgut for sewing the tissue together along the fascia is ensuring suture degradation and residue-free healing.

21. Suture the epidermis of the mouse with sewing silk.

22. Discontinue the anesthesia.

23. Administer analgesia (e.g., Buprenorphine) i.p. directly after surgery. For the next 3 days, analgesia is provided via drinking water (Tramadol) (Table 1, **Note 10**).

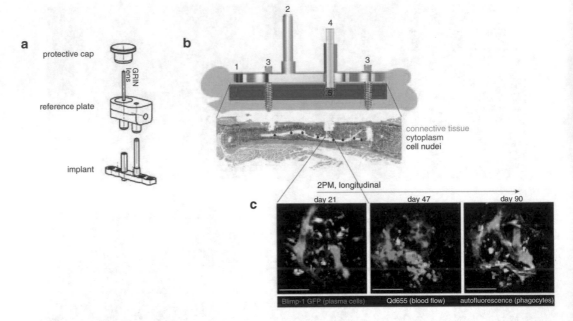

Fig. 2 (**a**) Schematic representation of the three major parts of the LIMB endoscope. (**b**) Schematic representation of the parts of the implant: (1) fixation plate, (2) bicortical screws, (3) positioner, (4) GRIN lens attached to tubing, (5) imaging volume. Masson-Goldner stain of mouse femur slice after explantation, the central sinus is marked with an asterisk. (**c**) Representative images of 90 day LIMB measurements with plasma cells visible in green, blood vessels in white and macrophages in yellow (autofluorescence signal overlap of multiple channels). Scale bar 50 um

24. After 7 days, the acute immune reaction following implantation should have sufficiently declined for a first observation with the endoscope.

3.5 Imaging

1. Insert the reference plate (Figs. 1d and 2a) into the specialized microscopic stage by sliding it in.

2. Adjust the focus first to the upper surface of the GRIN lens, then lower at low laser power by several hundred micrometers until the tissue becomes visible in the field of view.

3. Image at desired resolution, depth and length (Fig. 2b, c and **Note 11**).

4. After imaging, image correction procedures (due to wave front distortions of the GRIN lenses) might apply, for those please refer to our publication [21].

5. Proceed to analysis (e.g., using Imaris, time-resolved z-stack data in the format of .tif-files can be imported and rendered (*see* **Note 12**)).

4 Notes

1. To ensure that the same optical conditions are maintained at all times during longitudinal imaging and prevent crushing by mice bite or getting lost in the cage, it is necessary to fix the GRIN lens into the tubing of the implant using glue. The gluing process needs to be done delicately, in order not to interfere with the optical properties of the lens. To do so, the tubing is positioned opening facing downwards onto the sticky side of a stripe of tape. This seals the opening and simultaneously holds the tubing in position. The lens is then inserted and a small amount of two-component glue (e.g., EPO-TEK 301-2, John P. Kummer) is let into the exposed opening via a syringe. Let dry overnight, at 65 °C [21]. Make sure to inspect the visible surface of the lens before every imaging step and clean with a dry swab or cloth from dust or particles, if necessary.

2. 3D rendering software requires considerable graphics board equipment and processing unit(s). Carefully check system requirements before purchasing a license. There will be different options available, but (especially for time-resolved 3D analysis) avoid to go for a set up that has minimum-required processing power; rather choose middle-class or "gold standard" equipment, that will save time and money in the long run.

3. Labeling blood vessels with injectable fluorophores not only helps for orientation within the imaging volume (though position and structure may slightly change after some days, *see* [6]) but could also be used for interaction studies of plasma cells with blood vessels. Trained personnel for i.v. injection is indispensable. Furthermore, always provide some amounts of surgical tape, swabs, surface and skin disinfectant, syringes with different cannulas, saline and glucose solution (both injectable for assistance of hydration or energy during anesthesia).

4. Lethal irradiation is classically done with gamma rays, but there are also different radiation sources being used (e.g., X-ray). Research institutes working in immunological fields normally hold special facilities for irradiation purposes. Irradiation devices can be purchased from suppliers for laboratory medical technologies (e.g., Theratronics, RADsource). Several doses with reduced amount of radiation are recommended to attenuate side effects, for example 4–7 Gy twice within 3–4 h. It is absolutely necessary to keep animals under antibiosis for the following 2 weeks since they are immune-suppressed and prone to infections.

5. It is recommended to leave the purified cells untouched in order not to prestimulate them, that is, to choose negative selection protocols. In column-based negative selection, all unspecific cells are labelled and retained (this will be the bigger proportion, and you will have to choose the column capacity likewise) whereas all other cells are flushed out and collected. The purity of the desired cell type should be >95%, as checked by flow cytometry before proceeding with adoptive transfers.

6. Cell amounts might vary and would have to be determined individually for the experimental purpose. A commonly used number for instance is 3×10^6 antigen-specific B cells.

7. A typical amount to apply is 100 μg of a 0.5 mg/ml hapten-protein conjugate/Alum emulsion. So, for a total volume of 1 ml emulsion, mix 100 μl hapten-protein conjugate solution, 400 μl 1× PBS, and 500 μl Alum. This should be sufficient for approximately four mice (calculate huge dead volume of syringe, since emulsion is sticky).

8. Surgical depth of anesthesia is reached when hind legs plus one fore leg are negative for pinching reflex. One leg should still be positive in order to prevent overdosing.

9. Inhaled anesthesia (2% isoflurane) is easily adjustable by gas flow, easily reverted by revoking the inhalation device, and has less side effects. Additional pain management is necessary by injection of, for example, Metamizol or Buprenorphine, since isoflurane does not provide inherent analgesia. For anesthesia of small rodents, a transportable device with face mask as provided by Luigs & Neumann GmbH, Germany is well-suited.

10. A postsurgical clinical scoring approach to assess proper healing and animal welfare can be found in [6]. Activity measurements performed after surgery were able to confirm that the impairment of the animals is minimized after a few days and that carrying the implant is well tolerated over the course of several weeks.

11. The size of the circular field of view depends on the properties of the GRIN lens and of the tubing used and ranges from 150 to 500 μm, in diameter. For resolution purposes, it is suggested to acquire >1pixel per μm. Adjusting the scan frequency and line averaging has a beneficial effect on the later image quality since it improves the photon budget, but has to be in consensus with the acquisition speed. Typically, up to 21 z-planes with a distance of about 5 μm can be acquired over times up to 2 h, with a time resolution of 30 s. If lower time resolutions are acceptable, up to 41 z-planes in steps of 5 μm can be imaged, covering 200 μm z-depth—the total working

distance of the used GRIN lenses at both object and image side.

12. Parameters to export include, for example, fluorescence intensities of the imaged cells and structures (after segmentation), track properties of moving objects (length, duration, displacement, velocity, and so on). For the analysis of plasma cells within niche structures, the colocalization tool that calculates threshold-based overlap of two different fluorescent signals is of great value.

Acknowledgments

This work was funded by the Deutsche Forschungsgemeinschaft, DFG TRR130, TP17 to AEH and C01 to AEH and RAN, and SFB 1444, project 14, to AEH and RAN. This work was also supported by a grant from the Einstein Foundation Berlin (A-2019-559) to AEH and RN.

References

1. Lemke A, Kraft M, Roth K et al (2016) Long-lived plasma cells are generated in mucosal immune responses and contribute to the bone marrow plasma cell pool in mice. Mucosal Immunol 9:83–97. https://doi.org/10.1038/mi.2015.38

2. Manz RA, Thiel A, Radbruch A (1997) Lifetime of plasma cells in the bone marrow. Nature 388:133–134. https://doi.org/10.1038/40540

3. Slifka MK, Antia R, Whitmire JK, Ahmed R (1998) Humoral immunity due to long-lived plasma cells. Immunity 8:363–372. https://doi.org/10.1016/S1074-7613(00)80541-5

4. Holzwarth K, Köhler R, Philipsen L et al (2018) Multiplexed fluorescence microscopy reveals heterogeneity among stromal cells in mouse bone marrow sections. Cytom A 93:876–888. https://doi.org/10.1002/cyto.a.23526

5. Zehentmeier S, Roth K, Cseresnyes Z et al (2014) Static and dynamic components synergize to form a stable survival niche for bone marrow plasma cells. Eur J Immunol 44:2306–2317. https://doi.org/10.1002/eji.201344313

6. Reismann D, Stefanowski J, Günther R et al (2017) Longitudinal intravital imaging of the femoral bone marrow reveals plasticity within marrow vasculature. Nat Commun 8:2153. https://doi.org/10.1038/s41467-017-01538-9

7. Turner CA, Mack DH, Davis MM (1994) Blimp-1, a novel zinc finger-containing protein that can drive the maturation of B lymphocytes into immunoglobulin-secreting cells. Cell 77:297–306. https://doi.org/10.1016/0092-8674(94)90321-2

8. Reimold AM, Iwakoshi NN, Manis J et al (2001) Plasma cell differentiation requires the transcription factor XBP-1. Nature 412:300–307. https://doi.org/10.1038/35085509

9. Shapiro-Shelef M, Lin K-I, Savitsky D et al (2005) Blimp-1 is required for maintenance of long-lived plasma cells in the bone marrow. J Exp Med 202:1471–1476. https://doi.org/10.1084/jem.20051611

10. Klein U, Casola S, Cattoretti G et al (2006) Transcription factor IRF4 controls plasma cell differentiation and class-switch recombination. Nat Immunol 7:773–782. https://doi.org/10.1038/ni1357

11. Ulbricht C, Lindquist RL, Tech L, Hauser AE (2017) Tracking plasma cell differentiation in living mice with two-photon-microscopy. Methods Mol Biol 1623:37–50

12. Fooksman DR, Schwickert TA, Victora GD et al (2010) Development and migration of plasma cells in the mouse lymph node. Immunity 33:118–127. https://doi.org/10.1016/j.immuni.2010.06.015

13. Kallies A, Hasbold J, Tarlinton DM et al (2004) Plasma cell ontogeny defined by quantitative changes in Blimp-1 expression. J Exp Med 200:967–977. https://doi.org/10.1084/jem.20040973

14. Luche H, Weber O, Nageswara Rao T et al (2007) Faithful activation of an extra-bright red fluorescent protein in "knock-in" Cre-reporter mice ideally suited for lineage tracing studies. Eur J Immunol 37:43–53. https://doi.org/10.1002/eji.200636745

15. Srinivas S, Watanabe T, Lin C-S et al (2001) Cre reporter strains produced by targeted insertion of EYFP and ECFP into the ROSA26 locus. BMC Dev Biol 1:4. https://doi.org/10.1186/1471-213X-1-4

16. Brunsing R, Omori SA, Weber F et al (2008) B- and T-cell development both involve activity of the unfolded protein response pathway. J Biol Chem 283:17954–17961. https://doi.org/10.1074/jbc.M801395200

17. Ara T, Itoi M, Kawabata K et al (2003) A role of CXC chemokine ligand 12/stromal cell-derived factor-1/pre-B cell growth stimulating factor and its receptor CXCR4 in fetal and adult T cell development in vivo. J Immunol 170:4649–4655. https://doi.org/10.4049/jimmunol.170.9.4649

18. Logan M, Martin JF, Nagy A et al (2002) Expression of Cre recombinase in the developing mouse limb bud driven by a Prxl enhancer. Genesis 33:77–80. https://doi.org/10.1002/gene.10092

19. Ferreira FM, Palle P, vom Berg J et al (2019) Bone marrow chimeras—a vital tool in basic and translational research. J Mol Med 97:889–896. https://doi.org/10.1007/s00109-019-01783-z

20. Shih T-AY, Meffre E, Roederer M, Nussenzweig MC (2002) Role of BCR affinity in T cell dependent antibody responses in vivo. Nat Immunol 3:570–575. https://doi.org/10.1038/ni803

21. Stefanowski J, Fiedler AF, Köhler M et al (2020) Limbostomy: longitudinal intravital microendoscopy in murine osteotomies. Cytom A 97:1–13. https://doi.org/10.1002/cyto.a.23997

Chapter 15

Intravital Imaging of Bone Marrow Microenvironment in the Mouse Calvaria and Tibia

Changming Shih, Leonard Tan, Jackson Liang Yao Li, Yingrou Tan, Hui Cheng, and Lai Guan Ng

Abstract

The complex bone marrow microenvironment or niche is an important anatomical structure responsible for hematopoiesis and providing support to the immune cells function. Being the source of immune and blood cells, the interaction of these hematopoietic stem and progenitor cells with the cellular niches regulates their ability for self-renewal, proliferation, and differentiation. Dynamic imaging not only provides spatiotemporal information of cell motility but also the morphological changes due to cell–cell interactions in the bone marrow, providing insights into the ongoing physiological activities within the tissue. Here, we describe customized stages with compatible equipment best suited for the upright two-photon microscope, accompanied by detailed methods for both calvarial and tibial intravital imaging. We demonstrate a general protocol for calvarial imaging using a minimally invasive surgical approach, and introduce a bone shaving-based tibial imaging as a complementary method. To demonstrate the applicability of our method we used Lyz2-EGFP transgenic mice to track bone marrow neutrophil activities as an example.

Key words In vivo imaging, Mouse, Bone marrow, Calvaria, Skull, Long bone, Tibia, Hematopoiesis, Immune system

1 Introduction

The bone marrow (BM) is a spongy tissue which occupies both cancellous bones and long bone medullary cavities [1]. The BM is a complex microenvironment, which is composed of structures such as hematopoietic cells, adipose tissue, vasculature, stromal networks and nerve fibers [1]. Concomitant with its complicated structure, the BM plays versatile roles in different physiological processes, such as hematopoiesis and osteogenesis. Increasing evidence suggests that the cross-talk between hematopoietic cells and their supportive cellular niche is indispensable for sustaining

---

Changming Shih and Leonard Tan contributed equally to this work.

Marion Espéli and Karl Balabanian (eds.), *Bone Marrow Environment: Methods and Protocols*, Methods in Molecular Biology, vol. 2308, https://doi.org/10.1007/978-1-0716-1425-9_15, © Springer Science+Business Media, LLC, part of Springer Nature 2021

physiological homeostasis during steady and inflammatory states [2–4]. Specifically, this cross talk depends on the interactions between the hematopoietic stem and progenitors and the stroma, and the secretory signals from diversified niches as well. Thus, exemplifying the importance of spatiotemporal information to decipher the hematopoiesis. [2–4]

Two-photon microscopy provides high spatial resolution and minimizes tissue damage during prolonged temporal imaging, making it an optimal method to perform intravital imaging of BM microenvironment. In comparison to traditional histology which provides spatial information, intravital imaging offers insight into spatial and temporal information of ongoing cell dynamics and cell–niche interactions. Amongst the different BM niches, the calvarium has several unique characteristics, making it the most advantageous site for intravital BM imaging. For instance, the calvarium is sited just directly under the scalp, which makes it convenient to expose via a minimally invasive surgery [5, 6]. On the other hand, the tibia provides a good representation of the medullary BM cavity, but at the cost of more tissue damage due to the need of thinning the bone for imaging [7].

Therefore, we demonstrate our two-photon intravital imaging procedure into two parts here. For the first part, we offer calvarial BM imaging that is minimally invasive. As a complement, we provide a detailed tibial BM imaging method using a bone-shaving approach in the second part. We illustrate this using lysozyme-EGFP transgenic mice to capture neutrophil trafficking, so that our readers can choose a suitable procedure for their own bone marrow–related studies.

2   Materials & Equipment

2.1   Mouse

1. 6- to 12-week-old male or female Lysozyme-EGFP C57BL/6 mice ($Lyz2^{tm1.1Graf}$, Thomas Graf, Center for Genomic Regulation, Spain) [8] (*see* **Notes 1** and **2**).

2.2   Anesthesia

1. Digital weighing scale (weight range between 0.1 g and 500 g).

2. Digital timer.

3. Mouse body temperature homeothermic monitoring system and heating pad.

4. 30-gauge insulin syringe.

5. Ketamine–xylazine mixture: 150 mg/kg ketamine hydrochloride, 10 mg/kg xylazine hydrochloride in sterile water (*see* **Note 3**).

2.3 Hair Removal for Skull Imaging

1. Mouse body temperature homeothermic monitoring system and heating pad.
2. Precision hair trimmer (*see* **Note 4**).
3. Cotton-tipped applicators.
4. 70% ethanol.

2.4 Hair Removal for Tibial Imaging

1. Digital timer.
2. Mouse body temperature homeothermic monitoring system and heating pad.
3. Single edge razor blade.
4. Cotton-tipped applicators.
5. Disposable wipes (e.g., M-fold towels, Kimwipes).
6. Depilatory cream.
7. 70% ethanol.
8. PBS.

2.5 Intravenous Injection

1. 30-gauge insulin syringe.
2. Evans Blue: 10 mg/mL in PBS.
3. Phosphate-buffered Saline (PBS, without calcium and magnesium): potassium phosphate monobasic 0.2 g/L, potassium chloride 0.2 g/L, sodium chloride 8.0 g/L, sodium phosphate dibasic (anhydrous) 1.15 g/L in sterile water.

2.6 Skin Incision of the Calvarium

1. Mouse body temperature homeothermic monitoring system and heating pad.
2. Surgical straight scissors (130 mm).
3. Curved serrated forceps (130 mm).
4. Cotton-tipped applicators.
5. Disposable Pasteur pipettes.
6. 70% ethanol.
7. PBS.

2.7 Skin, Muscle, and Tibia Incision

1. Customized stage with attached base support, heating pad, and cuboid spacer (*see* **Note 5**) (Fig. 1).
2. Mouse body temperature homeothermic blanket and control system.
3. Rectal thermal probe.
4. Custom coverslip holder and stand (Fig. 2a–d).
5. Coverslip (22 × 32 mm).
6. High vacuum grease.
7. Blu-Tack (*see* **Note 6**).

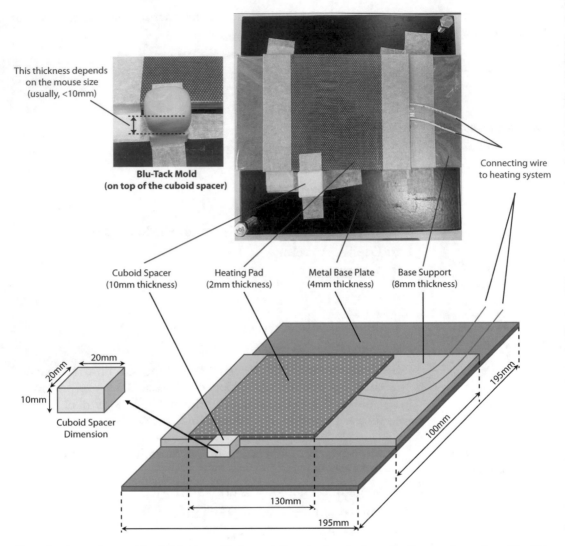

This thickness depends on the mouse size (usually, <10mm)

Blu-Tack Mold
(on top of the cuboid spacer)

Connecting wire to heating system

Cuboid Spacer (10mm thickness)

Heating Pad (2mm thickness)

Metal Base Plate (4mm thickness)

Base Support (8mm thickness)

20mm
20mm
10mm
Cuboid Spacer Dimension

195mm
100mm
130mm
195mm

Fig. 1 Design and assembly of tibia stage. Photograph (top view) and schematic diagram dimension of the tibia stage, with metal base plate, base support, heating pad, and cuboid spacer for the placement of Blu-tack mold

8. Dissecting stereomicroscope.

9. Micro motor drill, with a carbide round burr (diameter: 0.2 mm).

10. Scissors and forceps (*see* **Note 7**).

11. Cotton-tipped applicators.

12. Disposable Pasteur pipettes.

13. Micropipette (20–200 μL) and compatible micropipette tips.

14. 70% ethanol.

15. PBS.

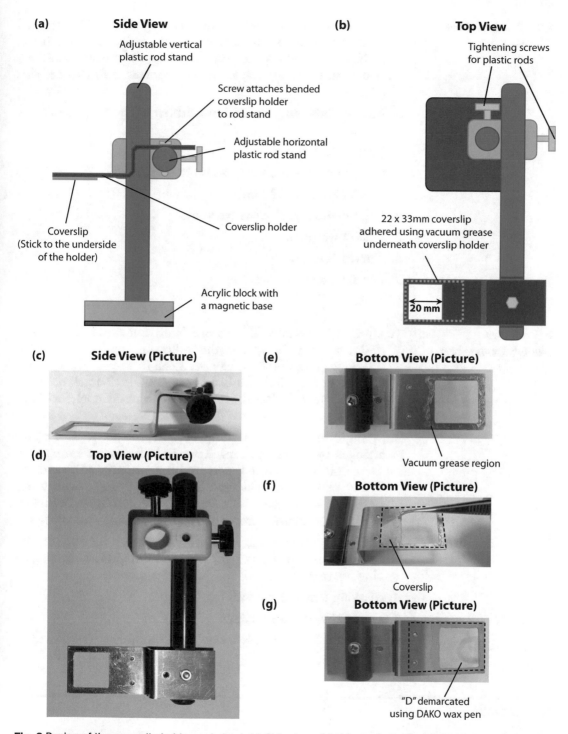

Fig. 2 Design of the coverslip holder and stand. (**a**) Side view of holder and stand, showing the bending of the coverslip holder. (**b**) Top view of holder and stand, showing the placement of the coverslip and the tightening screws for both the coverslip holder and the stand. (**c**) Photograph of side view of holder. (**d**) Photograph of top view of holder. (**e**) Photograph of the bottom view of the holder, showing that the vacuum grease was applied to the coverslip hole circumference, (**f**) the placement of the coverslip, and (**g**) the "D" demarcation drawn using the DAKO wax pen

2.8 Calvarial Bone Marrow Stage

1. Custom mouse calvarial bone marrow stage with attached heating pad (65 × 95 mm) and integrated stereotactic parts, including one adjustable nose bar clamp adaptor with nose bar clamp, and two adjustable ear bar adaptors with two ear bars (Fig. 3).

2. Mouse body temperature homeothermic blanket and control system.

3. Rectal thermal probe.

4. Custom coverslip holder and stand (Fig. 2a–d).

5. Coverslip (22 × 32 mm).

6. High vacuum grease (*see* **Note 8**).

7. DAKO wax pen.

8. Curved forceps.

9. Cotton-tipped applicators.

10. PBS.

2.9 In Vivo Multiphoton Imaging

1. Upright two photon microscope (with emission laser pulses at 950 nm capability), and the following optical filter sets: 495 long-pass, 560 long-pass, 475/42 band-pass (SHG channel, for bones), 525/50 band-pass (EGFP channel, for neutrophils) and 665/40 band-pass (Evans blue channel, for blood vessels).

2. Syringes attached with polypropylene tubing × 2. Prepare syringes by attaching 1 mL Luer-Lok disposable syringes to Luer-Lok adapters and polypropylene tubing (diameter of 1/16″ or smaller) of desired length. Fill syringes with water or diluted ketamine–xylazine mixture as accordingly. Ensure air bubbles are removed from the syringes, tubing, and needles.

3. Luer-Lok hypodermic 30G needle. Attach needle to the syringe with tubing that contains the diluted ketamine–xylazine mixture.

4. Masking tape (~2 cm wide).

5. Disposable Pasteur pipettes.

6. Diluted 1:1 ketamine–xylazine and PBS mixture (*see* **Note 9**).

7. Sterile water.

3 Methods

3.1 Preparation of Coverslip Holder for Skull Bone Marrow Imaging

1. At the underside of the coverslip holder, apply vacuum grease onto the perimeter of the coverslip holder hole (Fig. 2e). Carefully place a 22 × 32 mm coverslip onto the grease layer and apply pressure using the tip of the forceps on the perimeter of the coverslip that it is in contact with the grease (Fig. 2f).

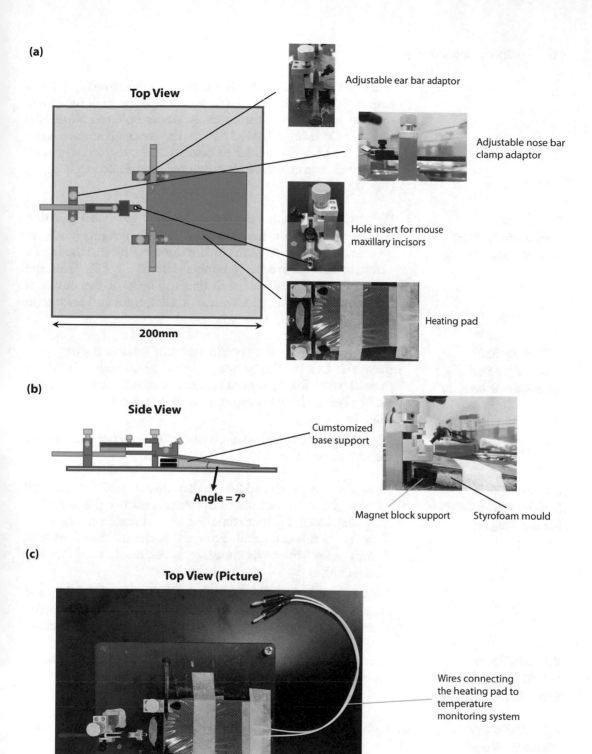

Fig. 3 Design of mouse calvarial bone marrow stage. (**a**) Top view of stage, with adjustable adaptors for ear and nose bar, and hole for the insertion of mouse maxillary incisors on the nose bar clamp. (**b**) Side view of stage, showing the custom base support made up of Styrofoam mold and magnets to elevate the heating pad at 7° angle. (**c**) Photograph of top view of stage, with heating pad and wires that can be connected to the temperature monitoring system

Remove excess grease that is not stuck to the coverslip using a cotton-tipped applicator. Avoid touching the area of the coverslip that is exposed to the coverslip holder hole (*see* **Note 10**). Draw a "D" using a DAKO pen on the coverslip at the underside of the coverslip holder to demarcate the region of interest which will be in contact with the mouse calvaria during imaging (Fig. 2g).

3.2 Mouse Anesthesia for Skull Bone Marrow Imaging

1. Weigh and anesthetize the mouse via intraperitoneal injection of ketamine–xylazine solution (8 µL/g body weight) (*see* **Note 11**). Start a timer to monitor the duration of anesthesia. To maintain the mouse body temperature at 37 °C, place the mouse on a heating pad such that the belly of the mouse is touching the heating pad, with its head upright and resting on its chin (*see* **Note 12**).

3.3 Scalp Hair Removal for Skull Bone Marrow Imaging

1. While maintaining the mouse position on the heating pad, shave the hair on the mouse scalp in the direction from the parietal to the frontal area using a hair trimmer. Ensure that the surface being shaved is approximately 1×1 cm (*see* **Note 13**) (Fig. 4a–c).

2. Dip one cotton tipped applicator into 70% ethanol and sterilize the shaved region.

3.4 Scalp Skin Incision for Skull Bone Marrow Imaging

1. Using forceps, pinch and lift the shaved scalp skin at the frontal location. Make an incision in the direction from the frontal to the parietal area. Further trimming of the incision can be made if the incision is too small. Ensure that the incision is at least 2 mm away from the unshaved region (*see* **Note 14**) (Fig. 4d–e).

2. Moisten the exposed region of the incision by applying one drop of PBS onto the incision area using a disposable pipette. Then, remove excess PBS using a dry cotton tipped applicator.

3.5 Blood Flow Labelling for Skull Bone Marrow Imaging

1. Inject Evans Blue dye (10 mg/mL) intravenously into the mouse (2 µL/g body weight) (*see* **Note 15**).

3.6 Positioning Mouse Skull onto Customized Stereotactic Frame Stage for Skull Bone Marrow Imaging

1. Push back the nose bar clamp. Then, insert the tip of the forceps into the side of the mouse's mouth and gently part the upper and lower jaws to expose the maxillary incisor (*see* **Note 16**) (Fig. 5a). Hook the mouse maxillary incisors onto the hole of the lower jaw clamp (*see* **Note 17**) (Fig. 5b). Move the nose bar clamp toward the mouse such that the clamp is resting approximately 2 mm away from the eyes (*see* **Note 18**) (Fig. 5c). Tighten the nose bar clamp knob. Do not overtighten the nose bar clamp knob as it will obstruct the mouse's

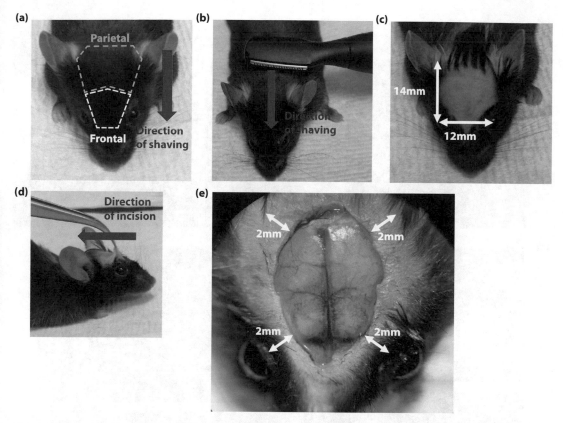

Fig. 4 Scalp hair shaving and skin incision of the region of interest on mouse calvaria. (**a**) The parietal and frontal location of the mouse skull. (**b**) Direction of shaving using a precision hair trimmer. (**c**) The approximate dimension of the shaved region. (**d**) Direction of the incision to expose mouse calvaria. (**e**) The 2 mm allowance between the incision and unshaved region

breathing. Check that the mouse breathing is regular (*see* **Note 19**). Ensure the mouse head is resting comfortably on the heating pad by tuning the height adjustment knob on the nose bar clamp adaptor (*see* **Note 20**) (Fig. 5d).

2. Check the position of the ear bars to adjust position of the mouse accordingly. Adjust the up-down axis of the mouse by shifting the nose bar clamp back and forth (Fig. 6a). Adjust the height of the ear bars accordingly by twisting the ear bar adjustment knobs to ensure they are aligned with the mouse ears (*see* **Note 21**) (Fig. 6b). Insert the tapered tip of each ear bar into the ear canal such that the ear bar rests on the canal wall located at the occipital of the skull (i.e., back of the skull) (Fig. 6c). Do not insert the ear bar too deep into the ear canal. Repeat for the other ear bar (*see* **Note 22**) (Fig. 6d). Ensure that the head is firmly immobilized by gently tapping the exposed calvarial region using forceps and checking for movement (Fig. 6e). The mouse skull should be firmly anchored to avoid drifts during imaging (*see* **Note 23**).

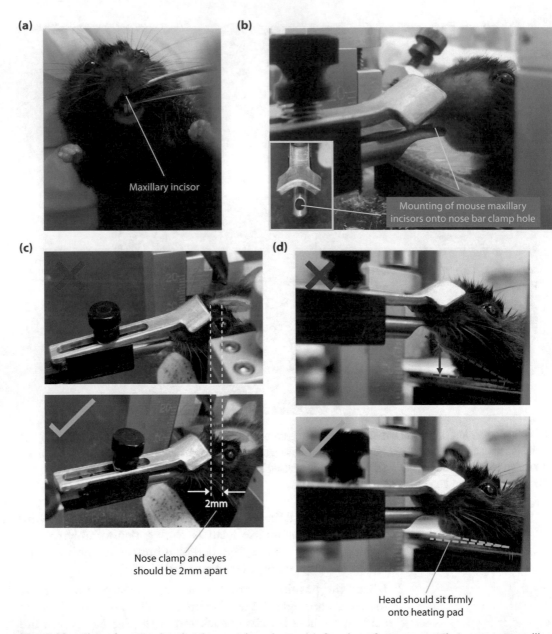

Fig. 5 Mounting of mouse head onto nose bar clamp. (**a**) Opening of mouse mouth to expose maxillary incisors. (**b**) Mounting of maxillary incisors onto the hole insertion of the nose bar clamp unit. (**c**) Positioning of the nose bar clamp with at least 2 mm apart from mouse eyes. (**d**) Resting of mouse head onto the heating pad

3. Add a drop of PBS on the incised region using a disposable pipette.

4. Position the coverslip over the calvaria such that the region of interest is within the DAKO wax boundary. Gently lower and anchor the coverslip holder such that the coverslip is resting on the calvaria, not on the ear bars (*see* **Note 24**) (Fig. 6f–g).

5. Tighten the screw on the coverslip holder and stand.

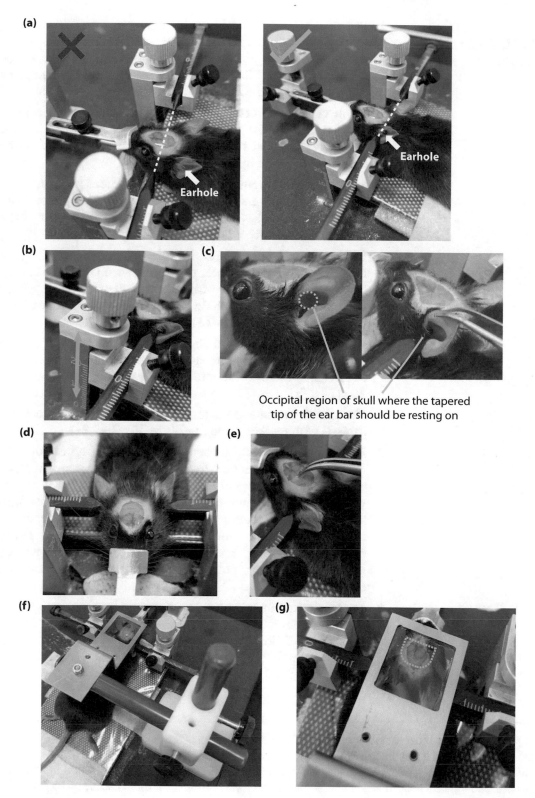

Fig. 6 Tuning of nose bar and ear bars for positioning the mouse calvaria before imaging. (**a**) Adjustment of the nose bar clamp for proper positioning of the ear bars onto the mouse ear holes. (**b**) Vertical adjustment of the

3.7 Imaging Setup for Skull Bone Marrow Imaging

1. Place the mouse calvarial bone marrow stage on to the microscope stage. Ensure that the area of interest ("D" demarcation) within coverslip is positioned directly beneath the microscope objective (*see* **Note 25**). Position the mouse and bone marrow stage to simulate the final imaging position (*see* **Note 26**).

2. Connect all wires necessary for the maintenance of mouse body temperature at 37 °C. This includes electrical wires of the mouse heating pad to the heating system, and inserting the temperature probe into the rectum.

3. Secure the rectal temperature probe and mouse tail onto the heating pad with masking tape to prevent the probe from sliding out, as well as to restrict any possible movement of the mouse tail (*see* **Note 27**).

4. Using masking tape, secure flexible polypropylene tubing of the syringe containing water to the microscope objective, such that when the plunger is pushed, the water will be dispensed from the mouth of the tubing directly to the microscope objective lens region (*see* **Note 28**) (Fig. 7a).

5. Insert the 30G needle of the syringe containing 1:1 diluted ketamine–xylazine mixture subcutaneously into the back skin. Secure the needle to the mouse using masking tape (*see* **Note 29**) (Fig. 7a).

6. Place both syringes outside the dark box, and secure loose tubings with masking tape (*see* **Note 30**) (Fig. 7a).

7. Following the manufacturer's instructions, set up the upright multiphoton microscope for imaging of the coronal vein, where bone marrow sinusoids are located (Fig. 7b). For previewing, use the titanium:sapphire (Ti-Sa) laser at wavelength 950 nm with an output power of about 30 mW. Gradually raise/lower the laser power for better image contrast. However, do not raise the laser power too high as heat will be generated, inducing potential thermal damage and causing the 'speckling' effect' [9] (*see* **Note 31**).

8. Locate desired imaging field of view and perform imaging of the mouse bone marrow. Capture 250 μm × 250 μm × 30 μm ($X \times Y \times Z$) image time-lapse z-stacks, with approximately 30 s time gaps between image acquisition (*see* **Note 32**), for the desired length of time (typically 1–3 h).

Fig. 6 (continued) ear bar using the adaptor. (**c**) The occipital region of the mouse skull in the ear hole where the tapered tip of the ear bar should be resting on. (**d**) Locking of the two ear bars in place. (**e**) Gently tapping of the central area within incised region, ensuring that the mouse head is secured. (**f**) Placement of the coverslip holder stand. (**g**) PBS trapped within the "D" demarcation when coverslip holder is gently pressed onto the region of interest

(a)

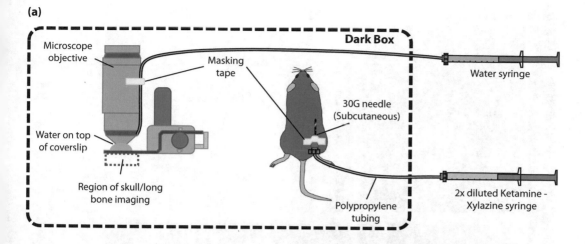

(b)

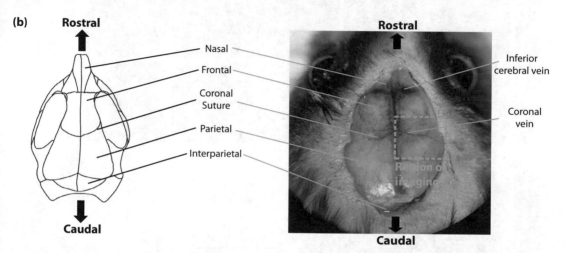

Fig. 7 Setup for continuous intravital imaging of mouse bone marrow within the microscope dark box. (**a**) Schematic diagram showing the positioning of the syringes to deliver respective fluids for continuous intravital imaging. (**b**) Mouse skull anatomy and region of imaging

9. Slowly deliver the diluted ketamine–xylazine mixture using the syringe located outside the dark box every half an hour to maintain the anesthesia of the mouse (after the first hour, 4 μL/g body weight) (*see* **Note 33**). Likewise, refill approximately 30 μL the water every hour under the microscope objective lens.

10. Once the imaging session has ended, disconnect all wirings and tubing accordingly. Remove the mouse from the skull bone marrow stage and sacrifice the mouse according to institutional guidelines.

11. Visualize and analyze the imaging results as appropriate using the relevant software (e.g., ImageJ). GFP-positive myeloid cells should be observed in the sinusoids of the skull bone marrow (Fig. 8a).

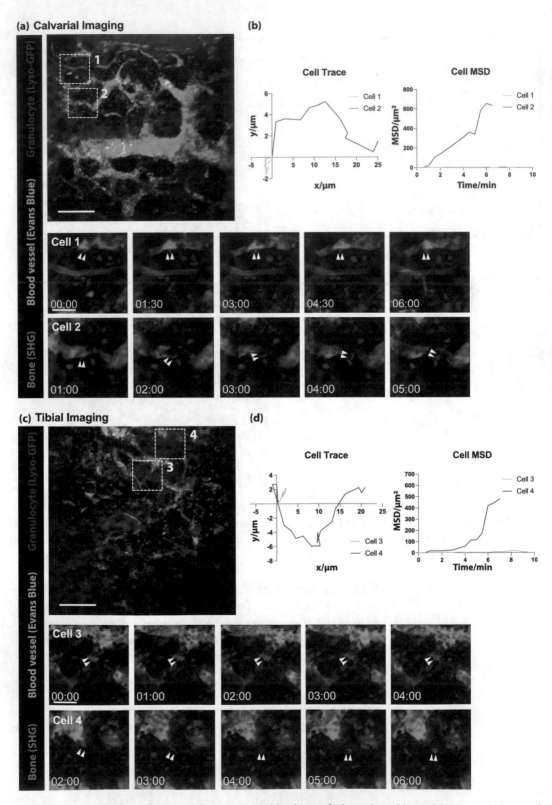

Fig. 8 Time-lapse images of neutrophils (magenta), blood vessels (green), and bone (blue, second harmonic generation signal generated from collagen) in the mouse calvarium (**a**) and tibia (**c**) at steady state. (**a**)

3.8 Surgical Stage Preparation for Tibial Bone Marrow Imaging

1. Secure the cuboid spacer next to the heating pad using masking tape. Shape the Blu-Tack mold into a cuboidal form roughly the size of the cuboid space, and place molded Blu-Tack on top of the cuboid spacer (Fig. 1).

2. Using the same coverslip holder for the skull bone marrow imaging, prepare the coverslip holder similar to Subheading 3 **for Skull Bone Marrow Imaging**, with the exception of the "D" demarcation (Fig. 2).

3.9 Mouse Anesthesia for Tibial Bone Marrow Imaging

1. Weigh and anesthetize the mouse via intraperitoneal injection of ketamine–xylazine solution (8 µL/g body weight) (*see* **Note 11**). Start a timer to monitor the duration of anesthesia. To maintain the mouse body temperature at 37 °C, place the mouse on its back on the heating pad (*see* **Note 12**).

3.10 Thigh Hair Removal for Tibial Bone Marrow Imaging

1. Gently stretch the right leg of the mouse, and moisten the medial thigh and shin region with 70% ethanol. Use the single edge razor blade to gently shave the hair around the desired leg region (Fig. 9a).

2. Dip one cotton tipped applicator into PBS and clean the shaved region. Apply depilatory cream on the shaved region with a new cotton tipped applicator (*see* **Note 34**) (Fig. 9b).

3. Wait for 2 min before wiping away the depilatory cream with dry disposable wipes. Remove residual cream by using a cotton tipped applicator dipped in PBS, and repeat this step for another time using a new cotton tipped applicator dipped in PBS. Gently wipe the shaved region with disposable wipes to remove excess PBS (*see* **Note 35**) (Fig. 9c).

3.11 Tibial Surgery for Tibial Bone Marrow Imaging

3.11.1 Mouse Placement to Surgical Stage

1. Use the flat side of a metal piece to shape the Blu-Tack mold such that it is able to fit the right leg of the mouse (Fig. 9d–e).

2. Gently transfer the mouse from the heat pad to the surgical stage, and still keep the mouse laying on its back, and then put the right leg (with the inner thigh facing upward) onto the Blu-Tack mold (Fig. 9f).

Fig. 8 (continued) Calvarial BM imaging and dynamic tracking of neutrophils. The static snapshot (**a**, upper panel) shows neutrophils around the sinusoids. The zoom-in time series images (**a**, lower panels) show morphology and mobility of two typical neutrophils (Cell 1 and Cell 2, double arrowheads); regions of interest are demarcated with dotted squares in the static snapshot. (**b**) Dynamics of cell movement illustrated by cell trace and mean squared displacement (MSD) in the regions of interest (**a**, lower panels) in calvarial BM. (**c**) Tibial BM imaging illustrated by static snapshot (**c**, upper panel) and corresponding zoom-in time series images (**c**, lower panel) of neutrophils in a manner similar to calvarial BM imaging in (**a**). (**d**) Dynamics of cell movement illustrated by cell trace and mean squared displacement (MSD) in the regions of interest (**c**, lower panels) in tibial BM. SHG: Second harmonic generation; scale bar: 100 µm for static snapshots, 20 µm for zoom-in dynamic time-lapse view; time = mm:ss for the time-lapse view

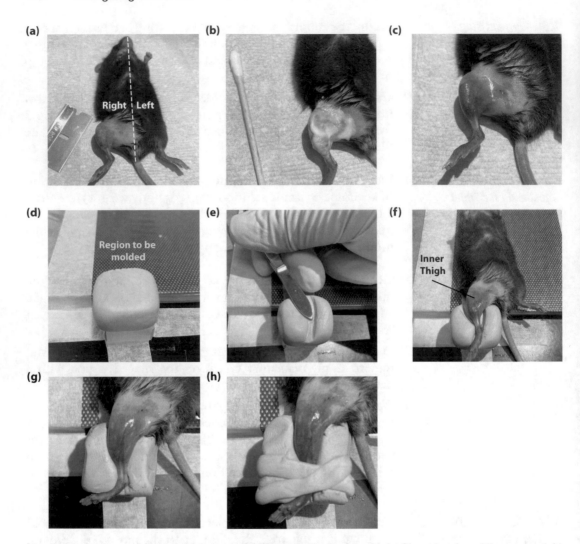

Fig. 9 Hair shaving and mounting of mouse right leg onto Blu-Tack mold. (**a**) Shaved region of the mouse right leg that was subsequently applied with (**b**) depilatory cream, and (**c**) washed with PBS and wiped dry. (**d**) Approximate region of the Blu-tack that is to be molded. (**e**) Shaping of the Blu-Tack mold, and (**f**) fitting of right inner thigh onto preshaped Blu-Tack mold. (**g**) Reshaping of Blu-Tack mold to immobilize leg. (**h**) Placement of two crisscrossed Blu-Tack strips onto mouse ankle

3. Reshape the Blu-Tack mold with hands or forceps, such that the leg is anchored and immobilized into the mold (Fig. 9g).

4. Make two strips of Blu-Tack (approximately 20 × 3 mm) and paste them onto the mold across the mouse ankle in a crisscross pattern for further stabilization of the leg (*see* **Note 36**) (Fig. 9h).

3.11.2 Skin Incision and Tissue Removal

1. Using regular surgical straight scissors and curved serrated forceps, make a tiny incision on the skin near the tibia (Fig. 10a). Then, gently enlarge the incision by cutting the skin along the tibia to expose the adjacent muscles and connective tissue (Fig. 10b).

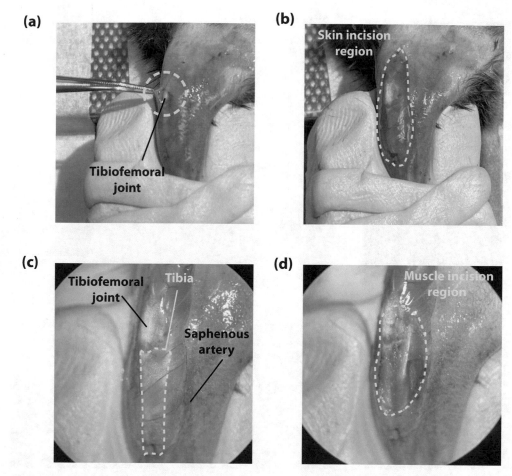

Fig. 10 The exposure of mouse tibia by the sequential surgical manipulation. (**a**) An initial tiny skin incision is made at the tibiofemoral joint, followed by the trim and enlargement of the skin incision (**b**). (**c**) Zoom-in view of the tibia beneath the thin layer of connective tissue and muscles. (**d**) Approximate region of the muscle incision to completely expose the tibia

2. Add a few drops of PBS using a disposable pipette to moisten the incised region. View the incision under the dissecting microscope. The tibia can be vaguely visualized under a thin layer of connective tissue (*see* **Note 37**) (Fig. 10c).

3. Using delicate straight surgical scissors and micro dissecting curved serrated forceps, carefully remove the connective tissue layer by layer, followed by the adjacent muscles, until the tibia is exposed (*see* **Note 38**) (Fig. 10d).

3.11.3 Tibia Shaving

1. Locate the region of interest on the incision (the upper field of the tibia below the mouse knee) using the dissecting microscope (Fig. 11a).

2. Gradually shave the bone surface of the region of interest (the upper part of the tibia, indicated in Fig. 11a) using the micro

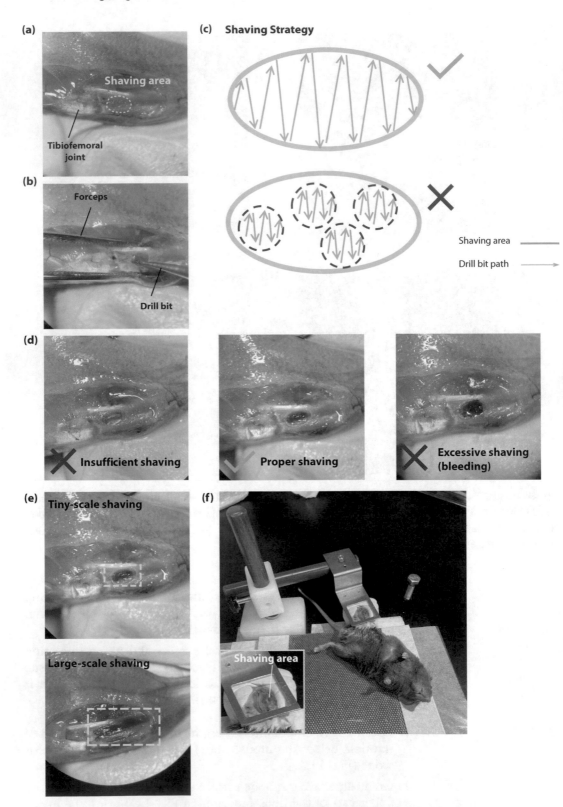

Fig. 11 Tibia bone shaving to expose the marrow region for imaging. (**a**) Demarcation of the region for bone shaving. (**b**) Stabilizing of the tibia using forceps while shaving with the micro motor drill. (**c**) Proper bone

motor drill. Use forceps to stabilize the tibia so that the shaving can be precise (Fig. 11b). Keep shaving until the bone color turns from white to pale red (*see* **Notes 39–41**) (Fig. 11c–e).

3. Add a drop of PBS on the tibia incision.

3.11.4   Blood Flow Labelling

1. Inject Evans Blue dye (10 mg/mL) intravenously into the mouse (2 µL/g body weight) (*see* **Note 15**).

3.11.5   Coverslip Positioning

1. Use disposable wipes to remove the excess PBS, so that the PBS will not overflow too much when the coverslip is placed.

2. Adjust the coverslip holder such that the shaved region (imaging area) is in the middle of the coverslip. Then, gently push the coverslip holder down toward the shaved region (imaging area). Make sure that there is PBS with no air bubbles in between (Fig. 11f).

3. Once the coverslip touches the tibia, push down a bit more to make sure the imaging site is close enough to the coverslip (*see* **Note 42**).

4. Tighten the screw on the coverslip holder and stand (*see* **Note 43**).

3.12   Imaging Setup for Tibial Bone Marrow Imaging

This step is the same as Subheading 3 **for Skull Bone Marrow Imaging**, except that we put the temperature probe right under the mouse body, as it is difficult to insert the probe into the rectum while the mouse is in a supine position.

After acquisition of the imaging, visualize and analyze the imaging results using the proper software (e.g., ImageJ), to achieve descriptive or quantitative information. GFP-positive myeloid cells should be observed around the sinusoids within the tibial bone marrow (Fig. 8b).

4   Notes

1. In terms of the mouse age, ~8-week-old mice are the best choice to do the surgery. Older or younger mice can also be used depending on the experiment design. However, the bone stiffness and thickness may vary according to ages or genders. Therefore, the imaging quality may vary with age, and the assay needs to be slightly modified if possible.

Fig. 11 (continued) shaving strategy within the demarcation. (**d**) Illustration of good and bad examples of the shaved region. (**e**) The difference scales of shaving for the region of interest depending on different experimental objectives. (**f**) Placement of the coverslip holder where the imaging area is at the center of the coverslip

2. Cells of the monomyelocytic lineage will express EGFP in the lysozyme-EGFP transgenic mice. These include granulocytes (higher EGFP expression levels), monocytes and macrophages (lower EGFP expression levels) [10].

3. The efficacy of the ketamine–xylazine solution will degrade over time. Thus, anesthesia solutions should be prepared fresh, which will allow better temporal control when administering anesthesia for continuous imaging sessions.

4. Ensure hair trimmers are properly maintained by brushing using a toothbrush or paintbrush after each shaving procedure to remove excess hair stuck in the device. Well-maintained hair trimmers can improve the required time for shaving. In addition, hair trimmers are preferable to single edge razor blades as shaving of the skull region requires delicate care. Shaving the skull using razor blades may cause accidental laceration of the eye of the mouse during shaving.

5. We use iron to make the metal base plate so that the magnetized coverslip holder and stand can be stabilized on it. Also, we use wood (or other thermal insulation materials) to make the base support, so that the heating pad will not overheat the whole stage but only warm the mouse. The cuboid spacer, together with Blu-Tack (depicted as below in **Note 8**), is used for immobilization of the mouse leg to prevent imaging drifting. In addition, putting the hard cuboid spacer beneath Blu-Tack provides support to the soft Blu-Tack, which makes the Blu-Tack easier to mold. Therefore, the cuboid spacer can be made from hard material such as acrylic plastic (need to be fixed on the stage by masking tapes) and magnets (easy to be fixed on the stage).

6. We use commercial Blu-Tack to make the mold that fits into the leg due to its pressure-sensitive ability to adhere onto the custom stage, and it is also suitable for being shaped into the support for the mouse tibia for further manipulation. Furthermore, it acts as a damping agent to reduce vibrations, which is advantageous for imaging. Plasticine or other substances with good adhesiveness and plasticity can also be utilized as substitutes.

7. A variety of scissors and forceps are used here. A set of regular surgical straight scissors (130 mm) and curved serrated forceps (130 mm) are used for skin incision, while a set of delicate surgical straight scissors (115 mm) and micro dissecting curved serrated forceps (100 mm) are used for more delicate surgeries such as the removal of connective tissue and muscle. Delicate equipment is highly recommended as larger equipment may cut into the saphenous artery which may result in massive bleeding.

8. To make the vacuum grease easy to use, we place the grease into a 10 mL syringe, and gently expel it out from the syringe tip when we apply it to surfaces.

9. The ketamine–xylazine solution is diluted with PBS so as to hydrate the mouse during the imaging session.

10. Fingerprint or dirt marks on the coverslip will affect the quality of images obtained. Use forceps to carefully press down the perimeter of the coverslip without touching the area of imaging.

11. Adequate dosing of the anesthesia is vital in any intravital imaging experiment, as incorrect dosing often leads to the premature termination of the experiment. Insufficient anesthesia may cause the mouse to fidget, which may result in the unstable images capture during imaging. On the other hand, excessive anesthesia may cause the mouse to die prematurely. The optimization of the dosage will be necessary as the drug efficacy may vary depending on many factors such as the manufacturer, mice strains and so on.

12. In our experience, setting a timer immediately after injecting the first dose of anesthesia can allow better control over subsequent dosage for continuous intravital imaging sessions.

13. Shaved area should be big enough for an incision to be made that could accommodate the "D" demarcation drawn onto the coverslip with the DAKO wax pen.

14. The incision must be smaller than the shaved area. This is to ensure that no hair will be caught in the "D" demarcation during imaging as it will result in unnecessary interference with image acquisition or leakage of PBS out of the demarcated area.

15. The binding of the albumins from the mouse blood plasma to the Evans blue dye is able to generate a very bright fluorescence signal which is easily excited over an array of wavelengths. If necessary, the injection volume can be increased by at least five times (or depending on the volume determined by your institutional guidelines) with no apparent toxicity to the mouse.

16. Ensure suitable forceps are used to open the mouth of the mouse. Large or sharp forceps will cause unnecessary distress to the mouse. Insert the forceps with the two grasping ends touching each other into the side of the mouse's mouth and gently release the two grasping ends of the forceps. This will allow the opening of the lower and upper jaws of the mouse, to expose the maxillary incisors.

17. Mouse maxillary incisors must fit vertically down into the hole of the lower jaw clamp. Fitting at a slanted angle, possibly caused by one maxillary incisor inserted more than the other,

will result in tilting of the exposed skull region. This is likely to cause the leaking of PBS out of the "D" demarcation, and uneven illumination of the imaging plane, contributing to a certain area being darker as compared to another area in an image.

18. If the nose bar clamp is very close to the eyes (<2 mm), the height of the exposed calvarial region might be lower than the top of the nose bar clamp, potentially restricting the coverslip from touching the exposed calvarial region. Also, the nose bar clamp might injure the eye of the mouse.

19. Overtightening of the nose bar clamp will obstruct the mouse airway and thus cause irregular breathing patterns. Before commencing the next step, observe the breathing pattern of the mouse for approximately 1 min. Loosen the knob for the nose bar clamp if necessary.

20. Resting of the mouse body and head onto the heating pad allows thermal monitoring and regulation. In addition, the resting of the mouse head onto the heating pad also allows a steady support when the coverslip is pressed onto the exposed calvarial region. If the height of the nose bar clamp is too high, the mouse head will not be in contact with the heat pad. This will result in the instability during imaging caused by the breathing or movement of the mouse.

21. The height of the ear bar must not be higher than the plane of the exposed calvarial region. This will potentially restrict the coverslip from touching the exposed calvarial region.

22. Ensure that when both ear bars are secured/tightened, the exposed calvarial region is not tilted. Tilted skull is usually caused by one side of the ear bar being inserted further into the mouse ear as compared to the other side of the ear bar. The differences in height between the two ears or two eyes are usually the indicator that the mouse head is being tilted to one side. Readjust the two ear bars so that they allow the skull to be leveled correctly.

23. Do not tap too hard on the exposed calvarial region using forceps or other blunt instruments as this may cause the mouse to fidget even when the mouse is anesthetized. Unnecessary movement will cause the misalignment to the calibrated adjustments, thus leading to more time needed to reset the adjustments. While tapping on the mouse calvaria, any movement of the skull indicates that the head is not immobilized sufficiently. Inadequate head immobilization will cause drifting during imaging. Readjust the ear bar or the nose bar clamp if necessary.

24. Leaking of excess PBS out of the demarcation is expected as the demarcation can accommodate only a thin layer of PBS within

it. Ensure that the coverslip holder is calibrated as such that it is parallel to a flat surface (i.e., benchtop) when it is tightened. Any deviation will cause drifting or uneven imaging.

25. The hydrophobic barrier demarcated by the wax pen can trap the PBS within the drawn region which prevents the loss of PBS during imaging.

26. With the position of the stage that is closed to the final imaging position, this reduces the likelihood of some wirings or tubing that are under undue tension or be in an awkward position when the stage setup is eventually moved into the final imaging position.

27. Any movement of the mouse tail during imaging may hit the rectal temperature probe, possibly releasing the probe from the mouse and thus, causing thermal dysregulation.

28. Doing so will prevent the water under the objective lens from drying out. However, this step is actually optional if the user is imaging for only a short period of time.

29. The mouse can be remotely anesthetized from the outside of the dark box without disrupting the imaging process. In addition, physical manipulation on the mouse can be avoided if subsequent dosage of anesthesia is required, which ensures a stable setup. This step is highly recommended even without using the dark box.

30. Record the timings for each injection, as their timely dosage of the anesthesia can ensure a higher chance of a successful continuous imaging session and provide help in future troubleshooting with the drug. Data may be affected and irrecoverable if anesthesia is not given at the right timings.

31. The 'Speckling' effect has been shown to trigger chemotactic responses by the neutrophils, which results in unnecessary imaging artifacts during imaging [9].

32. Increase the temporal resolution if the granulocytes are traveling too fast for cell tracking. To achieve that, you can either increase the image scanning frequency, decrease the thickness of the z-stack, or decrease the spatial resolution of the images.

33. Optimization of the drug injection timing is required, as drug efficacy may vary. To improve the consistency of the anesthesia, always maintain a consistent body temperature of the mouse throughout every imaging session, ensure the imaging room is consistently quiet (to prevent the mouse from fidgeting due to sudden loud noises), and invest in a remote monitoring system for the mouse (pulse oximeter, etc.). In our experience, a mouse that is in consistent deep anesthesia should not have constant movement of its whiskers and shallow breathing

pattern. Alternatively, a continuous drug delivery system such as an inhalational anesthetic system (e.g., isoflurane) can be used.

34. Although shaving using the razor blade can be very fast and clean, the chances of wounding or cutting the skin is also higher. Alternatively, the hair can be shaved briefly using the razor blade, and depilatory cream applied to remove excess hair completely. This will minimize the chances of wounding the skin unnecessarily.

35. Avoid leaving the depilatory cream on the skin for too long, as it may cause local inflammation to the skin.

36. The stabilization of the mouse leg is crucial for the subsequent surgery and imaging. Therefore, ensure that the entire tibia sits firmly onto the mold, and keep the leg flat, without protruding in an awkward angle.

37. It is important to keep the open incision moist when operating the surgery, otherwise the connective tissue and muscles are difficult to remove, and may cause local inflammation as well. Therefore, we need to drop PBS onto the incision once in a while during the whole surgery process.

38. Avoid cutting into any large blood vessels (e.g., saphenous artery in Fig. 10c) when doing both skin incision, and connective tissue and muscle removal, as it may cause massive bleeding.

39. This is a critical step which can determine the imaging quality. Excessive shaving of the tibial bone may create a deep hole in the bone marrow, causing it to bleed. This will also destroy the structural intactness. Conversely, insufficient shaving will lead to the poor quality of imaging, in which the signals will be precipitously attenuated along with depth (Fig. 11c). In general, the final thickness after shaving is roughly around 100 μm but often empirical and difficult to control. Therefore, we recommend doing rounds of pilot assays to familiarize and optimize the depth to shave. For beginners, we recommend the shaving within a small region (e.g., 2 mm × 2 mm square or smaller, the tiny-scale shaving in Fig. 11e, upper). After familiarizing the shaving technique, one can try to shave most of the tibia to expand the imaging fields (the large-scale shaving in Fig. 11e, lower).

40. For the mouse tibial shaving, we recommend the use of the carbide round burr with a low spinning rate of between 5000 and 8000 rpm. Before shaving, locate the region and size of the area to be shaved. Then, homogeneously shave across the whole area layer by layer with the drill bit drawing in an up and down motion (Fig. 11c, upper). This is to prevent the shaving area from being unevenly thinned. Do not shave

multiple small areas within the region of interest. It is difficult to maintain even shaved thickness throughout the whole region of interest (Fig. 11c, lower). In addition, the shaving surface should always be dry. Hence, when we start shaving, use a cotton-tipped applicator to absorb PBS in the incision. Then, shave for about 30–60 s (clean the shavings with a cotton-tipped applicator), and remoisten the incision. Repeat this dry–shave–moisten cycle until the thickness is appropriate for imaging (Fig. 11d, middle).

41. It is inevitable to break some capillaries and cause minor bleeding. If this happened, use PBS to wash the blood away if possible.

42. Avoid pushing down the coverslip too close and tight to the imaging site as this may cause ischemia for the whole leg.

43. When fastening and fixing the coverslip holder, keep the imaging surface flat and not tilted.

Acknowledgments

We would like to thank Chi Ching GOH for technical contributions on the development and optimization of the methods.
This research was funded by SIgN core funding, A*STAR. L. G. N is supported by SIgN core funding.

References

1. Porwit A, McCullough J, Erber WN (2011) Blood and bone marrow pathology, 2nd edn. Churchill Livingstone, London. https://doi.org/10.1016/C2009-0-52942-X

2. Mercier FE, Ragu C, Scadden DT (2011) The bone marrow at the crossroads of blood and immunity. Nat Rev Immunol 12(1):49–60. https://doi.org/10.1038/nri3132

3. Crane GM, Jeffery E, Morrison SJ (2017) Adult haematopoietic stem cell niches. Nat Rev Immunol 17(9):573–590. https://doi.org/10.1038/nri.2017.53

4. Pinho S, Frenette PS (2019) Haematopoietic stem cell activity and interactions with the niche. Nat Rev Mol Cell Biol 20(5):303–320. https://doi.org/10.1038/s41580-019-0103-9

5. Lo Celso C, Fleming HE, Wu JW, Zhao CX, Miake-Lye S, Fujisaki J, Cote D, Rowe DW, Lin CP, Scadden DT (2009) Live animal tracking of individual haematopoietic stem/progenitor cells in their niche. Nature 457 (7225):92–96. https://doi.org/10.1038/nature07434

6. Mazo IB, Gutierrez-Ramos JC, Frenette PS, Hynes RO, Wagner DD, von Andrian UH (1998) Hematopoietic progenitor cell rolling in bone marrow microvessels: parallel contributions by endothelial selectins and vascular cell adhesion molecule 1. J Exp Med 188 (3):465–474. https://doi.org/10.1084/jem.188.3.465

7. Kohler A, Geiger H, Gunzer M (2011) Imaging hematopoietic stem cells in the marrow of long bones in vivo. Methods Mol Biol 750:215–224. https://doi.org/10.1007/978-1-61779-145-1_15

8. Faust N, Varas F, Kelly LM, Heck S, Graf T (2000) Insertion of enhanced green fluorescent protein into the lysozyme gene creates mice with green fluorescent granulocytes and macrophages. Blood 96(2):719–726. https://doi.org/10.1182/blood.V96.2.719

9. Li JL, Goh CC, Keeble JL, Qin JS, Roediger B, Jain R, Wang Y, Chew WK, Weninger W, Ng LG (2012) Intravital multiphoton imaging of immune responses in the mouse ear skin. Nat

Protoc 7(2):221–234. https://doi.org/10.1038/nprot.2011.438

10. von Bruhl ML, Stark K, Steinhart A, Chandraratne S, Konrad I, Lorenz M, Khandoga A, Tirniceriu A, Coletti R, Kollnberger M, Byrne RA, Laitinen I, Walch A, Brill A, Pfeiler S, Manukyan D, Braun S, Lange P, Riegger J, Ware J, Eckart A, Haidari S, Rudelius M, Schulz C, Echtler K, Brinkmann V, Schwaiger M, Preissner KT, Wagner DD, Mackman N, Engelmann B, Massberg S (2012) Monocytes, neutrophils, and platelets cooperate to initiate and propagate venous thrombosis in mice in vivo. J Exp Med 209(4):819–835. https://doi.org/10.1084/jem.20112322

Chapter 16

Intravital Imaging of Bone Marrow Niches

Myriam L. R. Haltalli and Cristina Lo Celso

Abstract

Haematopoietic stem cells (HSCs) are instrumental in driving the generation of mature blood cells, essential for various functions including immune defense and tissue remodeling. They reside within a specialised bone marrow (BM) microenvironment, or niche, composed of cellular and chemical components that play key roles in regulating long-term HSC function and survival. While flow cytometry methods have significantly advanced studies of hematopoietic cells, enabling their quantification in steady-state and perturbed situations, we are still learning about the specific BM microenvironments that support distinct lineages and how their niches are altered under stress and with age. Major advances in imaging technology over the last decade have permitted in-depth studies of HSC niches in mice. Here, we describe our protocol for visualizing and analyzing the localization, morphology, and function of niche components in the mouse calvarium, using combined confocal and two-photon intravital microscopy, and we present the specific example of measuring vascular permeability.

Key words Bone marrow, Bone marrow microenvironment, Niche, Hematopoietic stem cell, Bone marrow stroma, Endothelial cells, Intravital imaging, Confocal and Multiphoton microscopy, Time-lapse imaging, Three-dimensional image analysis

1 Introduction

The primary site of adult hematopoiesis is the bone marrow (BM), from which billions of mature cells, arising from proliferating hematopoietic progenitors, are released into the circulation every day. Hematopoietic stem cells (HSCs) reside within this highly dynamic tissue and are characterized by their ability to self-renew and differentiate on demand to sustain blood cell production.

In 1987, it was proposed that a specialized microenvironment, or niche, is essential for the maintenance of stem cells in their quiescent state and influences HSC self-renewal and differentiation within the BM [1]. A series of studies followed, exploring this concept, and answering initial questions about the molecular and cellular nature of the HSC niche—further refining our understanding of the complexity of cellular interactions that regulate hematopoietic stem and progenitor cells (HSPCs). We now know that an

Marion Espéli and Karl Balabanian (eds.), *Bone Marrow Environment: Methods and Protocols*, Methods in Molecular Biology, vol. 2308, https://doi.org/10.1007/978-1-0716-1425-9_16, © Springer Science+Business Media, LLC, part of Springer Nature 2021

orchestra of non-hematopoietic cells—such as endothelial cells (ECs) [2–4], perivascular cells [5–8], and osteoblasts [9–12]—immune cells and both soluble and metabolic factors are associated with HSCs [13], and understanding their interactions is critical for the development of improved regenerative medicine approaches and to prevent disease.

There remains a need to decipher the spatial organization of the BM microenvironment, and the rapid advance of imaging technologies over the last years has made it possible to directly visualize the HSC niche. While two-dimensional (2D) histological analyses of BM sections provide valuable information, particularly when reconstructed into three-dimensional (3D) representations of the tissue [7, 14–17], preprocessing methods used to decalcify the bone can alter tissue morphology, protein antigenicity and may result in the loss of intrinsic fluorescence or increased autofluorescence. The main disadvantage to these methods is the lack of a temporal dimension, necessary to analyze cellular interactions and differentiation within the BM microenvironment. The use of in vivo confocal and multiphoton microscopy of the BM has permitted real-time imaging of cell dynamics [18–20] and various complex biological processes [21–25], both at steady-state and during disease development [26, 27], that cannot be detected in static tissue sections. Numerous genetically modified mouse models, as well as a vast selection of injectable fluorophore-conjugated antibodies, dyes, and regents, have enhanced the study of the HSC niche by intravital microscopy (IVM) (*see* Table 1). In contrast to the challenges faced by imaging the long bones—including mechanical stress caused by thinning of the bone by shaving prior to imaging—the calvarium (skull cap) bone plate is thin and the BM cavity can be imaged at single cell resolution after a minimally invasive surgery to implant a specially designed headpiece.

In this chapter, we provide a technical overview of the procedures we use to image various components of BM niches. We include tips for IVM of reporter mice to acquire a cellular map of the calvarium BM and image processing for data analysis. Furthermore, we describe a method developed to quantify vascular permeability using time-lapse imaging. We indicate alternative options that we know would not preclude the generation of results and may be relevant to addressing specific research questions.

2   Materials

2.1   Mice

The use of fluorescent reporter mice to detect different cell types in vivo has become a gold standard for imaging studies investigating interactions of different niche cells, as well as hematopoietic cells. An increasing number of transgenic mice are being developed, in which a fluorescence protein (e.g., commonly, green

Table 1
Summary of reporter animal models and other labeling methods for visualizing HSPCs and BM niche components by IVM

Cell type	Reporter mice available	Other labeling methods
HSCs	Hoxb5-tri-mCherry [28] α-catulin GFP [16] vWF-eGFP [29, 30] Tie2-GFP [31] Fgd5$^{mCherry/+}$ [32] Mds1$^{GFP/+}$ Flt3^{Cre} [19]	Lipophilic carbocyanine membrane dyes (e.g., Vybrant DiO, DiI, DiD, and DiR from life technologies or commercial sources) [25] FACS sort HSCs from mice expressing fluorescence in all tissues and cell types (e.g., mT/mG [33])
Haematopoietic cells	mT/mG [33] H2B-eGFP [34] LifeAct-GFP [35]	In vivo administration of anti-CD45 antibody (*see* **Note 1**) or other markers of interest BM chimeras in which hematopoietic cells and stroma express different fluorescent proteins [27] (*see* **Note 2**)
Vasculature	Flk1-GFP [36] Tie-2 Cre [4] VECad-Cre [37] Fgd5$^{mCherry/+}$	In vivo administration of anti-CD31, anti-Endomucin antibody (*see* **Note 1**) or fluorescently conjugated lectins Vascular dyes, for example FITC-, TRITC- or Cy5-dextran, non-targeted quantum dots or angiosense probes
Bone, osteoclasts, osteolineage and perivascular cells	Col2.3GFP [38] and CFP [39] Osterix-EGFPCre [40] Nestin-GFP [7] Prx1-Cre [41, 42] CXCL12-GFP [43] and dsRed [44] LepR-Cre;Rosa26-tdTomato [45] NG2-dsRed [46]	OsteoSense for newly formed bone [47] Calcium-binding reagents such as tetracycline or Calcein blue or alizarin red (sigma) [19] Cathepsin K for osteoclasts and osteoclasts [19, 48] Second harmonic generation (typically 840 nm excitation and 415–445 nm detection) [49, 50]

fluorescence protein, GFP but, more recently, RFP and mCherry too) is driven by, or fused to, the gene of interest therefore only cells that express the gene are fluorescent. *See* Table 1 for a comprehensive list of reporter mice and other labeling approaches currently available for these studies (*see* **Notes 1** and **2**). Depending on the specific investigation, care should be taken to minimize fluorescence signal overlap between reporters and additional antibodies, dyes or reagents used (*see* **Note 3**). Typically, mice aged 6–16 weeks are used, and it is important to remain consistent between experimental repeats. Mice older than 16 weeks may have larger skulls, however, may compromise depth of imaging due to increased bone thickness.

2.2 Surgery and Intravital Microscopy Reagents, Supplies, and Equipment

1. A biological safety cabinet that ensures sterility and the containment of allergens.

2. Anesthesia drugs, for example, isoflurane (*see* **Note 4**).

3. Gas anesthesia vaporizer and scavenging system (if using isoflurane).

4. Heatpad to keep the mouse warm during surgery as well as on the stage of the microscope, best equipped with body temperature probe (for example, the ThermoStar Homeothermic Monitoring System).

5. Sterile surgical tools (scissors and forceps).

6. Cotton buds.

7. Specially designed headpieces that attach to the skull of the mouse and fit into a lock and key mechanism on the microscope stage. This may vary depending on the specific IVM mouse holding set up available to you.

8. Dental cement (we use Diamond Carve from Kemdent).

9. Lacri-Lube or equivalent eye ointment to lubricate and protect the animal's eyes from drying while anesthetized.

10. Lubricant to facilitate inserting the rectal temperature probe connected to the heatpad.

11. Phosphate buffered saline (PBS, without calcium and magnesium) and plastic Pasteur pipettes.

12. Microscope—for example, we use a Zeiss LSM 780 upright confocal/two-photon hybrid microscope equipped with Argon (458, 488, and 514 nm), a diode-pumped solid-state 561 nm, a Helium-Neon 633 nm, and a tunable infrared multiphoton laser (Spectraphysics Mai Tai DeepSee 690–1200 nm), four non-descanned detectors (NDDs), and an internal spectral detector array (*see* **Note 5**). Signal is visualized using a Zeiss W Plan-Apochromat 20× DIC water immersion lens (1.0 NA).

13. Vascular dyes and other antibodies or injectable reagents required for the specific IVM experiment to be carried out. For the analysis and measurement of vasculature permeability, 65–80 kDa TRITC-dextran (Sigma) is preferred.

14. Disposable insulin syringes (ideally 29 G × 12.7 mm and 0.5 ml or 1 ml, as required).

15. Heat box and mouse restrainer to facilitate intravenous (i.v.) injections.

16. (*Optional*—for recovery imaging) Intrasite gel, plasters and analgesic for pain relief during recovery.

2.3 Image Analysis

1. A powerful workstation, for example (at minimum) an Intel-Core i5-3427U processor at 1.80 GHz and 4.0 GB RAM, running the 64-bit Windows 7 operating system.

2. FIJI/ImageJ.

3. (*Optional*) Further software including Imaris, Matlab, and R packages.

3 Methods

3.1 Microscope Considerations for In Vivo Imaging

1. Microscopes for in vivo imaging should include a stage and a holder that secures the animal in a steady but comfortable position, which will minimize movement from breathing and heartbeat during imaging. Multiple variations of this set up are now available, thus this protocol may need to be adjusted depending on your specific system. The microscope should also have a means of keeping the mouse warm, such as a heatpad and rectal probe thermometer for measuring and adjusting body temperature throughout. If using Isoflurane for anesthesia, the mechanism of delivery to the animal, as well as a scavenging system to protect users, would also be necessary.

2. When planning your experiments, ensure the microscope to be used will have the capacity to allow multichannel setup in order to capture and record information from the fluorescent probes labeling your cells of interest, as well as any reference structures within the BM space. Ideally, and to help with navigating around the calvarium, the bone can be imaged by the endogenous second harmonic generation (SHG) signal from bone collagen through two-photon excitation.

3. Autofluorescence from BM cells can cause false positives when identifying fluorescently labeled cells in vivo (*see* Fig. 1). The intensity of this signal varies from cell to cell and from mouse to mouse and can be particularly strong within the blue to red wavelengths. To deal with this, it is useful to know your sample and understand the spectra of the autofluorescence in your experiment, for example by using spectral lambda scanning. This will allow you to optimize your fluorophore choices and select those with spectra as far away as possible from the background noise. Selecting fluorophores with narrow excitation and emission spectra makes it easy to specifically acquire signal from your cell or structure of interest. If your experimental set up permits it, an autofluorescence channel can be included during in vivo imaging which does not overlap with those used to detect other fluorophores. During acquisition, this

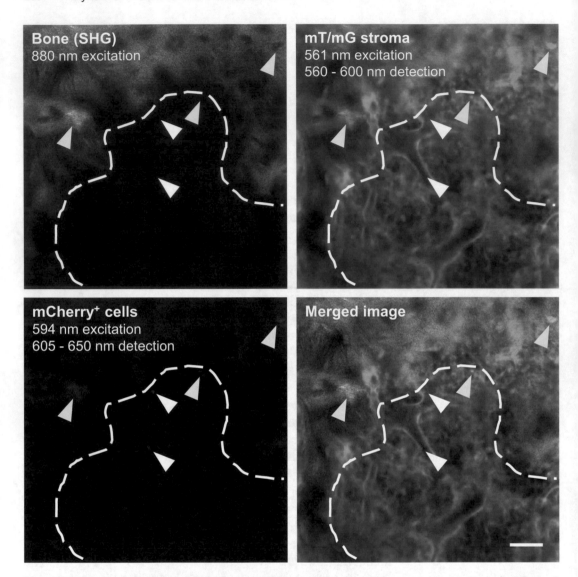

Fig. 1 Deciphering true sample signal from autofluorescence. An image acquired of mCherry⁺ cells (red) acquired from an mT/mG mouse calvarium where the BM stroma can be visualised (green). The bone was imaged using SHG (gray) and is outlined with the white dashes in each panel. The white arrows indicate true cells as they appear brightly only in the mCherry channel. The yellow arrows indicate examples of cells/ structures which appear bright in the mCherry channel but can also be seen in the tomato and/or SHG channel therefore are likely autofluorescence and can be excluded from analysis. Details of the set up for each channel is provided within the images. Scale bar represents 100μm

channel can share the same detection set up as any other channel of interest but with the excitation laser turned off. If your cell of interest is still bright compared to neighboring cells when the excitation laser is off, it can be considered to be another cell with high autofluorescence.

3.2 Surgery

1. If using fluorescent probes, such as antibodies, for in vivo labeling of BM components, inject these i.v. prior to inducing anesthesia and commencing surgery to allow time for labeling.

2. Anesthetize the animal, for example using isoflurane and Oxygen mixed in a vaporizer (4% in 4 L/min Oxygen for induction and 1–2.5% in 1 L/min Oxygen for maintenance) and delivered through a nose cone with an appropriate scavenging system in place. Monitor the breathing rate of the animal constantly throughout the imaging session in order to maintain it at approximately 1 breath per second. Adjust the anesthesia as necessary. The mouse should be placed on a heatpad for the duration of surgery and body temperature also monitored.

3. Expose the calvarium for imaging, starting by dampening the fur on the scalp with ethanol. Using a forceps, raise the skin between the ears and make a small incision using surgical scissors. Insert one side of the scissors under the skin to make two long incisions towards the nose of the mouse. Remove a rectangular flap of skin by cutting horizontally behind the nose, resulting in a rectangular window to the bone (*see* Fig. 2a, left panel). Care should be taken to avoid trimming the mouse's whiskers, especially if planning to recover it after imaging.

4. Use clean, dry cotton buds to consistently wipe away all fur from the exposed area of bone (hair is highly autofluorescent under the microscope, therefore should be eliminated as much as possible during preparation). The exposed area of the skull should be large enough for the metal headpiece to sit on the bone itself and not the skin and fur (the latter would likely lead to the detachment of the headpiece during imaging) (*see* Fig. 2a, right panel). Measure the opening made at this stage against the headpiece and adjust it accordingly by making further incisions and removing extra skin to widen the window, if required.

5. Attach the headpiece to the mouse using dental cement. Form a firm paste and apply it to all edges of the base of a clean headpiece. Place the headpiece on the head, using forceps and a steady hand, and ensure that it is as level as possible with the bone. It is also useful to leave the coronal suture uncovered and ensure that the central sinus can be seen running through the middle of the exposed area and parallel to the sides of the headpiece (*see* Fig. 2b). Allow the cement to fully dry.

6. If using fluorescent vascular dyes to visualize vasculature, inject these i.v. prior to mounting the mouse onto the microscope to ensure it will last through image setup and acquisition. For measuring vascular permeability, inject TRITC-dextran immediately prior to imaging, when the mouse is already on the stage and image acquisition parameters are set up (*see* Subheading 3.7 for a detailed protocol).

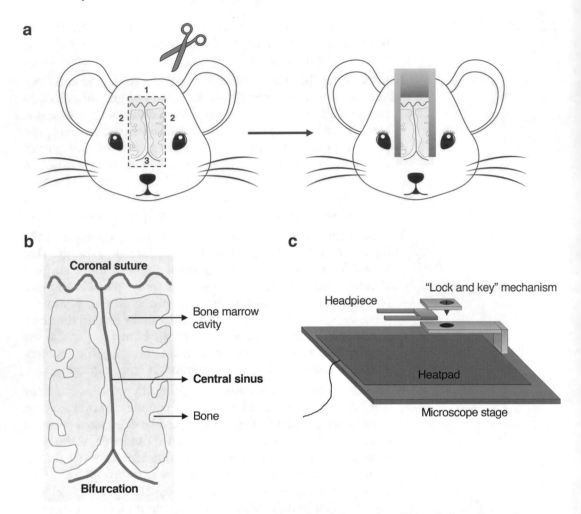

Fig. 2 Preparation of the calvarium for intravital imaging. (**a**) A schematic detailing an example of an optimal method for removing the scalp to create a sufficiently large imaging window so that the headpiece can be placed directly onto the bone of the skullcap, with the main landmarks of the calvarium clearly in view. These are highlighted in (**b**). (**c**) depicts an example of a "lock and key" mechanism that can be used to attach the headpiece to the microscope stage. The screw must be tightened well to avoid movement of the head during image acquisition. The mouse is laid onto a heatpad to control its body temperature throughout the procedure

3.3 Mounting the Mouse onto the Stage

1. Transfer the anaesthetized animal to the heatpad on the microscope stage and secure the headpiece into the "lock and key" positioning brace mechanism (*see* **Note 6** and Fig. 2c). Insert a rectal probe to monitor the animal's body temperature during imaging, using a lubricant to avoid causing discomfort to the mouse, and place Lacri-Lube on the eyes.

2. Apply PBS to the imaging window using a Pasteur pipette.

3. Using a cotton bud, clean away the membrane that covers the bone to increase image quality (*see* **Note 7**).

4. Lower the lens onto fresh PBS on the calvarium. Using the brightfield setup and shining a white light over the calvarium, locate the sagittal suture and move to locate its intersection with the coronal suture, which has a characteristic curvature and striations which can be seen in the bone. This is a prominent landmark of the calvarium which serves as a helpful orientation guide while imaging and for keeping each imaging session between animals as consistent as possible.

3.4 Image Acquisition Setup

1. Once the lens is positioned appropriately, turn on the excitation source for the bone (for example, using SHG) and adjust the Z-axis until the bone surface comes into focus in the imaging software.

2. Switch on the relevant excitation sources and detection channels for the cells of interest and carry out a quick scan of the whole imaging area to help with adjusting the power of your excitation source and the gain of the detection channels to ensure it is optimal for all the features you wish to capture.

3. A tilescan image (individual tiles stitched together to form a composite) of the calvarium, which provides you with a detailed 3D overview of the BM cavity in your experiment, can be acquired (*see* Fig. 3a). A number of BM niche components can be visualized simultaneously by setting up the appropriate excitation and detection parameters. A typical tilescan captures the area between the coronal suture and the bifurcation of the central sinus (*see* **Note 8**). Using the motorized, precision stage control of the microscope, select the area where overlapping Z-stacks are to be captured (*see* **Note 9**). These can be stitched together postacquisition to create one image of the calvarium BM. It is recommended, in order to get a good stitching effect, to have at least 10% tile overlap.

4. A tilescan provides a static, 3D overview of the BM cavity. It can also be used as a map to select positions of interest and set Z-stacks for time-lapse imaging of BM components of interest within the calvarium (*see* **Note 10** and Fig. 3b). In the latter case, a low-resolution image would reduce acquisition time and reduce "set up" time prior to starting time-lapse acquisition. The time interval should be selected based on the cell type you are imaging, the specific experimental question and the potential for bleaching the fluorophore in question. In our hands, we have found that taking an image every 3 min is sufficient for capturing mobilization, interactions between cells and niche components and cell division events (*see* **Note 11**).

5. The main limitations to the in vivo imaging sessions are time constraints imposed due to animal welfare concerns and the gradual loss of fluorescence signals. The mice may remain

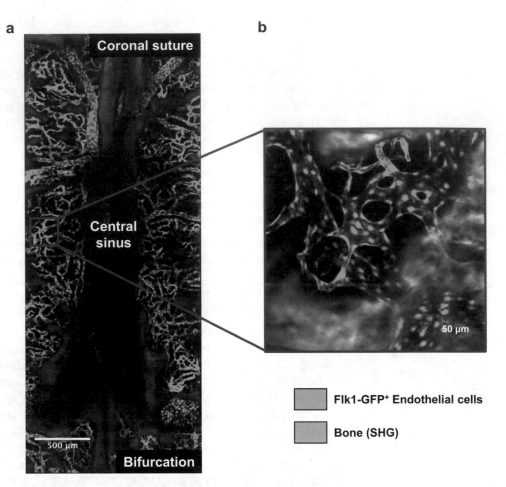

Fig. 3 Generating an overview of the calvarium BM facilitates selection of specific positions for further analysis. (**a**) Individual 3D z-stacks are acquired and stitched together to build a 3D tile scan (Z projection shown here), allowing for whole-tissue visualization of the calvarium BM with single cell resolution as seen in (**b**). Positions for subsequent time-lapse imaging can be selected from the tilescan overview. These images were obtained from a Flk1-GFP mouse, in which the endothelial cells express GFP (green). The bone was imaged using SHG

continuously anesthetized however will, over time, suffer from dehydration. Hydration and nutrients should be supplied for imaging sessions lasting longer than 8 h and longer than 6 h if followed by recovery. The gradual loss of fluorescence is unavoidable as fluorescent probes will be cleared from the animal's circulation over time, or by photobleaching of the fluorophore themselves, but can be managed by reinjecting some dyes and careful set up of excitation parameters.

3.5 Recovery of the Mouse for Reimaging

For analyzing various cell types over the course of a few days, for example, the mice can be recovered and reimaged when necessary (*see* **Note 12**). The calvarium will need to be covered postimaging each time to reduce the buildup of scar tissue and dehydration of the bone. Local expertise from veterinary surgeons should be sought for devising the most appropriate experimental protocol that will ensure animal welfare.

1. With the mouse still under anesthesia, dry the headpiece and imaging window thoroughly with a cotton bud, clean the area and evenly apply enough intrasite gel to cover the entirety of the exposed calvarium in the imaging window.

2. Once the gel is dry, cut a piece of plaster to the size of the headpiece, lay it over the gel and secure it in place.

3. Place the mouse in a heated recovery box with an appropriate analgesic. Once the animal begins to exhibit normal behavior, place it in its cage, ensure that any cage enrichment that the headpiece could get entwined with is removed. Extra analgesic for pain relief can be placed in the cage or in the water, as required. Monitor animals frequently over the following days.

4. For subsequent imaging sessions, induce anesthesia, remove the plaster and clean away the gel. The headpiece is immediately ready to be attached to the microscope and the imaging session can readily commence. As long as the headpiece has been attached well and does not fall off postrecovery, the main advantage of this method is that the exact positions imaged before can be imported from the metadata of previous imaging sessions and reimaged again.

3.6 Image Processing and Analysis

For stitching 3D BM tilescans, we generally use ZEN black software (Zeiss, Germany) as it is the same used for image acquisition. However, stitching can also be done using FIJI/ImageJ. We use ImageJ to visualize and process raw data using various tools including bleach correction and maximum projections (*see* **Note 13**). Tilescans will often contain a large amount of autofluorescent signal obtained during acquisition and it helps with visualization and quantification of these images to manually crop that signal out using ImageJ (*see* Fig. 4). Macros can be designed to aid with the process; however, a full automation of this step is challenging. Often, time-lapse images may shift or tilt due to movement of the animal or of the headpiece during image acquisition. The registration tool in ImageJ [51] can be used to rectify this. Cell tracking can be performed in ImageJ using plugins such as TrackMate or in other analysis software such as Imaris or R packages. Ultimately, the processing and analysis of images obtained by IVM will mainly be driven by the specific research questions. Despite the challenging and lengthy protocol, the power of IVM lies in the vast amount of information that can be extracted from each image obtained, not forgetting how impactful it is to achieve direct visualization of cells within the BM.

Maximum projection
before **cropping**

Example slice:
30 out of 56

1. Draw around Nestin+ signal
2. Clear all signal outside of the dotted lines
3. Repeat on next slice

Example slice:
50 out of 56

Maximum projection
after **cropping**

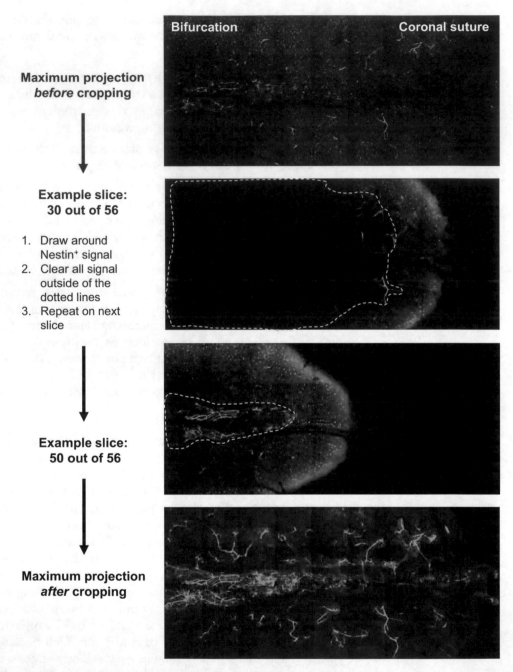

Fig. 4 Processing to remove autofluorescence drastically improves the clarity of images. An example of the impact that manually cropping tilescans acquired by IVM can have on the final image. This image represents the calvarium of a Nestin-GFP mouse, in which a proportion of perivascular cells are identifiable through GFP expression. High levels of autofluorescence can be seen. Cropping was carried out on each slice of the tilescan in ImageJ following the steps detailed above and the striking difference can be seen in the final maximum projection, clearly depicting the distribution of Nestin+ cells in the calvarium BM

3.7 Application of the Protocol for Measuring Vascular Permeability

Blood vessels are lined by specialized ECs in the BM that establish an instructive and nurturing vascular niche, responsible for orchestrating homeostasis and regeneration of cells within the tissue. By producing specific angiocrine factors, ECs are able to regulate HSPC fate decisions, including self-renewal, proliferation, differentiation, and intra- and extravasation [52]. It has been shown that less permeable arterial vessels maintain HSPCs in a metabolically quiescent state, indicated by undetectable levels of reactive oxygen species (ROS) levels, whereas more permeable sinusoids, for example, are able to sustain ROS^{high} HSPCs, promoting their differentiation and trafficking [18]. Interesting questions about the mechanisms by which various insults impact BM vascular structural integrity have been raised and already some work has shown that leukemic cells disrupt BM vascular architecture and permeability in order to sustain malignant cells which disrupt normal hematopoiesis [53]. Here we present a protocol, adapted by our research group from previously published protocols [18, 53] to test vascular leakiness at steady-state and under stress.

1. Carry out surgery and mounting of the mouse onto the stage as described in Subheadings 3.2 and 3.3. At this stage, vascular dyes should *not* be injected into the animal as this will perturb the acquisition of the readout of vascular permeability by the microscope (*see* **Note 14**).

2. Once the lens has been positioned appropriately and excitation sources and detection channels for the ECs, any other cell type or structure to be imaged and the specific dextran to be used have been set up, acquire a quick scan of the BM cavity. This can be used to select appropriate positions for measuring vascular leakiness (*see* **Note 15**). This protocol requires an image of each position to be taken every 1 min for up to 15 min immediately after injecting the dextran (this timing can be altered based on your specific experiment and research question). Therefore, the number of positions, Z-stacks and averaging should be optimized to ensure that all positions can feasibly be acquired within this set time period.

3. Confirm that all positions to be imaged are selected and all imaging parameters have been adjusted accordingly within the microscope software, including the time interval set up for time-lapse imaging. Check the temperature and breathing of the mouse and that there is sufficient PBS on the calvarium (avoid cleaning the bone with a cotton bud at this point and if it is necessary to add PBS to the imaging window, be gentle and try not to touch the head as this can disrupt the positions that have been selected for imaging).

4. Prepare an insulin syringe with 60–80 μl of low molecular weight TRITC dextran (3 mg/ml, 65–80 kDa). Gently support the tail of the mouse in one hand and with the other, inject

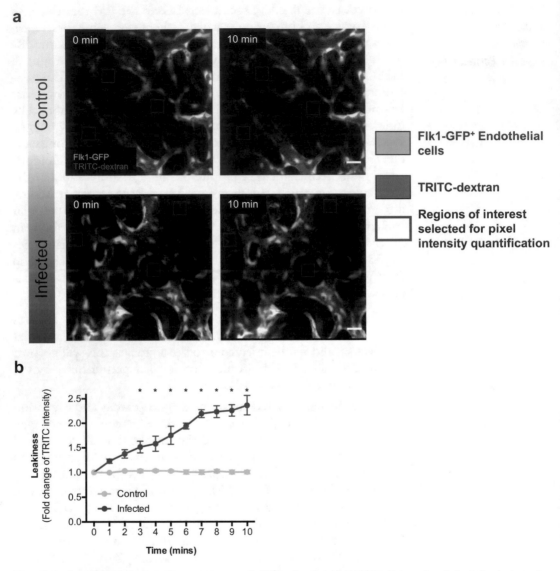

Fig. 5 Applying IVM to quantify vascular permeability in vivo. (**a**) Using the protocol described, vascular leakiness was assessed by time-lapse imaging of randomly selected regions of interest (red boxes) within the calvarium of mice for 10 min following the administration of TRITC-dextran. Here we provide representative frames from control as well as malaria infected mice, in which vessels become significantly permeable due to the inflammatory response generated. These images were obtained from a Flk1-GFP mouse, in which the endothelial cells express GFP (green), injected with TRITC-dextran (magenta). (**b**) An example of the quantification that is obtained after analyzing the time-lapse images from this experiment

the dextran into the tail vein (*see* **Note 16**). As soon as the dextran has been administered, immediately start the time-lapse recording, and you should already see the dextran within the vasculature (*see* Fig. 5a). If you cannot see any signal, the dextran may have been injected subcutaneously. In this case, stop the image acquisition and readminister the vascular dye to the mouse before restarting the time-lapse.

5. Once the videos have been acquired and exported, the data can be analyzed using ImageJ. Ensure that the videos have been registered and make a maximum projection of each one to be analyzed. Once all parameters have been adjusted and applied, including brightness, contrast and bleach correction, calculate the pixel intensity within three equally sized (usually square) and randomly placed regions of interest in the BM parenchyma per time frame (*see* Fig. 5a). This can be done using the "measure" option on ImageJ, making sure that the software is measuring the pixel intensity of the correct channel for your dextran. Record these values, take an average and calculate the fold change increase in intensity. These values can be plotted in a graph to assess how permeability changes over time or to compare between different conditions (*see* Fig. 5b). It is advisable to take measurements for all control and test animal groups in your experiment within the same imaging session to control for any technical variability that may occur. If this is not possible, try to split the groups in a way that allows for control mice to be imaged each time and these can help you to normalize the data if variation is observed.

4 Notes

1. For in vivo labeling of BM components with antibodies, we typically inject 5–10μg i.v. via the tail vein before starting surgery to give the antibody at least 30 min to circulate and label the appropriate cells. This should be optimized based on the specific antibody used. As an example, for labeling ECs in vivo, we have found anti-CD31, clone 390, to work best.

2. BM chimeras, in which hematopoietic cells and stroma express different fluorescent proteins, can be made, for example, by transplanting whole BM (approx. 1×10^6 cells) from mT/mG donor mice into lethally irradiated C57BL/6 wild-type recipients (11 Gy total, usually split into two doses administered at least 3 h apart, up to 24 h prior to transplantation). In this specific case, hematopoietic cells will be labeled with mTomato fluorescence protein and all stroma will remain dark. The opposite combination can be performed which will allow IVM of stromal components (which will in this case be mTomato$^+$) and their interactions in vivo.

3. GFP is a popular fluorophore and derivatives of it include yellow fluorescence protein (YFP) and cyan fluorescence protein (CFP). The emission maxima of these proteins are limited in the red fluorescence; therefore, red fluorescence protein (RFP) and RFP-derived fluorophores have been developed which complement GFP fluorophores very well, for example when used in tandem to tag multiple proteins of interest and

visualize interactions between them. Furthermore, given that the native autofluorescence of cells and tissues is typically in the green-yellow range, RFP suffers less from spectral overlap with autofluorescent signal, resulting in better visualization of the signal than GFP which does emit in the autofluorescence range. More recently, far-red shifted fluorescence proteins, such as mCherry, have been generated with the aim to increase the palette of fluorophores available for imaging studies. It is important to carefully consider the choice of fluorophores for intravital imaging to get the best results. Freely available spectral plotting and analysis tools can be found online (for example: https://searchlight.semrock.com), allowing researchers to model and evaluate the spectral performance of fluorophores, filter sets, light sources and detectors in order to quantify a number of metrics including signal-to-noise ratio and bleed-through of fluorophores.

4. In our experience, isoflurane is more labor intensive than injectable anesthetics as it requires continuous monitoring of the animal's breathing and adjusting anesthesia and oxygen levels throughout surgery and imaging. However, it is advantageous in that it is more controllable and better in the long-term for the mouse—especially if the experiment involves recovery and reimaging at a later timepoint.

5. Single two-photon or single confocal microscopes are also suitable for carrying out IVM. We find that two-photon microscopy is ideal to detect signal from the bone, therefore allowing visualization of the BM cavity edges, however it is limited in identifying very weak signals. Confocal microscopy provides generally less penetration and resolution but allows the maximum signal collection. For example, weakly labeled cells can be identified using the confocal modality with a wide pinhole and then imaged optimally with the two-photon modality.

6. It is extremely important to secure the mouse well at this stage as any small movements due to loosening of the headpiece from the brace can have drastic (and often unsalvageable) effects on image acquisition. Ensure that the screw holding the headpiece in place (if applicable in your setup) is tightened well. It may also be beneficial to change the screw between imaging sessions to avoid blunting of the head which will impede tightening of the screw.

7. If, while cleaning away the membrane, the mouse starts to bleed, stop wiping. Instead, gently press with the cotton bud and allow the bleed to airdry and clot before wiping away the clotted blood with PBS and a fresh cotton bud prior to imaging. If blood is present on the calvarium during imaging, it will not allow the lasers to penetrate through to the tissue and will obstruct image acquisition.

8. When setting up tilescans for a cohort of mice within an experiment, it is useful to keep the images as consistent as possible and aiming for a similar number of Z-stacks horizontally and vertically. This becomes important at the stage of image quantification to ensure that bias is not introduced due to variability in the surface area of BM imaged each time. Optimal positioning of the imaging window contributes to maximizing reproducibility of the images.

9. The interval between each image acquired in each Z-stack should be set to ensure that there are no gaps in your acquisition while avoiding oversampling which could lead to bleaching due to excessive illumination. Generally, intervals of up to 5μm are fine. A 3–5μm step is usually sufficient for capturing the locations of your cell of interest. Smaller steps can allow acquisition of the fine details of the BM microarchitecture, if required.

10. Averaging helps to improve the image by increasing the signal-to-noise ratio. The higher the averaging, the more time image acquisition takes and the higher the risk of photobleaching. Therefore, a balance between the number of positions to image, the number of channels to be acquired and averaging should be found to ensure you are able to image all cells of interest in the calvarium for the desired time.

11. If a particular cell you are tracking moves deeper into the marrow or out of frame, you can stop the current time-lapse at the end of a cycle, save the movie, readjust the Z-axis or position and restart the movie. It is important to keep track of time to maintain the selected time intervals between each frame. These resultant movies can be concatenated using ImageJ postacquisition.

12. If you are planning to carry out recovery imaging, it is best to work as quickly and efficiently as possible to minimize the time that the mouse is under anesthesia in order to ensure effective recovery each time.

13. When generating projections of acquired tilescans or time-lapse positions, you may want to consider using different projection types for different channels, depending on the data you are trying to represent. For example, maximum projection is preferable for viewing structures and cells deeper within the calvarium BM. However, if this is applied to the channel used to acquire the bone, for example, it can mask all other channels as the signal intensity is extremely bright at the highest point of the calvarium. To overcome this, split the channels and do not include the bone channel when merging them again to create a maximum projection of your channels of interest. For the bone, you can create a *median* projection which will result in

a clear outline of the central bone structures as seen in Fig. 3. These two projections can subsequently be merged together to create a final composite image.

14. To visualize ECs in the calvarium, we use Flk1-GFP reporter mice, in which CD45$^-$ Ter119$^-$ CD31$^+$ phenotypic ECs express GFP. This greatly aids the selection of positions to image and to determine the appropriate Z-stack set up in order to effectively capture the vascular dye extravasating from the vessels. Other reporter mice for visualizing the vasculature are available (*see* Table 1). Otherwise, anti-CD31 antibody, anti-Endomucin antibody, or other fluorescently conjugated lectins can be administered to wild-type mice prior to surgery to label ECs. Ensure that the fluorophore of antibody chosen will not interfere with the emission spectra of the dextran used to quantify vascular permeability.

15. Ensure that the positions selected allow you to clearly see the blood vessels as well as substantially large areas of the BM parenchyma around the vessels. This will allow you to reliably quantify the dextran as it extravasates from the vasculature in ImageJ. SHG signal can be used to visualize the bone pillars in the calvarium and to avoid confounding those areas for BM parenchyma, as there will of course be minimal dextran signal there and this could bias the final quantification.

16. Placing the tail directly on the heatpad during image acquisition set up and prior to injecting the dextran, as well as shining a light behind the tail using a lamp or torch can help to emphasize the veins making them easier to see for aiming the needle. Be very gentle and try not to pull the tail too taught to avoid moving the mouse or headpiece, thus avoiding changes to the positions selected for time-lapse imaging. Keep a steady hand and be careful not to injure yourself with the needle.

Acknowledgments

We are grateful to the Imperial College London Central Biomedical Services and Sir Francis Crick Institute Biological Research Facility for their support with in vivo experimental models. Thanks to Andreas Bruckbauer and Steve Rothery from the Imperial College Facility for Imaging by Light Microscopy for their advice on image acquisition and particularly analysis. This work was funded by the Wellcome Trust (Investigator award 212304/Z/18/Z to C.L.C. and PhD studentship 105398/Z/14/Z to M.H.), Cancer Research UK (Programme Foundation award C36195/A26770 to C.L.C.), the ERC (Starting grant 337066 to C.L.C.), and BBSRC (grant BB/L023776/1 to C.L.C.).

References

1. Schofield R (1978) The relationship between the spleen colony-forming cell and the haemopoietic stem cell. Blood Cells 4:7–25

2. Kiel MJ, Yilmaz ÖH, Iwashita T et al (2005) SLAM family receptors distinguish hematopoietic stem and progenitor cells and reveal endothelial niches for stem cells. Cell 121:1109–1121

3. Kiel MJ, Radice GL, Morrison SJ (2007) Lack of evidence that hematopoietic stem cells depend on N-cadherin-mediated adhesion to osteoblasts for their maintenance. Cell Stem Cell 1:204–217

4. Ding L, Saunders TL, Enikolopov G, Morrison SJ (2012) Endothelial and perivascular cells maintain haematopoietic stem cells. Nature 481:457

5. Katayama Y, Battista M, Kao W-M et al (2006) Signals from the sympathetic nervous system regulate hematopoietic stem cell egress from bone marrow. Cell 124:407–421

6. Méndez-Ferrer S, Lucas D, Battista M, Frenette PS (2008) Haematopoietic stem cell release is regulated by circadian oscillations. Nature 452:442

7. Méndez-Ferrer S, Michurina TV, Ferraro F et al (2010) Mesenchymal and haematopoietic stem cells form a unique bone marrow niche. Nature 466:829

8. Pinho S, Lacombe J, Hanoun M et al (2013) PDGFRα and CD51 mark human Nestin+ sphere-forming mesenchymal stem cells capable of hematopoietic progenitor cell expansion. J Exp Med 210:1351–1367

9. Calvi LM, Adams GB, Weibrecht KW et al (2003) Osteoblastic cells regulate the haematopoietic stem cell niche. Nature 425:841

10. Zhang J, Niu C, Ye L et al (2003) Identification of the haematopoietic stem cell niche and control of the niche size. Nature 425:836

11. Arai F, Hirao A, Ohmura M et al (2004) Tie2/ Angiopoietin-1 signaling regulates hematopoietic stem cell quiescence in the bone marrow niche. Cell 118:149–161

12. Lo Celso C, Fleming HE, Wu JW et al (2009) Live-animal tracking of individual haematopoietic stem/progenitor cells in their niche. Nature 457:92

13. Pinho S, Frenette PS (2019) Haematopoietic stem cell activity and interactions with the niche. Nat Rev Mol Cell Biol 20(5):303–320

14. Nombela-Arrieta C, Pivarnik G, Winkel B et al (2013) Quantitative imaging of haematopoietic stem and progenitor cell localization and hypoxic status in the bone marrow microenvironment. Nat Cell Biol 15:533

15. Kusumbe AP, Ramasamy SK, Adams RH (2014) Coupling of angiogenesis and osteogenesis by a specific vessel subtype in bone. Nature 507:323

16. Acar M, Kocherlakota KS, Murphy MM et al (2015) Deep imaging of bone marrow shows non-dividing stem cells are mainly perisinusoidal. Nature 526:126

17. Coutu DL, Kokkaliaris KD, Kunz L, Schroeder T (2017) Multicolor quantitative confocal imaging cytometry. Nat Methods 15:39–46

18. Itkin T, Gur-Cohen S, Spencer JA et al (2016) Distinct bone marrow blood vessels differentially regulate haematopoiesis. Nature 532:323

19. Christodoulou C, Spencer JA, S-CA Y et al (2020) Live-animal imaging of native haematopoietic stem and progenitor cells. Nature 578:278–283

20. Upadhaya S, Krichevsky O, Akhmetzyanova I et al (2020) Intravital imaging reveals motility of adult hematopoietic stem cells in the bone marrow niche. Cell Stem Cell 27(2): P336–P345.e4

21. Sipkins DA, Wei X, Wu JW et al (2005) In vivo imaging of specialized bone marrow endothelial microdomains for tumour engraftment. Nature 435:969–973

22. Adams GB, Alley IR, Chung U-I et al (2009) Haematopoietic stem cells depend on Galpha (s)-mediated signalling to engraft bone marrow. Nature 459:103–107

23. Köhler A, Filippo KD, Hasenberg M et al (2011) G-CSF-mediated thrombopoietin release triggers neutrophil motility and mobilization from bone marrow via induction of Cxcr2 ligands. Blood 117:4349–4357

24. Beck TC, Gomes A, Cyster JG, Pereira JP (2014) CXCR4 and a cell-extrinsic mechanism control immature B lymphocyte egress from bone marrow. J Exp Med 211:2567–2581

25. Rashidi NM, Scott MK, Scherf N et al (2014) In vivo time-lapse imaging shows diverse niche engagement by quiescent and naturally activated hematopoietic stem cells. Blood 124:79–83

26. Hawkins ED, Duarte D, Akinduro O et al (2016) T-cell acute leukaemia exhibits dynamic interactions with bone marrow microenvironments. Nature 538:518

27. Duarte D, Hawkins ED, Akinduro O et al (2018) Inhibition of Endosteal vascular niche remodeling rescues hematopoietic stem cell loss in AML. Cell Stem Cell 22:64–77.e6

28. Chen JY, Miyanishi M, Wang SK et al (2016) Hoxb5 marks long-term haematopoietic stem

cells and reveals a homogenous perivascular niche. Nature 530:223

29. Sanjuan-Pla A, Macaulay IC, Jensen CT et al (2013) Platelet-biased stem cells reside at the apex of the haematopoietic stem-cell hierarchy. Nature 502:232–236

30. Pinho S, Marchand T, Yang E et al (2018) Lineage-biased hematopoietic stem cells are regulated by distinct niches. Dev Cell 44:634–641.e4

31. Ito K, Hirao A, Arai F et al (2006) Reactive oxygen species act through p38 MAPK to limit the lifespan of hematopoietic stem cells. Nat Med 12:446–451

32. Gazit R, Mandal PK, Ebina W et al (2014) Fgd5 identifies hematopoietic stem cells in the murine bone marrow. J Exp Med 211:1315–1331

33. Muzumdar M, Tasic B, Miyamichi K et al (2007) A global double-fluorescent Cre reporter mouse. Genesis 45:593–605

34. Hadjantonakis A-K, Papaioannou VE (2004) Dynamic in vivo imaging and cell tracking using a histone fluorescent protein fusion in mice. BMC Biotechnol 4:33

35. Schachtner H, Li A, Stevenson D et al (2012) Tissue inducible Lifeact expression allows visualization of actin dynamics in vivo and ex vivo. Eur J Cell Biol 91:923–929

36. Ishitobi H, Matsumoto K, Azami T et al (2010) Flk1-GFP BAC Tg mice: an animal model for the study of blood vessel development. Exp Anim 59:615–622

37. Alva JA, Zovein AC, Monvoisin A et al (2006) VE-cadherin-Cre-recombinase transgenic mouse: a tool for lineage analysis and gene deletion in endothelial cells. Dev Dynam 235:759–767

38. Kalajzic Z, Liu P, Kalajzic I et al (2002) Directing the expression of a green fluorescent protein transgene in differentiated osteoblasts: comparison between rat type I collagen and rat osteocalcin promoters. Bone 31:654–660

39. Paic F, Igwe JC, Nori R et al (2009) Identification of differentially expressed genes between osteoblasts and osteocytes. Bone 45:682–692

40. Liu Y, Strecker S, Wang L et al (2013) Osterix-Cre labeled progenitor cells contribute to the formation and maintenance of the bone marrow stroma. PLoS One 8:e71318

41. Logan M, Martin JF, Nagy A et al (2002) Expression of Cre recombinase in the developing mouse limb bud driven by aPrxl enhancer. Genesis 33:77–80

42. Greenbaum A, Hsu Y-MS, Day RB et al (2013) CXCL12 in early mesenchymal progenitors is required for haematopoietic stem-cell maintenance. Nature 495:227

43. Sugiyama T, Kohara H, Noda M, Nagasawa T (2006) Maintenance of the hematopoietic stem cell Pool by CXCL12-CXCR4 chemokine signaling in bone marrow stromal cell niches. Immunity 25:977–988

44. Ding L, Morrison SJ (2013) Haematopoietic stem cells and early lymphoid progenitors occupy distinct bone marrow niches. Nature 495:231

45. Comazzetto S, Murphy MM, Berto S et al (2019) Restricted hematopoietic progenitors and erythropoiesis require SCF from leptin receptor+ niche cells in the bone marrow. Cell Stem Cell 24:477–486.e6

46. Zhu X, Bergles DE, Nishiyama A (2008) NG2 cells generate both oligodendrocytes and gray matter astrocytes. Development 135:145–157

47. Lassailly F, Foster K, Lopez-Onieva L et al (2013) Multimodal imaging reveals structural and functional heterogeneity in different bone marrow compartments: functional implications on hematopoietic stem cells. Blood 122:1730–1740

48. Kozloff KM, Quinti L, Patntirapong S et al (2009) Non-invasive optical detection of cathepsin K-mediated fluorescence reveals osteoclast activity in vitro and in vivo. Bone 44:190–198

49. Ambekar R, Chittenden M, Jasiuk I, Toussaint KC (2012) Quantitative second-harmonic generation microscopy for imaging porcine cortical bone: comparison to SEM and its potential to investigate age-related changes. Bone 50:643–650

50. Houle M-A, Couture C-A, Bancelin S et al (2015) Analysis of forward and backward second harmonic generation images to probe the nanoscale structure of collagen within bone and cartilage. J Biophotonics 8:993–1001

51. Preibisch S, Saalfeld S, Schindelin J, Tomancak P (2010) Software for bead-based registration of selective plane illumination microscopy data. Nat Methods 7:418

52. Rafii S, Butler JM, Ding B-S (2016) Angiocrine functions of organ-specific endothelial cells. Nature 529:316–325

53. Passaro D, Tullio A, Abarrategi A et al (2017) Increased vascular permeability in the bone marrow microenvironment contributes to disease progression and drug response in acute myeloid leukemia. Cancer Cell 32:324–341.e6

Part IV

In Vivo and Ex Vivo Modelling of the Bone Marrow Ecosystem

Chapter 17

Modeling Human Fetal Hematopoiesis in Humanized Mice

Seydou Keita, Bruno Canque, and Kutaiba Alhaj Hussen

Abstract

Due to difficulties to access primary human bone marrow samples and age or donor effects, human hematopoiesis has long remained far less well characterized than in the mouse. Despite recent progresses in single-cell RNA profiling only little is known as to phenotype, function and developmental trajectories of human lymphomyeloid progenitors and precursors. This is especially true regarding the developmental architecture of the lymphoid lineage which has been the subject of persistent controversies over the past decades. Here, we describe an original approach of in vivo modeling of human fetal hematopoiesis immunodeficient NSG mice engrafted with neonatal CD34$^+$ hematopoietic progenitor cells (HPCs) allowing for rapid identification and isolation of lymphomyeloid developmental intermediates.

Key words Immunodeficient mice, CD34$^+$ hematopoietic progenitor cells, Bone marrow, Immunophenotypic stratification, multiparameter flow cytometry

1 Introduction

Much of our understanding of the cellular bases of lymphoid development comes from mouse studies showing that lymphoid specification is initiated downstream of HSCs in lymphomyeloid-primed progenitors (LMPPs) that subsequently upregulate the interleukin 7 (IL-7) receptor alpha-chain defining the stage of common lymphoid progenitors (CLPs) [1, 2]. Works from the last few years have established that the CLP subset is subjected to continuous diversification and that, beside multilymphoid progenitors, it comprises a wealth of already T, B, or NK/ILC lineage-specified precursors whose phenotype and functions are now well documented [3].

Despite the technical limitations inherent to human studies [4], it has now been established that lymphoid fate cosegregates with monocyte and dendritic cell (DC) potentials in CD38$^{lo/}$$^+$CD45RA$^+$ hematopoietic progenitor cells (HPC) [5–7]. However, despite these progresses, the phenotype, developmental status and

Marion Espéli and Karl Balabanian (eds.), *Bone Marrow Environment: Methods and Protocols*, Methods in Molecular Biology, vol. 2308, https://doi.org/10.1007/978-1-0716-1425-9_17, © Springer Science+Business Media, LLC, part of Springer Nature 2021

function of lymphoid precursors has long been hotly debated [8–10]. To resolve longstanding controversies and bypass limitations of conventional approaches based on analysis of cellular subsets isolated from the BM of primary fetal or adult donors, we developed a modeling approach in humanized mice engrafted with neonatal $CD34^+$ HPCs isolated from the umbilical cord blood (UCB). This model is referred to as the NSG-UCB model. Kinetic follow-up combined to in-depth immunophenotypic and functional characterization of human $CD45^+$ cells found that by week 3 after grafting the BM of NSG-UCB mice supports a fully diversified xenogeneic hematopoiesis reminiscent of the mid-fetal period [11, 12]. This allowed us to develop optimized antibody cocktails to capture lymphomyeloid precursors, follow their developmental trajectory from early emergence from multipotent $CD34^{hi}CD45RA^+$ Lympho-Mono-Dendritic cell Progenitors (LMDPs) to the stage mature or semi-mature immune effectors leaving the BM to seed secondary lymphoid organs, and propose an updated two-family model of lymphoid architecture.

Immunophenotypic stratification and in-depth functional characterization of lymphoid precursors isolated directly from the bone marrow of NSG-UCB mice led to finding that human lymphopoiesis actually displays an original bipartite organization stemming from $CD127^-$ and $CD127^+$ early lymphoid precursors (ELPs) that possess both common and specific differentiation potentials and display family-specific gene signatures which also determine their lineage restriction modes. Whereas the $CD127^-$ ELPs retain a myeloid trait and can still diverge toward the monocyte-DC lineage, their $CD127^+$ counterparts are almost devoid of myeloid potential and B lineage-biased. Consistent with a two-family model, we could demonstrate that the $CD127^-$ lymphoid family includes prototypic T-cell precursors, NK/ILC precursors (NKIP) particularly efficient at generating NK cells in the sole presence of IL-15, as well as early B cell precursors (EBP) prone to generation of marginal zone B lymphocytes. On the other hand, the $CD127^+$ lymphoid family, reminiscent of mouse CLPs, displays a strict dependency upon IL-7 for growth and survival, retains only a marginal capacity to enter the T-cell pathway, and subdivides into NKIPs that require Notch1 and IL-15 for optimal differentiation into ILC3s, and EBPs generating follicular B cells.

The NSG-UCB is very flexible and can be easily optimized to study monocyte and dendritic cell differentiation or carry out molecular studies using lentivirally transduced $CD34^+$ HPCs for characterizing candidate regulators of lymphomyeloid differentiation.

2 Materials

2.1 Cell Preparation and Culture Plates

1. Blunt ended sterile scissors.
2. Forceps.
3. 70% ethanol.
4. Insulin syringes.
5. 23-gauge needle.
6. 70 µM nylon cell strainers.
7. Staining buffer: 2% FBS in PBS.
8. 5-, 10-, and 25-mL sterile pipettes.
9. 15- and 50-mL polypropylene conical centrifuge tubes.
10. 5 mL round bottom polystyrene tubes.
11. 24- or 96-well plates.
12. Cryovials.

2.2 Magnetic Cell Separation Kits

1. Human CD34 Microbead kit (Miltenyi Biotec) (*see* **Note 1**).
2. Mouse cell depletion kit (Miltenyi Biotec).
3. Magnetic field (Miltenyi Biotec).
4. LS columns (Miltenyi Biotec).

2.3 Buffers and Media

1. Staining buffer: 2% FBS in PBS.
2. Bovine serum albumin (BSA).
3. Density gradient medium such as Ficoll-Hypaque.
4. Freezing medium: FCS, 10% DMSO.
5. Thawing medium: IMDM, 10% FCS, 1% L-glutamine, 1% penicillin–streptomycin.
6. Culture medium: IMDM medium 10% BIT 1% Glu 1% PS supplemented with SCF (10 ng/mL), Flt3L (10 ng/mL), TPO (4 ng/mL), and IL-3 (2 ng/mL).
7. MACS buffer: Bovine serum albumin, EDTA, PBS, 0.09% Azide (pH 7.2) (Miltenyi Biotec).
8. Injection medium: RPMI 1640 + 2% FCS.

2.4 Handling of Immunodeficient Mice

1. NOD scid gamma (NSG) mice (NOD.Cg-Prkdc^{scid}IL2RG^{tm1wjl}/SzJ; Jackson Laboratory, Bar Harbor, MI).
2. Pathogen-free animal facility.
3. Accreditation by local ethical committees.
4. X-ray irradiator.

2.5 Antibody Staining

1. Fc receptor-binding inhibitor (Fc Block).

2. Viability dye: Zombie Violet Fixable Viability Kit for exclusion of dead cells.

3. Fluorophore-conjugated antibodies: CD45 AF700 (clone HI30); CD34 pacific blue (clone 581); CD45RA PE (clone HIT100); CD7 FITC (clone 8H8.1); CD2 PerCPCy5.5 (clone RPA2.10); CD115 APC (clone 9-4D2-1E4); CD116 APC-vio770 (clone REA211); CD123 BV786 (clone 7G3); CD127 PC5 (clone A019D5); ITGB7 PC7 (clone FIB504); CD10 BV650 (clone HI10a); CD19 BV711 (clone HIB19); CD24 PE-CF594 (clone ML5).

2.6 Data Collection and Analysis

1. Flow cytometer. We use a BD Fortessa analyzer.

2. Flow cytometry acquisition and analysis. For example, FlowJo software version 10.6.2 (BD Biosciences).

3 Methods

This is a basic protocol for the identification of human CD34$^+$ lymphomyeloid precursors in the BM of xenografted mice using a cocktail of 13 fluorescent antibodies.

3.1 Purification and Storage of CD34$^+$ HPC from Umbilical Cord Blood (UCB) (See Note 2)

1. To isolate PBMCs from UCB dilute the blood 1:3 with PBS (*see* **Note 3**).

2. Add 15 mL of Ficoll-Hypaque solution in a 50-mL tube.

3. Layer 30 mL of the blood/PBS mixed solution onto the surface of the Ficoll-Hyaque solution and centrifuge for 20 min at $400 \times g$ (acceleration = 2, brake = 0) at 20 °C (*see* **Note 4**).

4. After centrifugation, using 5 mL pipette, aspirate gently the lymphocyte ring at the interface between plasma and Ficoll-Hypaque containing erythrocytes and immediately transfer the umbilical cord blood mononuclear cells (CBMCs) into a 50-mL tube. Add 30-40 mL of PBS and wash twice 10 min at $400 \times g$.

5. Resuspend the cells in PBS 2% FCS and take an aliquot for counting.

6. To isolate CD34$^+$ HPCs resuspend CBMCs in 300 μL of MACS buffer and add 100 μL of CD34 microbeads and 100 μL of FcR Blocking Reagent (Human IgG) per 10^8 CBMCs following the manufacturer's instructions. Mix well and incubate for 30 min at 2–8 °C (*see* **Note 5**).

7. Resuspend the cells in 5–10 mL Macs buffer for up to 10^8 cells and wash 10 min at $300 \times g$.

8. Resuspend the cell pellet in 500 μL of buffer per 10^8 cells.

9. Place a suitable MACS separator column (LS) in the magnetic field and rinse the column with 3 mL of MACS buffer (*see* **Note 6**).

10. Apply cell suspension onto the column and collect flow-through containing unlabeled cells.

11. Rinse (three times) the LS column with 3 mL of MACS buffer and collect unlabeled cells (*see* **Note 7**).

12. Remove column from the separator, place it on a 15 mL collection tube, Pipette 5 mL of buffer onto the column and immediately flush out the magnetically labelled cells pushing the plunger into the column.

13. Spin down and resuspend the cells in PBS.

14. Take an aliquot to evaluate the purity of the cells by flow cytometry (purity should be >90% CD34$^+$ cells).

15. Spin down the cells and resuspend them in 1 mL (up to 2×10^6 cells) of Freezing Medium and put them in cryovials (*see* **Note 8**).

16. Store immediately in −80 °C and the day after, transfer in liquid nitrogen until use.

3.2 CD34$^+$ Cell Activation and Xenotransplantation

1. Irradiate NSG mice (2.25 Gy) the day before intravenous injection of the CD34+ cells.

2. The day before injection to irradiated mice, also thaw frozen aliquots of CD34$^+$ HPCs covered by a watertight protective plastic overwrap in a water bath at 37 °C until you see the last piece of ice in cryovial.

3. Dilute rapidly in 10 mL of Thawing Medium and wash the cells twice 10 min at $300 \times g$.

4. Resuspend the cells at a concentration of 10^6 cell/mL in Culture medium.

5. Transfer the CD34$^+$ HPCs in 24-well plates (1 mL/well) and keep them in culture overnight.

6. After overnight incubation in culture medium, collect, spin down and count the cells.

7. Resuspend the cells at a final concentration of 2.5×10^6/mL in Injection medium.

8. Inject 100 µL of the final cell suspension intravenously into each recipient mouse by tail-vein using a 1-cm^3 insulin syringe (2.5×10^5 CD34$^+$ HPCs per mouse).

3.3 Isolation of Human Cells from the BM of Xenografted Mice

1. Sacrifice recipient mice 3 weeks after grafting according to institutional guidelines and dissect femurs and tibias.

2. Position the mouse in a supine position and affix by pinning all four legs through the mouse paw pads below the ankle joint, then spray leg with 70% ethanol.

3. Open the abdominal cavity with blunt-end sterile scissors and remove the surface muscles and find the pelvic-hip joint.

4. Cut off the foots, tear off the skin, Remove the muscles and residue tissues surrounding the tibia and femur with sterile forceps and scissors. Cut off the hind leg above the pelvic-hip joint and the tibia from the hind leg below the knee joint with sharp sterile scissors. Remove the rest of muscles on the bones (*see* **Note 9**).

5. Cut the femurs and tibias at both ends with scissors. Use a 23-gauge needle and a 10-cm^3 syringe filled with PBS to flush the bone marrow out onto a 70 μm nylon cell strainer placed in a 50 mL Falcon conical tube. Repeat until the flow through turns white.

6. Smash the bone marrow through the cell strainer.

7. Wash the strainer 2–3 times with another 5–10 mL PBS.

8. Centrifuge cells at $300 \times g$ for 10 min at 4 °C.

9. Discard the supernatant and resuspend cell pellet in 80 μL of MACS buffer (including red blood cells).

10. Add 20 μL of Mouse Cell Depletion Cocktail (Mouse cell depletion kit).

11. Mix well and incubate for 15 min at 2–8 °C.

12. Adjust volume to 500 μL by adding PBS 0.5% BSA and proceed to magnetic separation.

13. Place a suitable MACS separator column (LS) in the magnetic field and rinse the column with 3 mL of MACS buffer.

14. Apply cell suspension onto the column and collect the flow-through containing unlabeled cells enriched in human cells.

15. Rinse (three times) the LS column with 1 mL of MACS buffer and collect unlabeled cells.

16. Dilute the cell suspension in 10 mL of staining buffer and wash them minutes at $300 \times g$.

17. Resuspend the pellet in 10 mL of staining buffer and filter the cell suspension through the cell strainer to remove cell aggregates and debris.

18. Count the cells and adjust concentration at 10^8 cell/mL and proceed for immunolabeling.

3.4 Antibody Staining

1. Incubate the single-cell suspension with Fc Block for 10 min at 4 °C.

2. Without washing the cell suspension add the Zombie Violet Fixable Viability Kit and the fluorescent antibodies and incubate for 30 min at 4 °C in the dark.

3. Add 5–10 mL of staining buffer and centrifuge the cells at $400 \times g$ to wash.

4. Discard the supernatant, resuspend the pellet with 1–5 mL of staining buffer and keep them on ice before flow cytometry.

3.5 Flow Cytometry Gating Strategy (Fig. 1)

1. Gate on the white blood cell population.

2. Gate out cell doublets.

3. Gate on human CD45+ cells.

4. Gate on CD34^{hi}CD45RA^{hi} (CD115$^-$CD116$^-$CD123$^-$) cells to isolate early lymphoid progenitors.

4 Notes

1. The CD34 Microbead Kit from Miltenyi Biotec is recommended for CD34-positive cell purification as we obtain a high level of purity with this method. Note also that purity above 90% is essential to ensure efficient engraftment of NSG mice.

2. Purification of CD34$^+$ HPCs from UCB must be carried out under hood to ensure sterility. Bacterial contamination of the CD34$^+$ cell preparation could be fatal for engrafted mice.

3. It is advisable to dilute the blood before performing Ficoll. This allows for recovery of a clearer CBMCs ring. When UCB is not diluted it frequently happens that erythrocytes contaminate the leukocyte ring.

4. To deposit blood gently over the Ficoll, it is recommended to keep the tube tilted while holding the pipette straight. Then simply run the diluted blood slowly on the wall of the tube.

5. Recommended incubation temperature for CD34 purification is 2–8 °C. Higher temperature and/or longer times may lead to nonspecific labeling and lower purity.

6. For CD34$^+$ cell separation it is recommended to add preseparator filter (30 μm) on the column to avoid clumping.

7. Always wait until the column reservoir is empty before proceeding to the next step.

8. The freezing solution containing DMSO should be used cold. Do not take it out of the fridge until the last moment.

9. During mouse dissection, to effectively remove muscles use a small scissor slightly open and then scrape the bone doing back and forth movements. A scalpel can also be used.

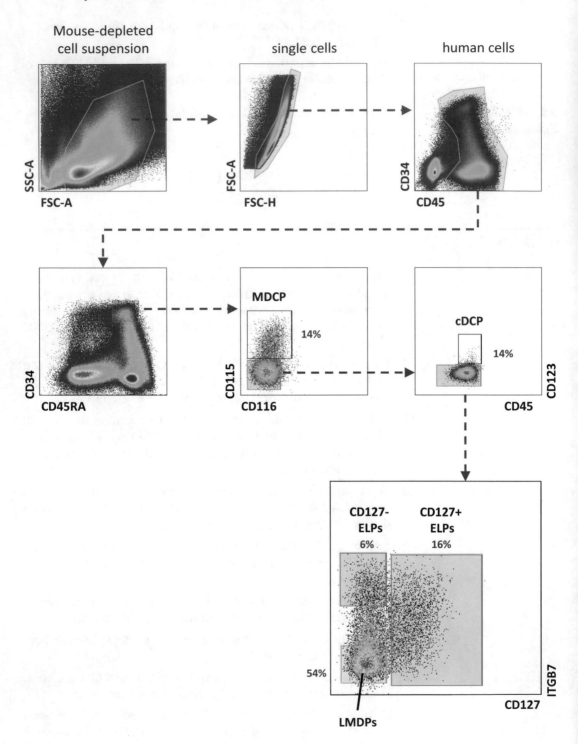

Fig. 1 Gating strategy to isolate human lymphoid progenitors

References

1. Adolfsson J, Mansson R, Buza-Vidas N, Hultquist A, Liuba K, Jensen CT et al (2005) Identification of Flt3+ lympho-myeloid stem cells lacking erythro-megakaryocytic potential a revised road map for adult blood lineage commitment. Cell 121(2):295–306. https://doi.org/10.1016/j.cell.2005.02.013.

2. Kondo M, Weissman IL, Akashi K (1997) Identification of clonogenic common lymphoid progenitors in mouse bone marrow. Cell 91(5):661–672

3. Cumano A, Berthault C, Ramond C, Petit M, Golub R, Bandeira A et al (2019) New molecular insights into immune cell development. Annu Rev Immunol 37:497–519. https://doi.org/10.1146/annurev-immunol-042718-041319

4. Doulatov S, Notta F, Laurenti E, Dick JE (2012) Hematopoiesis: a human perspective. Cell Stem Cell 10(2):120–136. https://doi.org/10.1016/j.stem.2012.01.006

5. Canque B, Camus S, Dalloul A, Kahn E, Yagello M, Dezutter-Dambuyant C et al (2000) Characterization of dendritic cell differentiation pathways from cord blood CD34(+)CD7(+)CD45RA(+) hematopoietic progenitor cells. Blood 96(12):3748–3756

6. Doulatov S, Notta F, Eppert K, Nguyen LT, Ohashi PS, Dick JE (2010) Revised map of the human progenitor hierarchy shows the origin of macrophages and dendritic cells in early lymphoid development. Nat Immunol 11(7):585–593. https://doi.org/10.1038/ni.1889

7. Lee J, Breton G, Oliveira TY, Zhou YJ, Aljoufi A, Puhr S et al (2015) Restricted dendritic cell and monocyte progenitors in human cord blood and bone marrow. J Exp Med 212(3):385–399. https://doi.org/10.1084/jem.20141442

8. Haddad R, Guardiola P, Izac B, Thibault C, Radich J, Delezoide AL et al (2004) Molecular characterization of early human T/NK and B-lymphoid progenitor cells in umbilical cord blood. Blood 104(13):3918–3926. https://doi.org/10.1182/blood-2004-05-1845

9. Hao QL, Zhu J, Price MA, Payne KJ, Barsky LW, Crooks GM (2001) Identification of a novel, human multilymphoid progenitor in cord blood. Blood 97(12):3683–3690

10. Kohn LA, Hao QL, Sasidharan R, Parekh C, Ge S, Zhu Y et al (2012) Lymphoid priming in human bone marrow begins before expression of CD10 with upregulation of L-selectin. Nat Immunol 13(10):963–971. https://doi.org/10.1038/ni.2405

11. Alhaj Hussen K, Vu Manh TP, Guimiot F, Nelson E, Chabaane E, Delord M et al (2017) Molecular and functional characterization of lymphoid progenitor subsets reveals a bipartite architecture of human Lymphopoiesis. Immunity 47(4):680–96 e8. https://doi.org/10.1016/j.immuni.2017.09.009

12. Parietti V, Nelson E, Telliam G, Le Noir S, Pla M, Delord M et al (2012) Dynamics of human prothymocytes and xenogeneic thymopoiesis in hematopoietic stem cell-engrafted nonobese diabetic-SCID/IL-2rgammanull mice. J Immunol 189(4):1648–1660. https://doi.org/10.4049/jimmunol.1201251

Chapter 18

Engraftment of Human Hematopoietic Cells in Biomaterials Implanted in Immunodeficient Mouse Models

Syed A. Mian and Dominique Bonnet

Abstract

Over the last 20 years, significant progress has been made in the development of immunodeficient mouse models that now represents the gold standard tool in stem cell biology research. The latest major improvement has been the use of biomaterials in these xenogeneic mouse models to generate human "bone marrow like" tissues, which not only provides a more relevant xenograft model but can also potentially enable us to delineate the interactions that are specific between human bone marrow cells. There are a number of biomaterials and strategies to create humanized niches in immunodeficient mouse models, and the methods can also differ significantly among various research institutes. Here, we describe a protocol to create a humanized 3D collagen-based scaffold human niche in immunodeficient mouse model(s). This humanized in vivo model provides a powerful technique for understanding the human BM microenvironment and the role it plays in the regulation of normal as well as malignant hematopoiesis.

Key words Ossicles, Xenotransplantation, Bioengineer scaffold, Bone marrow niche

1 Introduction

Human bone marrow is a spongy complex tissue, highly vascular that resides in the confines of bone and houses the vital hematopoietic stem and progenitor cells (HSPCs). Our understanding of the human bone marrow (BM) has changed immensely during the past few decades with increasing evidence that suggests an interdependency of HSPCs and their surrounding bone marrow microenvironment [1–5]. This increase in understanding the biology of the hematopoietic stem cell biology has been mainly driven by observations gathered from murine studies as well as the use of xenotransplantation in immunodeficient mouse models [6–12]. These immunodeficient mouse models have greatly improved our understanding of not only the normal human HSCs but has also helped in gaining the insight into the concept of cancer stem cells [13–18]. Even though these models have been useful, they still present various challenges. One of the biggest caveats is cross-species

Marion Espéli and Karl Balabanian (eds.), *Bone Marrow Environment: Methods and Protocols*, Methods in Molecular Biology, vol. 2308, https://doi.org/10.1007/978-1-0716-1425-9_18, © Springer Science+Business Media, LLC, part of Springer Nature 2021

differences that exist between the human and mouse microenvironment, and how this may influence the cell–cell (human HSPCs and mouse MM) interactions as well as the subsequent biological functions of the cells in question.

Advances in the integration of biomaterial engineering into developmental biology research has led to a worldwide scientific endeavor that has redefined the field of regenerative medicine [19]. Biomaterials provide a versatile tool to generate highly sophisticated architectural support for cell attachment and subsequent tissue development, therefore enabling key biological communications between the interacting cells. Some of the naturally occurring biomaterials such as collagen, laminin, fibronectin, vitronectin and reconstituted basement membrane (such as Matrigel) are the most commonly used materials in the bioengineering field because of their inherent cytocompatibility, intrinsic cell adhesion properties, and their ability to be remodeled by extrinsic factors (such as cells and cytokines) [20]. Therefore, combining biomaterials with cell-implantation techniques have yielded a promising strategy for engineering tissues that are capable of mimicking the human BM microenvironment. These technological advances have opened the possibility of using such bioengineering techniques along with xenotransplantation in mouse models to study human hematopoiesis in normal as well as in malignant state [21, 22].

Here, we describe a protocol to generate 3D human "BM-like" tissue (or human–mouse chimera) using collagen-based scaffold in humanized mouse models. The protocol is broadly divided into four main sections that includes preparation of the human cells, seeding of human cells into the collagen-based scaffolds, implantation of scaffolds into immunodeficient mice and finally retrieval of the human scaffold tissue following xenotransplantation.

2 Materials

2.1 Human Sample Preparation

1. Human cord blood (or bone marrow) derived HSPCs.
2. Human bone marrow derived Mesenchymal Stromal Cells (hMSCs).
3. Human umbilical vein endothelial cells (HUVEC).
4. Ficoll-Paque (GE Healthcare).
5. Red Blood Cell (RBC) lysis buffer.
6. Dimethyl sulfoxide (DMSO) (Sigma-Aldrich).
7. Fetal bovine serum (FBS) (Life Technologies).
8. MEM Alpha Medium (1×) + GlutaMAX-1 (Gibco).
9. Myelocult H5100 (Stem Cell Technologies).

10. Medium 199 ($1\times$) + GlutaMAX-1 (Gibco).

11. Human MSC-FBS, heat-inactivated (Gibco).

12. Penicillin–Streptomycin $100\times$ (P/S) (Sigma-Aldrich).

13. Endothelial cell growth supplement (ECGS) (Millipore).

14. 4-(2-hydroxyethyl)-1-piperazineethanesulfonic acid (HEPES) (Sigma-Aldrich).

15. Heparin (Sigma-Aldrich).

16. Glutamine (Gibco).

17. Trypsin–EDTA $10\times$ solution (ThermoFisher).

18. Sterile Phosphate-Buffered Saline (PBS).

19. Deoxyribonuclease (DNase 1) (Sigma-Aldrich).

20. Trypan blue (Sigma-Aldrich).

21. EasySep Human CD34 Positive Selection Kit II (Stemcell Technologies).

22. Cell separation magnet.

23. Hemocytometer or automatic cell counter.

24. Pipette dispenser (Complete set 1–1000μL).

25. Serological pipette.

26. 5 mL round-bottom polystyrene tubes.

27. 15 mL and 50 mL conical tubes.

28. Sterile tissue culture polystyrene flasks or multiwell plates.

29. 1.5 mL microcentrifuge tubes.

2.2 Seeding of Collagen-Based Scaffolds

1. 1 mL Insulin syringe with 25-G needle.

2. Sterile tissues.

3. Sterile forceps.

4. Ultra-Low Attachment Multiple Well Plates (Corning Inc.).

5. Gelfoam, Size 12–7 mm (Pfizer).

6. Sterile cell strainers.

7. Disposable scalpels with handle.

2.3 Scaffold Implantation

1. Ethanol (EtOH) (Sigma-Aldrich).

2. Distel.

3. Analgesic drugs (buprenorphine, meloxicam, and carprofen).

4. Isoflurane (Abbott).

5. Carbomer (polyacrylic acid) as ophthalmic gel (Novartis).

6. Chlorhexidine (G9).

7. Sterile surgical instruments ($2\times$ small forceps, $2\times$ small straight surgical scissors, $2\times$ small straight surgical scissors

blunt-end, surgical stapler, surgical staples, sterile tissues, staple remover).

8. Electric trimmer with 40 mm wide blade.

9. Heating pads.

10. Sterile surgical drapes.

11. Sterile surgery gloves.

12. 50 mL and 5 mL tube holder racks.

13. 0.5–1 mL insulin syringes.

14. Small petri dish.

15. Sterile nonwoven swabs 5 × 5 cm (or equivalent).

16. Sterile water.

17. Immunodeficient NOD-Prkdc^{scid} IL2rg^{Tm1Wjl} (NSG) mice between 8 and 12 weeks of age.

2.4 Retrieval of Cells After Xeno-transplantation

1. Sterile surgical instruments (2× forceps, 1× straight surgical scissors, disposable scalpels with handle).

2. Collagenase Type1 (Stemcell Technologies).

3. Dispase (Sigma).

4. Antibodies: anti-mouse CD45, anti-human CD45, anti-human CD33, anti-human CD19, human CD3.

5. Fc Block reagent (BD biosciences).

6. 4′,6-diamidino-2-phenylindole (DAPI) (Sigma-Aldrich).

2.5 Solutions and Culture Medium

1. 70% EtOH Solution: 70% EtOH absolute, 30% Sterile Water.

2. Washing buffer: Sterile PBS, 2% FBS, 1% P/S.

3. Tissue digestion solution: Dispase, 2 mg/mL Collagen Type I, 1 mg/mL DNase I, 10% FBS.

4. Cell Freezing medium: 90% FBS and 10% DMSO.

5. MSC expansion media: MEM-α Medium, 1% P/S and 10% human MSC-FBS.

6. HUVEC culture media: Medium M199, 20% FBS, 1% P/S, 10 mM HEPES, 50μg/mL heparin, 2 mM glutamine, and 50μg/mL ECGS.

7. HSC culture media: H5100 and 1% P/S.

3 Methods

3.1 hMSC and HUVEC Culture

Stromal cells are recommended to be seeded 7–10 days prior to setting up the scaffold experiment. hMSCs can be used alone or in combination with HUVEC cells to generate the stroma layer in the scaffolds. Do not use HUVECs alone in this protocol. All

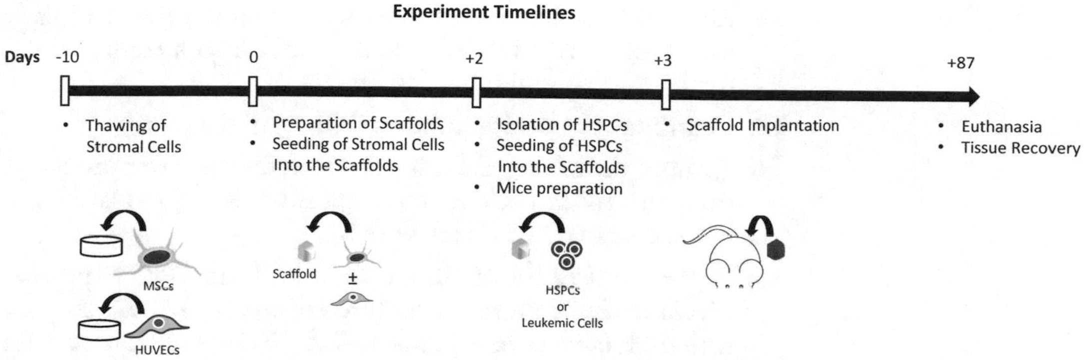

Experiment Timelines

Fig. 1 Schematics of the timeline for setting up the experiment. *HSPCs* hematopoietic stem and progenitor cells, *MSCs* Mesenchymal stromal cells, *HUVECs* Human umbilical vein endothelial cells

procedures must be done in sterile conditions in a Class II biological safety cabinet. *See* Fig. 1 for a timeline of preparation of experiments.

1. Remove the cryovials of hMSCs and HUVEC from liquid nitrogen and thaw immediately by incubating the cryovials in a 37 °C water bath (*see* **Note 1**).

2. Disinfect the outside of the cryovials with 70% ethanol.

3. Using a 1 mL pipette, transfer the hMSCs and HUVECs to two sterile 15 mL conical tubes.

4. Using 10 mL serological pipette, slowly add dropwise 4 mL of FBS (or a suitable alternative of choice), prewarmed to 37 °C, to the 15 mL conical tubes (*see* **Note 2**).

5. Centrifuge the 15 mL conical tubes at $300 \times g$ for 6 min to pellet the cells.

6. Discard the supernatant and break the cell pellet by slowly tapping the tube.

7. Add 5 mL of washing buffer and repeat **steps 5** and **6** once (*see* **Note 3**).

8. Resuspend the MSCs and HUVECs in a total volume of 1 mL of MSC expansion media or EC Culture media, respectively.

9. Count the number of cells using a hemocytometer (or an alternative method of choice).

10. Seed the cell suspension onto an appropriate tissue culture flask (or plate), with a seeding density of 5000–6000 cells/cm^2/ 0.2–0.3 mL of culture media.

11. Maintain the cells at 37 °C in a humidified incubator equilibrated with 5% CO_2.

12. After 24 h, collect the culture media from the flask (or plate) with hMSCs and HUVEC, and transfer it to a sterile 50 mL (or 15 mL) conical tubes (*see* **Note 4**).

13. Centrifuge the conical tubes at $300 \times g$ for 6 min.

14. Without disturbing the cell pellet, carefully transfer the media from the tubes back to the respective flask (or Plate) with hMSCs and HUVEC (*see* **Note 5**).

15. Replace 70% of the culture media with fresh MSC Expansion Media or EC culture media (prewarmed to 37 °C). Replace with fresh media every 3 days for HUVEC and once a week for hMSCs (*see* **Note 5**).

16. Subculture hMSCs and HUVEC once they have reached approximately 80% confluency (*see* **Note 6**).

3.2 Bioengineering Collagen-Based 3D Scaffolds with Human HSPCs and Stromal Cells

3.2.1 Prepare hMSCs and HUVECs

1. Carefully remove and discard the cell culture media from the flasks (or plates) containing up to 80% confluent layer of cells.

2. Wash the adherent cells in the flasks (or plates) twice with sterile PBS (*see* **Note 7**).

3. Apply $1\times$ trypsin-EDTA solution ($20\mu L$ per cm^2) and incubate in an incubator maintained at 37 °C and 5% CO_2 for 3–5 min (*see* **Note 8**).

4. Gently tap the flasks (or plate) with the palm of the hand and then inspect the flasks (or plate) under the microscope to confirm the complete detachment of adherent cells (*see* **Note 9**).

5. Add appropriate volume of MSC expansion media or EC Culture media (Final FBS ratio 10% of the total volume) to the flasks (or plates). Disperse the media over the entire surface area of the flask to aid in the detachment process.

6. Gently agitate the flasks (or plates) to mix the cell suspension.

7. Transfer the dissociated cells to a 15 mL conical tube.

8. Centrifuge the 15 mL conical tube at $300 \times g$ for 6 min to pellet the cells.

9. Discard the supernatant without disturbing the cell pellet and then break the cell pellet by slowly tapping the tube.

10. Add 1 mL of MSC Expansion media to hMSCs and EC culture media (prewarmed to 37 °C) to HUVECs and resuspend the cells thoroughly.

11. Count the number of cells using a hemocytometer (or an alternative method of choice).

12. If both cell types (hMSCs and HUVECs) are used together dilute each cell type at up to 4×10^6/mL. Alternatively, if only hMSCs are used for seeding into the scaffolds, dilute hMSCs at 2×10^6/mL in culture media.

13. Mix hMSC and HUVECs in a 1:1 ratio.

14. Up to 1×10^5 stromal cells (in 0.05 mL volume) need to be injected into each scaffold in the next step.

15. Calculate the number of scaffolds that are needed and transfer required cell numbers into a 1.5 mL microcentrifuge tube.

3.2.2 Preparation of Collagen-Based Scaffolds

1. Using a sterile scalpel, sterile forceps and sterile plate lid (or appropriate sterile hard surface), cut the sterile collagen-based sponge (e.g., Gelfoam, dimensions 20 mm × 60 mm × 7 mm) into 24 pieces of similar size (6.6 mm × 7.5 mm × 7 mm) Fig. 2a (*see* **Note 10**).

2. Transfer the excised scaffolds into a 50 mL conical tube that is prefilled with 70% EtOH. Incubate the scaffolds at room temperature for 5 min.

3. Transfer the scaffolds from the conical tube with 70% EtOH to a sterile tissue to remove excess 70% EtOH (*see* **Note 11**). Use a sterile forceps to perform **steps 3–6**.

4. Transfer the scaffolds to a new 50 mL conical tube that is prefilled with sterile PBS. Incubate the scaffolds at room temperature for 5 min (*see* **Note 11**).

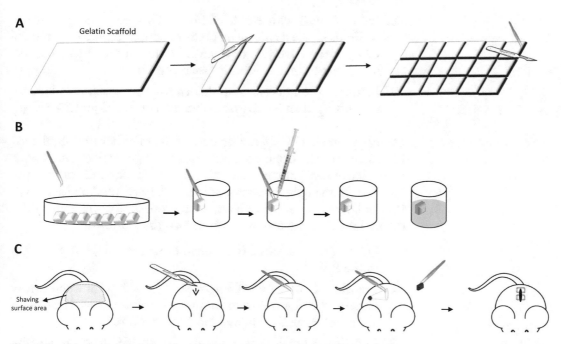

Fig. 2 Graphical representation depicting the processing of the collagen-based scaffolds. (**a**) Gelatin based scaffolds that is a derivative of collagen biomaterial was used in this protocol. Scaffold material is first cut vertically into 6 equal parts. Then each section is further cut horizontally into 3 parts. (**b**) Scaffolds are then injected with human stromal cells, followed by human HSPCs using a sterile forceps and syringe. (**c**) Surgical area on the back of the mouse is shaved and then scaffolds (preseeded with human cells) are implanted by following aseptic techniques

5. Repeat **step 3** (*see* **Note 11**).

6. Transfer the scaffolds to a new 50 mL conical tube that is prefilled with sterile PBS.

3.2.3 Seeding of Collagen-Based Scaffolds with Stromal Cells

1. Transfer the hMSC and HUVEC mix suspension (or MSCs only) to a 1 mL insulin syringe (*see* **Note 12**).

2. Using a sterile forceps, transfer the scaffold from PBS to a sterile tissue to remove excess PBS (*see* **Note 13**) (Fig. 2b).

3. Using the same sterile forceps, transfer the scaffolds to one well of an Ultra-Low attachment multiple well plate (preferably 24 or 48-well plate).

4. Using the insulin syringe containing stromal cells (or MSCs alone), inject 0.05 mL into the partially dried scaffold and leave the scaffolds hanging on the wall of the well (*see* **Note 14**).

5. Repeat **steps 2–4** with each scaffold until all required scaffolds are seeded with stromal cells (or MSCs alone).

6. Fill the outer wells of the plate with sterile PBS (*see* **Note 15**).

7. Incubate the injected scaffold for at least 1 h (up to a maximum of 2 h) inside a cell culture incubator maintained at 37 °C and 5% CO_2.

8. Fill each well with 3 mL of HUVECs media and return the scaffolds to a cell culture incubator maintained at 37 °C and 5% CO_2. Alternatively, if only hMSCs are used in this step, use MSC expansion media only (*see* **Note 16**).

9. Incubate the injected scaffolds overnight (maximum incubation 48 h) in an incubator maintained at 37 °C and 5% CO_2.

3.2.4 Isolation of Human HSPCs

Healthy human HSPCs can be derived from either umbilical cord blood or human bone marrow. Alternatively, malignant HSPCs from leukemic patients can also be used instead of healthy HSPCs. If malignant hematopoietic cells are used in this step, then follow an appropriate alternative protocol for the preparation of the HSPCs prior to the injection into the scaffolds.

1. Dilute the cord blood (or human bone marrow) sample 1:3 in sterile PBS.

2. Transfer 15 mL of Ficoll into a 50 mL conical tube and slowly dispense (tilting tube and running the blood down the side of the tube) 35 mL of diluted cord blood on top.

3. Centrifuge the 50 mL conical tube at $450 \times g$ for 30 min (at room temperature) with deacceleration set to 0.

4. Remove and discard half of the top layer (containing Plasma) (*see* **Note 17**).

5. Carefully collect the "cloudy thin" interface layer (Buffy coat), which contains the relevant mononuclear cells and transfer it into a new sterile 50 mL conical tube.

6. Add washing buffer to the collected cells to make a final volume of 50 mL.

7. Centrifuge the conical tube at $300 \times g$ for 6 min.

8. Repeat **steps 6** and **7** once.

9. Lyse the red blood cells by resuspending the cell pellet in 25 mL of cold $1\times$ red cell lysis buffer.

10. Incubate the cells at 4 °C for 10 min.

11. Deactivate the red cell lysis buffer by adding FBS (10% of the final volume).

12. Centrifuge the conical tube at $300 \times g$ for 6 min. Remove and discard the supernatant without disturbing the pellet.

13. Repeat **step 12**. Resuspend the cell pellet in 1 mL of washing buffer.

14. Using an appropriate commercially available kit, such as Easy-Sep™ Human CD34 Positive Selection Kit II in this case, isolate CD34$^+$ HSPCs by following the protocol available with the kit.

15. After the CD34$^+$ HSPC isolation, centrifuge the tube at $300 \times g$ for 6 min.

16. Resuspend the cell pellet in an appropriate volume of HSC culture media.

17. Count the number of cells using a hemocytometer or appropriate alternative method.

18. Dilute the purified CD34$^+$ HSPCs to 2×10^6 cells per mL in HSC culture media.

19. Up to 1×10^5 (in 0.05 mL volume) purified CD34$^+$ HSPCs are needed to be injected into each scaffold in the next step.

3.2.5 Seeding of Collagen-Based Scaffolds with Human HSPCs

1. Transfer the purified CD34$^+$ HSPCs to a 1 mL insulin syringe.

2. Using a sterile forceps, transfer the scaffold (seeded with MSC and HUVECs or MSCs only) from the culture well to a sterile tissue to remove excess media (*see* **Note 18**).

3. Using the same sterile forceps, transfer the scaffolds to a well of a new Ultra-Low attachment multiple well plate (preferably 24 or 48 wells plate, Fig. 2b).

4. Using the insulin syringe containing HSPCs, inject 0.05 mL into the partially dried scaffold and leave the scaffolds hanging to the wall of the well (*see* **Note 14**).

5. Repeat **steps 2–4** with each scaffold until all the required scaffolds are seeded with HSPCs.

6. Fill the outer wells of the plate with sterile PBS (*see* **Note 15**).

7. Incubate the injected scaffold for at least 1 h (up to a maximum of 2 h) inside a cell culture incubator maintained at 37 °C and 5% CO_2.

8. Fill each well with 3 mL of HSC culture medium and return the scaffolds to a cell culture incubator maintained at 37 °C and 5% CO_2 (*see* **Note 16**).

9. Incubate the injected scaffolds overnight in an incubator maintained at 37 °C and 5% CO_2 (maximum incubation 24 h).

3.3 Surgical Implantation of 3D Scaffolds in Immunodeficient Mice

All the materials, surgical instruments, and substances used in surgery and/or placed inside the animals must be either sterile single-use (disposable) or sterilized prior to each use. Follow the local named veterinary surgeon (NVS) guidelines for aseptic techniques. A dedicated surgical area must be used for all surgical procedures. Maintain the body temperature of the mouse at 37 °C throughout the surgical procedure. *See* Figs. 2c and 3 for experimental setup.

3.3.1 Animal Preparation

1. Remove the hair from the surgical site (parallel to the spine) by shaving the back of the mouse using an electric trimmer (Fig. 2c). This needs to be performed 24 h prior to the surgery (*see* **Note 19**).

2. Administer the mouse with appropriate analgesic (for example Carprofen in drinking water, 0.1 mg/mL of water) 24 h before the surgery (*see* **Note 20**).

3.3.2 Surgical Implantation

1. Prepare the surgical area according to the local NVS guidelines (As depicted in Fig. 3).

2. Anesthetize the mouse in a chamber filled with 0.5% isoflurane and 2 L/min O_2.

3. Apply a sterile drape over the working surface (surgery area) where the surgery will be performed.

4. Transfer the mouse to the surgical area in prone position in order to have easy access to the site of surgery. Keep the animal under anesthesia using a nose cone supplying 1.5% isoflurane and 2 L/min O_2. Keep the mouse under anesthesia during the surgical procedure and frequently check the animal state.

5. Apply ophthalmic ointment to both eyes to prevent corneal desiccation.

6. Administer appropriate analgesia (buprenorphine, 0.1 mg/kg and meloxicam, 10 mg/kg) via the subcutaneous route (*see* **Note 21**).

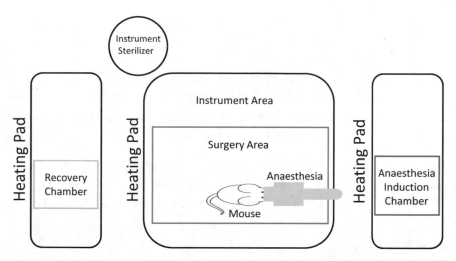

Fig. 3 Graphical illustration of the appropriate surgical set-up for performing scaffold implantation in immunodeficient mice. A dedicated recovery area, surgery area and induction area must be used for performing surgical implantation of the scaffolds. Body temperature of the mice must be always maintained at 37 °C using appropriate heating pads. If multiple surgeries are performed, surgical instruments must be sterilized in-between the animals using appropriate instrument (e.g., bead-based sterilizer)

7. Apply 10% chlorhexidine solution with clean swab in a circular fashion starting at the surgical incision site and rotating outward (*see* **Note 22**).

8. Repeat a minimum of three times discarding swab after each use.

9. Cover the mouse with a sterile drape (with an opening enough to see the surgery area; *see* **Note 23**).

10. Using a sterile forceps and a scalpel (or scissors), make a 0.5- to 0.7-cm anterior-to-posterior incision of the skin.

11. Hold the skin on one side of the incision side with the forceps and slowly insert round ended scissors (or forceps) to create a subcutaneous pocket away from the incision site toward the side of the animal.

12. Insert the scaffold subcutaneously, making sure that it is placed deep inside the pocket (*see* **Note 24**).

13. Close the incision site with surgical staples.

14. Transfer the mouse to the recovery chamber that is maintained at 37 °C for up to 2 h (*see* **Notes 25** and **26**).

15. If surgeries are done on multiple mice at the same time, sterilize all the instruments after performing the surgery on each mouse (*see* **Note 27**).

16. Transfer the second mouse to the induction chamber and perform the surgery by following **steps 3–14**.

17. Mice should be maintained on appropriate analgesia (Carprofen in drinking water, 0.1 mg/mL of water) for 48 h postsurgery.

18. Check the mouse and the wound frequently during the first 48 h following the surgery for possible adverse effects.

19. Remove the surgical staples after 7 days (no more than 10 days) of the surgical procedure.

3.4 Mouse Euthanasia, Tissue Retrieval, and Tissue Digestion

1. Euthanize the mouse using an appropriate schedule 1 method (preferably exposure to carbon dioxide).

2. Transfer the mouse to a sterile surface, preferably on a sterile tissue.

3.4.1 Mouse Euthanasia

3.4.2 Tissue Retrieval

1. Spray the back (surgical area) of the mouse with 70% EtOH.

2. Using a sterile forceps and sharp ended scissors, make a longitudinal skin incision on the back of the mouse, near the original implantation site.

3. With the help of a sterile forceps and scissors or scalpel, carefully separate the skin from the subcutaneous pocket where the scaffold had been implanted.

4. Hold the scaffold with a sterile forceps and gently explant it from the skin by cutting the residual membrane and tissue surrounding the scaffold using sterile scissors (*see* **Note 28**).

5. Fill a 1.5 mL microcentrifuge tube with 1 mL of sterile PBS and transfer the scaffold(s) to the tube.

3.4.3 Tissue Digestion

1. Excise the scaffold into small pieces using a sterile forceps and scalpel (*see* **Note 29**).

2. Add up to 1 mL of tissue digestion solution into a 1.5 mL microcentrifuge tube and transfer the minced tissue into this tube.

3. Incubate the 1.5 mL microcentrifuge tube in a water bath at 37 °C for up to 1 h.

4. Vortex the 1.5 mL microcentrifuge tube every 15 min to aid in the tissue digestion (*see* **Note 30**).

5. Filter the cell suspension through a sterile 5 mL tube with a cell strainer cap.

6. Add 4 mL of washing buffer.

7. Centrifuge the tube at $300 \times g$ for 6 min.

8. Discard the supernatant and repeat **steps 6** and **7**.

9. Resuspend the cells in 0.5 mL of washing buffer.

10. Count the number of cells using a hemocytometer or appropriate alternative method.

3.4.4 Sample Processing for Direct Flow Cytometry

1. Prepare the appropriate controls needed for flow cytometry analysis.

2. Transfer up to 5×10^4 cells to a new 5 mL round-bottom polystyrene tubes.

3. Add appropriate volume of washing buffer to make the final volume to 0.1 mL.

4. (Optional) To block nonspecific Fc-mediated interaction, add 2.5µg of Fc Block reagent per 10^6 cells and incubate for 10 min at room temperature.

5. Add 1 mL of washing buffer.

6. Centrifuge the tube at $300 \times g$ for 6 min.

7. Prepare an appropriate fluorochrome-labeled primary antibody cocktail (for example, anti-mouse CD45, anti-human CD45, anti-human CD33, anti-human CD19, anti-human CD3).

8. Add the antibody cocktail to cells and incubate for 20 min at 4 °C, protected from light.

9. Add 1 mL of washing buffer.

10. Centrifuge the tube at $300 \times g$ for 6 min.

11. Discard the supernatant and repeat **steps 9** and **10**.

12. Resuspend the cells in 0.5 mL of washing buffer.

13. Just before analysis, add appropriate viability dye (for example DAPI) to the cells.

14. Analyze the stained cells by flow cytometry, according to the manufacturer's instructions of the FACS instrument.

4 Conclusion

Our humanized collagen-based scaffold approach using immunodeficient mouse models provides a relevant xenograft model and will not only enable the scientific community to gain a better understanding of the behavior of HSPCs in a native niche but has also numerous other applications in stem cell research and drug screening, amongst others. This approach will also help us to understand the clonal evolution of various hematological cancers and their interactions with the human niche and dissect, how stroma cells contribute to the chemoresistance of leukemic stem cells.

5   Notes

1. Monitor the cells until they are completely thawed and this process should not take longer than 1 min. Do not vortex the cells.

2. (Optional) Add DNase I solution (1 mg/mL) to the FBS prior to adding to the cells. Do not add the entire volume of FBS all at once to the cells. This may result in decreased cell viability due to osmotic shock.

3. (Optional) Add DNase I solution (1 mg/mL) to the washing buffer prior to adding to the cells. Adding the DNase I solution in the washing buffer will minimize the presence of free-floating DNA fragments and cell clumps.

4. This step will help in removing all the floating nonadherent dead cells from the culture.

5. Dispense appropriate culture media down the side of the flask so as not to disrupt the adherent cells.

6. Do not let the cells become totally confluent during this log phase of growth as they will start to die off and may not be recoverable.

7. Do not dispense the PBS directly onto the adherent cells during the washing step, so as not to disrupt and detach the cells.

8. Make sure to agitate the flask (or plate) gently to disperse the Trypsin-EDTA solution evenly throughout the surface area of the flask (or plate).

9. Stromal cells will start to appear as round after the detachment process. Incubate the cells in trypsin-EDTA solution for longer time (up to 10 min), if cells are not detaching quickly.

10. Gently hold the scaffolds with a sterile forceps. While cutting the scaffolds, it is important to move the scalpel in a longitudinal direction (toward you) without applying too much force on top of the collagen-based scaffold material. Gelatin scaffold, that is, a derivative of collagen biomaterial, is used in this protocol.

11. Place the scaffolds on one side of the sterile tissue and flip the other side of the tissue over the scaffolds, and gently press on top of the tissue to remove the residual EtOH from the collagen-based scaffolds. Do not leave the scaffolds to dry out.

12. Using an appropriate pipette, gently mix cells prior to transferring them into an insulin syringe.

13. Remove around 95% of the PBS that is absorbed by the scaffolds. It is important not to let the collagen-based scaffolds dry completely.

14. Injected cells at multiple locations of the scaffolds to spread the stromal cells evenly throughout the scaffold. Leave the injected scaffolds hanging on to the wall of the well.

15. It is essential to maintain high levels of humidity environment in order to avoid drying of the scaffolds.

16. Add the appropriate media slowly in a dropwise manner along the wall of the wells. This is essential to avoid disturbing the stromal cells that are seeded inside the scaffolds.

17. Carefully remove the top layer without disturbing other layers. You can remove up to 80% of the plasma layer.

18. Place the scaffolds on the sterile tissue and wait for up to 10 s for liquid to egress out of the scaffolds. Do not let these scaffolds dry out.

19. While preparing the mouse for surgery, shave an additional minimum of 2 cm away from the site of the actual surgery. Remove shaved excess hair from the body of the mouse.

20. Mice must be maintained on drinking water supplemented with Carprieve (or equivalent analgesia) for 24 h before the surgery.

21. Prepare all the drugs that need to be injected into the mice prior to starting the procedure. If performing surgeries on multiple mice, prepare the syringes with the drugs for all the mice in your experiment.

22. Avoid excessive wetting of nonsurgical areas of the mouse with chlorhexidine as this can exacerbate hypothermia.

23. It is recommended to use clear drapes to facilitate monitoring of the mouse.

24. Insert the scaffold with the help of a sterile forceps. Use a second forceps to keep the scaffold at the appropriate location from outside the skin while the first forceps is removed from the subcutaneous pocket. Avoid touching the edge of the forceps with the nonsterile part of the mouse skin.

25. Remove the drape from the top of the mouse and fold it backward, keeping in mind that the sterile part of the drape is facing upward.

26. Place the mouse on its side in a prewarmed cage and leave it to recover until walking normally.

27. Place all the instruments (scissors, forceps) into a conical tube with sterile H_2O (Tube 1) and then transfer them into a hot bead sterilizer. Leave the instruments in the hot bead sterilizer for up to 30 s. Next, place the instruments into a different conical tube with sterile H_2O (Tube 2) before using them for subsequent surgery.

28. Remove all excess nonscaffold tissues that could be attached to the scaffold before moving on to the next step.

29. If histopathological examination needs to be performed on the scaffold tissue, find alternative protocols for further analysis.

30. Alternatively, break the tissue by pipetting up and down multiple times using a 1 mL pipette.

Acknowledgments

The authors acknowledge the funding received from Wellcome Trust (FC0010045), MRC (FC0010045), and CRUK (FC0010045) through The Francis Crick Institute; and Blood Cancer UK for supporting this work.

References

1. Velten L et al (2017) Human haematopoietic stem cell lineage commitment is a continuous process. Nat Cell Biol 19(4):271–281

2. Ramasamy SK et al (2016) Regulation of hematopoiesis and osteogenesis by blood vessel-derived signals. Annu Rev Cell Dev Biol 32:649–675

3. Raynaud CM et al (2013) Endothelial cells provide a niche for placental hematopoietic stem/progenitor cell expansion through broad transcriptomic modification. Stem Cell Res 11(3):1074–1090

4. Yu VW, Scadden DT (2016) Heterogeneity of the bone marrow niche. Curr Opin Hematol 23(4):331–338

5. Batsivari A et al (2020) Dynamic responses of the haematopoietic stem cell niche to diverse stresses. Nat Cell Biol 22(1):7–17

6. Hogan CJ et al (1997) Engraftment and development of human CD34+-enriched cells from umbilical cord blood in NOD/LtSz-scid/scid mice. Blood 90(1):85–96

7. Nicolini FE et al (2004) NOD/SCID mice engineered to express human IL-3, GM-CSF and steel factor constitutively mobilize engrafted human progenitors and compromise human stem cell regeneration. Leukemia 18 (2):341–347

8. Ito M et al (2002) NOD/SCID/gamma(c) (null) mouse: an excellent recipient mouse model for engraftment of human cells. Blood 100(9):3175–3182

9. Traggiai E et al (2004) Development of a human adaptive immune system in cord blood cell-transplanted mice. Science 304 (5667):104–107

10. Ishikawa F et al (2005) Development of functional human blood and immune systems in NOD/SCID/IL2 receptor {gamma} chain (null) mice. Blood 106(5):1565–1573

11. Rongvaux A et al (2014) Development and function of human innate immune cells in a humanized mouse model. Nat Biotechnol 32 (4):364–372

12. Cosgun KN et al (2014) Kit regulates HSC engraftment across the human-mouse species barrier. Cell Stem Cell 15(2):227–238

13. Bonnet D, Dick JE (1997) Human acute myeloid leukemia is organized as a hierarchy that originates from a primitive hematopoietic cell. Nat Med 3(7):730–737

14. Rouault-Pierre K et al (2017) Preclinical modeling of myelodysplastic syndromes. Leukemia 31(12):2702–2708

15. Mian SA et al (2015) SF3B1 mutant MDS-initiating cells may arise from the haematopoietic stem cell compartment. Nat Commun 6:10004

16. Anjos-Afonso F et al (2013) CD34(−) cells at the apex of the human hematopoietic stem cell hierarchy have distinctive cellular and molecular signatures. Cell Stem Cell 13(2):161–174

17. Goardon N et al (2011) Coexistence of LMPP-like and GMP-like leukemia stem cells in acute myeloid leukemia. Cancer Cell 19(1):138–152

18. Clappier E et al (2011) Clonal selection in xenografted human T cell acute lymphoblastic leukemia recapitulates gain of malignancy at relapse. J Exp Med 208(4):653–661

19. Place ES, Evans ND, Stevens MM (2009) Complexity in biomaterials for tissue engineering. Nat Mater 8(6):457–470

20. Ratner BD (2019) Biomaterials: been there, done that, and evolving into the future. Annu Rev Biomed Eng 21:171–191

21. Abarrategi A et al (2017) Versatile humanized niche model enables study of normal and malignant human hematopoiesis. J Clin Invest 127(2):543

22. Reinisch A et al (2016) A humanized bone marrow ossicle xenotransplantation model enables improved engraftment of healthy and leukemic human hematopoietic cells. Nat Med 22(7):812–821

Chapter 19

3D Engineering of Human Hematopoietic Niches in Perfusion Bioreactor

Steven J. Dupard and Paul E. Bourgine

Abstract

The hematopoietic microenvironment, also referred to as hematopoietic niche, is a functional three-dimensional (3D) unit of the bone marrow (BM) that planar culture systems cannot recapitulate. Existing limitations of 2D protocols are driving the development of advanced 3D methodologies, capable of superior modeling of the native organization and interactions between hematopoietic cells and their niche.

Hereafter we describe the use of a 3D perfusion bioreactor for in vitro generation of human hematopoietic niches. The approach enables the recapitulation of the interactions between hematopoietic stem and progenitor cells (HSPCs), mesenchymal cells (MSCs), and their extracellular matrix in a 3D relevant setting. This was shown to support the functional maintenance of blood populations, self-distributing in the system compartments depending on their differentiation status. Such 3D niche modeling represents an advanced tool toward uncovering human hematopoiesis in relation to its host microenvironment, for both fundamental hematopoiesis and personalized medicine applications.

Key words Human hematopoiesis, Hematopoietic niche, Perfusion bioreactor, 3D culture, Mesenchymal stromal cells, Bone marrow organoid

1 Introduction

The study of the human hematopoietic system is principally limited by the challenging functional maintenance of human HSCs ex vivo. New culture systems converge toward replicating the human niche organization, with the assumption that a superior biological fidelity will better support HSPC properties and allow their study in controllable conditions.

Key parameters need to be considered when attempting to mimic the human hematopoietic niche in vitro [1]. Within the bone marrow microenvironment, MSCs are essential HSPCs partners, capable of regulating their self-renewal/differentiation activities via paracrine or direct cell-to-cell interactions [2]. For this reason, MSCs have been exploited in coculture systems where

Marion Espéli and Karl Balabanian (eds.), *Bone Marrow Environment: Methods and Protocols*, Methods in Molecular Biology, vol. 2308, https://doi.org/10.1007/978-1-0716-1425-9_19, © Springer Science+Business Media, LLC, part of Springer Nature 2021

their capacity to secrete an extracellular matrix (ECM) was further shown to prevent HSPCs exhaustion [3].

From standard 2D approaches, the field progressively evolved toward the adoption of 3D protocols better replicating the native spatial tissue environment. This can be achieved through the use of scaffolding materials, which physical parameters (e.g., stiffness, degradability, elasticity) can directly impact the fate decision of seeded populations [4, 5]. In that regard, collagens and ceramic scaffolds consist in relevant candidates by analogy with the bone marrow composition. During culture, materials are further functionalized through deposition of an extracellular matrix by seeded MSCs. These 3D niches ultimately provide complex cell-ECM interactions and spatiotemporal factors presentation capable of conferring functional superiority over their 2D counterparts [6].

Finally, in the last decade the emergence of bioreactor device has provided the capacity of transitioning from 3D static to 3D dynamic culture. Bioreactors offer the control over a wider range of parameters, toward mimicking the native interstitial flow and associated shear stress within the modeled microenvironment [7]. In particular, perfusion bioreactors ensure the continuous medium delivery through the cell-seeded material, in an oscillating fashion. By providing uniform cell seeding, nutrient supply and waste removal [8], the dynamic culture assure a homogeneous tissue formation while providing scaled-up opportunities.

In this method article, we propose the combination of primary human MSCs, cord-blood HSPCs, a hydroxyapatite scaffold, and a perfusion bioreactor system in order to engineer 3D hematopoietic niches. This is achieved by the functionalization of the scaffolding material through human MSC seeding and their ECM-deposition during osteoblastic differentiation in perfusion bioreactor. The HSPC fraction is subsequently introduced in the system, leading to its self-distribution and establishment of interactions within the engineered osteoblastic niche. Our approach was shown to recapitulate key structural and functional features of the human bone marrow—including HSPCs maintenance—providing a tunable model to investigate human hematopoiesis [9]. In the following sections we will detail the methodology used from the cell isolation and 3D hematopoietic niche engineering, to the sample preparation for flow cytometry and histology.

2   Materials

Unless otherwise stated, all solution must be prepared under sterile conditions and, if applied, with deionized sterile water.

2.1 Isolation of Primary Human Bone Marrow MSCs and Cord Blood CD34⁺ HSPCs

1. T-175 flasks.

2. Bone marrow dilution buffer (BMDB): phosphate buffered saline (PBS) without Ca^{2+} or Mg^{2+}, supplemented with 3% fetal bovine serum (FBS).

3. Diluted bone marrow aspirate (dBMA): dilute the BMA with BMDB to a BMA:BMDB ratio of 1:2 (*see* **Note 1**).

4. Density gradient solution: Histopaque®-1077 from Sigma-Aldrich.

5. Isolation buffer (IB): 500 mL PBS with 10 mL FBS and 2 mL of 0.5 M ethylenediaminetetraacetic acid (EDTA) (pH 8) (*see* **Note 2**).

6. Complete medium (CM): α-minimal essential medium (α-MEM) with 10% FBS, 1% HEPES of a 1 M stock solution, 1% sodium pyruvate of a 100 mM stock solution, and 1% of Penicillin–Streptomycin–Glutamine (100×) solution.

7. Freezing medium 1 (FM1): Iscove's modified Dulbecco's media (IMDM) with 40% FBS.

8. Freezing medium 2 (FM2): FM1 with 20% dimethyl sulfoxide (DMSO).

9. Freezing medium 3 (FM3): FBS with 10% DMSO.

10. Negative selection: human CD34 MicroBead Kit from Miltenyi Biotec (described hereafter) or similar kit from other providers.

2.2 3D Perfusion Bioreactor

1. For in-process sterilization: 70% ethanol in spray bottle and absorbent tissue.

2. Bioreactor: U-CUP perfusion bioreactor from Cellec Biotek AG with infuse/withdraw PHD ULTRA™ syringe pump from Harvard Apparatus (*see* **Note 3**).

3. Scaffold: hydroxyapatite scaffold from (Finceramica, Engipore) of thickness 4 mm and diameter 8 mm (*see* **Note 4**).

4. Proliferative medium (PM): CM supplemented with 100 nM dexamethasone, 0.1 mM ascorbic acid-2-phosphate, and 5 ng/mL fibroblast growth factor 2 (FGF-2), all from Sigma-Aldrich (*see* **Note 5**).

5. Osteogenic medium (OM): CM supplemented with 100 nM dexamethasone, 10 mM β-glycerophosphate, and 0.1 mM ascorbic acid-2-phosphate, all from Sigma-Aldrich (*see* **Note 5**).

6. Coculture medium (CoCM): Serum Free Expansion Medium (SFEM, Stemcell Technologies: StemSpan™), supplemented with 10 ng/mL of stem cell factor (SCF), 10 ng/mL of Fms-like tyrosine kinase 3 (FLT3-ligand), and 10 ng/mL of thrombopoietin (TPO), all from Miltenyi Biotec (*see* **Note 5**).

2.3 Preparation for Flow Cytometry and Histology

Unless harvested cells are used for further culture or transplantation, the following does not require sterility.

1. Collection of cells from the engineered hematopoietic tissue: 0.3% collagenase and 0.05% Trypsin-EDTA.

2. FACS buffer (equivalent to IB): 500 mL PBS with 10 mL FBS and 2 mL of 0.5 M EDTA (pH 8).

3. Fixation buffer: 4% formalin.

4. Decalcification buffer: 15% EDTA (pH 7).

3 Methods

Unless otherwise stated the manipulation are performed at room temperature (RT) and in sterile conditions. Unless stated otherwise, the medium temperature needs to be prewarmed at 37 °C prior to usage.

3.1 Primary Human Bone Marrow MSC Isolation

1. In a 50 mL conical tube, carefully layer 30 mL dBMA over 15 mL of Histopaque®-1077, do not mix the two solutions (*see* **Note 6**).

2. Centrifuge at 480 × g for 30 min, without breaks. Collect the buffy coat—consisting of a ring of cells between the plasma and the Histopaque®-1077 fractions—with a 3 mL syringe and long needle (maximum gauge 21) into a new 50 mL conical tube and fill up with BMDB to 50 mL.

3. Centrifuge at 480 × g, at 4 °C for 7 min. Discard most of the supernatants and combine cell pellets into one 50 mL tube.

4. Repeat wash by adding BMDB up to 50 mL and spinning down 7 min, 480 × g, at 4 °C.

5. Resuspend cells in 10 mL of ice-cold IB and filter the suspension through 70 μM filter.

6. Count nucleated cells.

7. Centrifuge at 480 × g, at 4 °C for 7 min. Discard the supernatant.

8. Resuspend in prewarmed CM supplemented with 5 ng/mL of FGF-2. Seed a maximum of 15 million cells per T-175 Flasks and cultured in a humidified 37 °C/5% CO_2 incubator.

9. Change the media twice per week. When cells reach subconfluency (typically after 2 weeks of culture), proceed with trypsinization to harvest human MSCs (*see* **Note 7**). Cells can then be frozen (passage 0) until use. In experiment, cells need to be preferentially below passage 4 to prevent loss of proliferation or differentiation potential.

3.2 Cord Blood CD34+ HSPC Isolation

1. Using the harvested cord blood sample, proceed similarly as for the previously described bone marrow MSC isolation from **steps 1** to **7** included.

2. Resuspend in up to 300 μL of ice-cold IB (*see* **Note 8**), with a maximum 300 million cells per 100 μL of CD34 microbeads.

3. Add 100 μL of blocking reagent, provided by the Miltenyi Biotec kit, and mix well. Add 100 μL of human CD34 microbeads to the cell suspension and mix thoroughly once more.

4. Incubate 30 min at 4 °C, mix every 10 min by pipetting up and down gently (*see* **Note 9**).

5. Place MACS column on magnetic holder, and wash the column with 3 mL of ice-cold IB.

6. Add the 500 μL cell suspension to the column, and wash three times with 3 mL of ice-cold IB.

7. Place the column over a 15 mL conical tube. Add 5 mL of ice-cold IB and elute the cells using supplied plunger (*see* **Note 10**).

8. Count and centrifuge at $480 \times g$, at 4 °C for 7 min. Discard the supernatant.

9. Add 500 μL FM2 with 500 μL FM3 to the cell pellet (*see* **Note 11**), transfer the cell suspension into a labelled cryo-tube.

10. Place cryotubes in a freezing container and transfer immediately to −80 °C (*see* **Note 12**).

11. After 48 h, transfer to −150 °C.

3.3 Establishment of an Osteoblastic Niche in the 3D Perfusion Bioreactor

The establishment of the osteoblastic niche tissue is performed in two culture steps within the perfusion bioreactor. During the first week, MSCs are seeded and cultured using PM, promoting their proliferation thus ensuring scaffold colonization. Subsequently, MSCs are exposed to OM for 3 weeks in order to prime their osteoblastic differentiation and ECM deposition.

1. Open the sealed U-CUP perfusion bioreactor and the sealed hydroxyapatite scaffold under a hood and place the scaffold inside the bioreactor chamber before sealing it according to manufacturer instructions.

2. Place the mounted bioreactor on its support (*see* Fig. 1a).

3. Place a 70% ethanol spray bottle in the hood with absorbent tissue completely soaked by 70% ethanol for later sterilization.

4. Resuspend a fresh or frozen MSCs suspension in PM at a concentration of 1×10^5 cells/mL. Make 7 mL of cell suspension per bioreactor (*see* **Note 13**).

5. Make sure that the stopcocks are positioned to allow injection within the tubing compartments, and not the chamber.

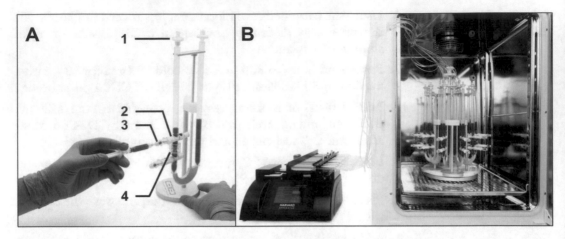

Fig. 1 (**a**) Illustration of the 3D perfusion bioreactor during medium loading. (**b**) Implementation of the bioreactors in a standard incubator, and connection to the syringe pump. The syringe pump is placed on top of the incubator during culture. (1) filter; (2) stopcock; (3) septum; (4) chamber. Images were provided by CELLEC Biotek AG

6. Spray 70% ethanol on the bioreactor septa prior to cell injection into the system to ensure sterilization. Place the tissue under the stopcock to absorb most of the dripping ethanol.

7. Collect 7 mL of cell suspension with a 10 mL syringe and a long needle (maximum gauge 21 recommended) (*see* **Note 14**).

8. Carefully remove the needle and inject 2 mL in the upper septum and 5 mL in the lower septum (*see* Fig. 1a).

9. Spray 70% ethanol on the bioreactor septa once more.

10. Open the stopcocks to allow connection of the tubing with the chamber.

11. Using an empty 10 mL syringe, gently inject air into the most distal filter relative to the chamber of the bioreactor to remove any air bubbles possibly remaining in the chamber.

12. Let the medium equilibrate for a few seconds.

13. Place the seeded 3D perfusion bioreactor in a humidified 37 °C/5% CO_2 incubator and connect to the syringe pump via the filter opposite to the chamber (*see* **Note 15**) (*see* Fig. 1b).

14. Initiate the following cell seeding perfusion program cycle: (i) infuse 2.8 mL/min with a 2 mL target, (ii) 30 s delay, (iii) withdraw 2.8 mL/min with a 2 mL target, (iv) 30 s delay, (v) repeat from step (i) (*see* **Notes 15** and **16**).

15. Let the cell seeding perfusion program cycle run overnight. Subsequently reduce the superficial velocity of the program from 2.8 mL/min to 0.28 mL/min (*see* **Note 17**).

16. PM is changed twice a week. Close the stopcocks and disconnect the bioreactor from the pump and place it from the incubator to a cell culture hood.

17. Spray 70% ethanol on the bioreactor septa. Using a 10 mL syringe, remove the PM from both sides of the chamber.

18. Notice that ≈1.5 mL of PM is inaccessible in the chamber. This dead-volume will need to be considered for the final concentration of factors for the PM change.

19. Spray 70% ethanol on the bioreactor septa once more.

20. Collect 6 mL of PM with a 10 mL syringe and a long needle (maximum gauge 21) (*see* **Note 14**).

21. Carefully remove the needle and inject 2 mL in the upper septum and 4 mL in the lower septum.

22. Spray 70% ethanol on the bioreactor septa once more.

23. Place the seeded 3D perfusion bioreactor in a humidified 37 °C/5% CO_2 incubator and connect the filter opposite to the chamber, to the syringe pump. Then open the stopcocks.

24. Initiate the following perfusion program cycle: (i) infuse 0.28 mL/min with a 2 mL target, (ii) 30 s delay, (iii) withdraw 0.28 mL/min with a 2 mL target, (iv) 30 s delay, (v) repeat from step (i).

25. After 1 week, proceed similarly by replacing PM by OM for the following 3 weeks of osteoblastic differentiation.

3.4 Seeding and Coculture of Cord Blood CD34+ HSPCs on Engineered Osteoblastic Niches

1. Resuspend a fresh or frozen CD34+ HSPCs suspension in CoCM to obtain 750.000 cells per 6 mL for each bioreactor.

2. Close the stopcocks and disconnect the bioreactor from the pump and place it from the incubator to a cell culture hood.

3. Spray 70% ethanol on the bioreactor septa. Using a10 mL syringe, remove the OM from both sides of the chamber.

4. Collect 6 mL of cell suspension with a 10 mL syringe and a long needle (maximum 21 gauge) (*see* **Note 14**).

5. Carefully remove the needle and inject 2 mL in the upper septum and 4 mL in the lower septum.

6. Spray 70% ethanol on the bioreactor septa once more.

7. Place the seeded 3D perfusion bioreactor in a humidified 37 °C/5% CO_2 incubator and connect the most distal filter relative to the chamber to the syringe pump (*see* **Note 15**).

8. Open the stopcocks to allow connection of the tubing with the chamber.

9. Initiate the following cell seeding perfusion program cycle: (i) infuse 2.8 mL/min with a 2 mL target, (ii) 30 s delay, (iii) withdraw 2.8 mL/min with a 2 mL target, (iv) 30 s delay, (v) repeat from step (i).

10. Let the cell seeding perfusion program cycle run overnight. Subsequently reduce the superficial velocity of the program from 2.8 mL/min to 0.28 mL/min.

11. Change the CoCM twice a week for 1 week. During CoCM change, the HSPCs fraction contained in it needs to be retrieved and reinjected in the system ultimately. As such, the medium from each bioreactor is harvested and spun down at $300 \times g$ for 10 min. The pelleted cells are then resuspended in fresh CoCM and injected back in the corresponding bioreactor.

3.5 Sample Harvesting for Flow Cytometry

From this point on, if no subsequent cell culture with isolated cells is planned, the sterility is no longer required.

1. For floating and loosely attached cells, harvest by collecting the CoCM present in the bioreactor in addition to a subsequent perfused wash with 6 mL PBS. Centrifuge the suspension for 10 min at $300 \times g$ and resuspend in ice-cold FACS buffer for immediate use.

2. For cells present on the scaffolding material, harvest by 5 min perfusion with 6 mL of collagenase solution at cell seeding velocity (2.8 mL/min) followed by 5 min perfusion with 6 mL of trypsin solution at the same velocity. In addition, two washes with 6 mL PBS are pooled with harvested cells. Centrifuge the suspension for 10 min at $300 \times g$ and resuspend in ice-cold FACS buffer for immediate use.

3.6 Sample Harvesting for Histology

1. Remove the scaffold from the bioreactor and fix in 4% formalin for 24 h with rotation at 4 °C.

2. Decalcify the scaffold with 50 mL of 15% EDTA (pH 7) at 4 °C under stirring for a minimum of 1 week, by changing the solution at least twice a week (*see* **Note 18**).

4 Notes

1. The bone marrow aspirate (BMA) is collected in the clinic from healthy donors in a tube containing Heparin (10,000–15,000 IU). Usually the BMA volume is around 20 mL. If visible clots are present in the BMA, remove them with a pipette before proceeding with the dilution. BMDB must be brought to RT prior to use.

2. IB must remain on ice during the whole procedure.

3. Other infuse/withdraw pumps with equivalent performance for superficial velocity between 0.28 and 2.8 mL/min can also be used.

4. The size of scaffolds to be used can also vary in thickness (2–4 mm) and diameter (6–10 mm) but the number of MSCs to be seeded should then be adapted accordingly. Other porous scaffolding materials of different composition can also be exploited (e.g., collagen-based) [10].

5. Add factors immediately prior to use. Thawed aliquots can be kept at 4 °C according to manufacturer instructions.

6. For maximum ease, pipettor must be set on low speed and the 50 mL conical tube be gradually held from a 45° angle to its upright position from the beginning to the end of loading.

7. Human MSCs are selected based on adhesion and proliferation on the plastic substrate for up to 2 weeks after seeding.

8. Take pellet volume into account.

9. Do not use a rotating wheel for mixing.

10. To increase purity, collected cells can be added to a second equilibrated column. Purity of the CD34$^+$ cell isolation can be assessed by flow cytometry.

11. This step needs to be done as quickly as possible for optimum cell survival.

12. Using the CoolCell® Cell Freezing Containers from Corning provides higher cell viability upon freezing.

13. One might consider the preparation of a 10–20% higher volume to anticipate potential loading errors and loss of material due to dead volumes.

14. To avoid cross-bioreactor contamination, ensure that a new syringe is used for each bioreactor.

15. For better performance, the syringe plug must be aligned with the 5 mL graduation. Also make sure that the syringe is properly lubricated.

16. To prevent air perfusion while the bioreactor is running, make sure during the first cycles that the media surface at equilibrium is circa 2 cm above the upper stopcock. If not the case, adjust manually by using the infuse/withdraw pump function, and restart the cycle.

17. A higher speed allows a more dynamic seeding for homogeneous cell distribution.

18. Assess the correct decalcification by applying a gentle pressure on the scaffold using forceps. If a certain physical resistance persists, extend the decalcification period until the material becomes spongy.

References

1. Bourgine PE, Martin I, Schroeder T (2018) Engineering human bone marrow proxies. Cell Stem Cell 22:298–301
2. Kfoury Y, Scadden DT (2015) Mesenchymal cell contributions to the stem cell niche. Cell Stem Cell 16(3):239–253
3. Muth CA, Steinl C, Klein G et al (2013) Regulation of hematopoietic stem cell behavior by the nanostructured presentation of extracellular matrix components. PLoS One 8:e54778
4. Holst J, Watson S, Lord MS et al (2010) Substrate elasticity provides mechanical signals for the expansion of hemopoietic stem and progenitor cells. Nat Biotechnol 28:1123–1128
5. Engler AJ, Sen S, Sweeney HL et al (2006) Matrix elasticity directs stem cell lineage specification. Cell 126:677–689
6. Raic A, Rödling L, Kalbacher H et al (2014) Biomimetic macroporous PEG hydrogels as 3D scaffolds for the multiplication of human hematopoietic stem and progenitor cells. Biomaterials 35:929–940
7. Rödling L, Schwedhelm I, Kraus S et al (2017) 3D models of the hematopoietic stem cell niche under steady-state and active conditions. Sci Rep 7:4625
8. Wendt D, Marsano A, Jakob M et al (2003) Oscillating perfusion of cell suspensions through three-dimensional scaffolds enhances cell seeding efficiency and uniformity. Biotechnol Bioeng 84:205–214
9. Bourgine PE, Klein T, Paczulla AM et al (2018) In vitro biomimetic engineering of a human hematopoietic niche with functional properties. Proc Natl Acad Sci U S A 115: E5688–E5695
10. Hoffmann W, Feliciano S, Martin I et al (2015) Novel perfused compression bioreactor system as an in vitro model to investigate fracture healing. Front Bioeng Biotechnol 3:1–6

Chapter 20

Manufacturing a Bone Marrow-On-A-Chip Using Maskless Photolithography

Benoit Souquet, Matthieu Opitz, Benoit Vianay, Stéphane Brunet, and Manuel Théry

Abstract

The bone marrow (BM) is a complex microenvironment in which hematopoietic stem and progenitor cells (HSPCs) interact with multiple cell types that regulate their quiescence, growth, and differentiation. These cells constitute local niches where HSPCs are confined and subjected to specific set of physical and biochemical cues. Endothelial cells forming the walls of blood capillaries have been shown to establish a vascular niche, whereas osteoblasts lying along the bone matrix organize the endosteal niche with distinct and specific impact on HSPC fate. The observation of the interaction of HSPCs with niche cells, and the investigation of its impact on HSPCs behavior in vivo is hindered by the opacity of the bone matrix. Therefore, various experimental strategies have been devised to reconstitute in vitro the interaction of HSPCs with distinct sets of BM-derived cells. In this chapter, we present a method to manufacture a pseudo BM-on-a-chip with separated compartments mimicking the vascular and the endosteal niches. Such a configuration with connected but distant compartments allowed the investigation of the specific contribution of each niche to the regulation of HSPC behavior. We describe the microfabrication of the chip with a maskless photolithography method that allows the iterative improvement of the geometric design of the chip in order to optimize the adaptation of the multicellular architecture to the specific aim of the study. We also describe the loading and culture of the various cell types in each compartment.

Key words Bone marrow-on-a-chip, Maskless photolithography, Microfabrication, Hematopoietic stem cells, 3D cell culture, Hydrogel, Organ-on-a-chip

1 Introduction

Hematopoietic stem and progenitor cells (HSPCs) are essential to maintain hematopoietic lineages homeostasis throughout life [1]. HSPCs home within the bone marrow (BM), where they interact with complex multicellular microenvironments that include HSC progenies and multiple non-hematopoietic cell types. These microenvironments, or niches, are potent regulators of HSPC proliferation and differentiation through the action of multiple diffusible factors, physical cues, or direct cell–cell interactions between HSPCs and the niche cells. It is generally admitted

Marion Espéli and Karl Balabanian (eds.), *Bone Marrow Environment: Methods and Protocols*, Methods in Molecular Biology, vol. 2308, https://doi.org/10.1007/978-1-0716-1425-9_20, © Springer Science+Business Media, LLC, part of Springer Nature 2021

that the major—and functionally distinct—niches for HSCs are on the first hand the sinusoidal and endothelial niches vascularizing the marrow and, on the other hand the peripheric endosteal niche in close contact with rigid bone wall [2, 3]. Finally, alterations of the niches components, or bidirectional interactions between leukemic cells and their niches, are now recognized as playing a critical role in the initiation and development of hematological malignancies [4, 5].

The molecular and cellular mechanisms at play when HSPCs interact with their complex environment, as well as the impact on HSPC behavior, and the implications for HSPC maintenance and expansion are still poorly understood, in physiological and malignant contexts. Investigating such interactions in vivo is indeed a real challenge. First, imaging the processes at play is obviously tremendously complicated [6, 7]. Most of the studies have been conducted in mouse by characterizing the impact of genetic ablation or amplification of specific stromal cell populations on HSPCs and on their regenerative capacities assessed by transplantation experiments [8–10]. However, these seminal works did not allow deciphering the cellular mechanisms at play.

In order to circumvent these limitations, we and others have developed an in vitro system based on microfabrication technologies to reconstitute some aspects of the physiological microenvironments and investigate HSPC interactions with stromal cells of the niche ranging from minimalistic models [11] to self-organizing organoids [12].

Our BM-on-a-chip [13] was first inspired by the pioneering work of Li Jeon and coll., who described the set-up for the microchannel geometry and the culture conditions necessary for inducing endothelial cells self-organization into hollow and perfusable three-dimensional (3D) networks [14]. From this first design, we used maskless photolithography with a digital micromirror (DMD)-based device (Fig. 1) to prototype different evolutions of the chip. We just had to modify our virtual mask (8-bit image) to test a new design. First, we added a channel to create an endosteal niche (Fig. 2a, channel 2), separated from the vascular niche (Fig. 2a, channel 4) by a channel allowing the loading of HSPCs (Fig. 2a, channel 3). This niche was made of osteoblasts cultured in a 3D matrix made of collagen-I and fibrin to model a minimal version of the endosteal niche [15]. We added two big channels for medium change (Fig. 2a, channels 1 and 6), and finally another channel for the cytokine-secreting fibroblasts necessary for the vascular network formation (Fig. 2a, channel 5) that were embedded in a fibrin hydrogel. All the channels of the chip were separated by pillars which allow compartmentalization but also nutrient diffusion and cell migration (Fig. 2a, b), inspired by those developed by Li Jeon and colleagues [14]. For the endosteal niche, we had to

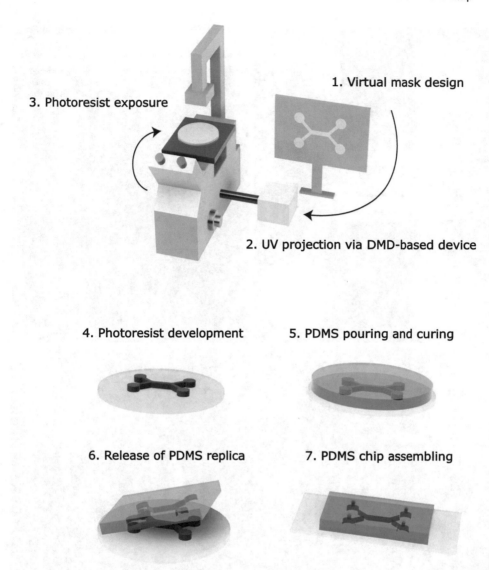

Fig. 1 Fabrication of the PDMS chip using maskless photolithography. A virtual mask is designed (1) and the corresponding UV image is projected via a DMD-based device (2) onto a photoresist (3). The exposed photoresist is then developed in a solvent, washed and baked, creating a master mold (4). Some PDMS is poured and cured onto the mold (5). The PDMS layer is then removed (6), punched to create inlets and bonded to a glass slide (7)

make another improvement. Indeed, the osteoblast imposed high constraints on the gel and made it collapse. To prevent this, we added some anchorage pillars in the channel, allowing the gel to resist cells' constraints (Fig. 2a, channel 2).

The endothelial cells and the HSPCs were cultured in the same hydrogel made of collagen-I and fibrin as for the osteoblasts. To confirm that this hydrogel was compatible with the self-organization of the endothelial cells into a hollow network, we

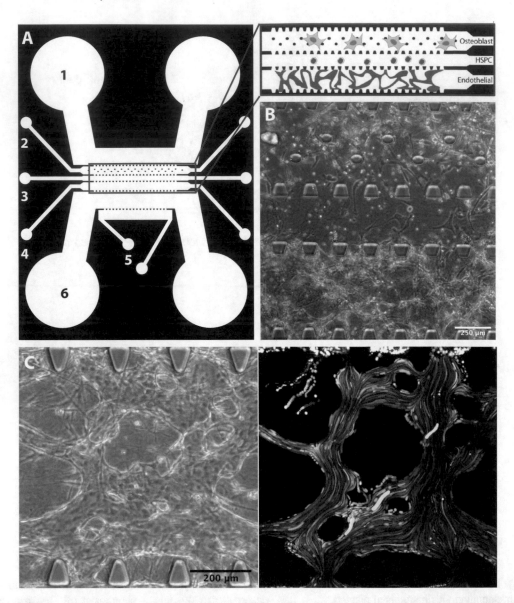

Fig. 2 BM-on-a-chip allowing the interaction of HSPCs with endosteal and vascular compartments in 3D hydrogels. (**a**) Illustration of the virtual mask (PDF file) loaded on the software of the DMD-based device. The chip comprises channels for medium circulation (no. 1 and no. 6), an endosteal compartment (no. 2), a HSPC injection channel (no. 3), a vascular compartment (no. 4), and a channel for cytokine-secreting fibroblasts (no. 5). The inset on the right (red rectangle) describes the organization of the three central channels. (**b**) Image of the three central channels of the chip in phase contrast (average projection of 5 z-slices with 5 μm spacing). (**c**) (left) Transmitted light image of the vascular network and (right) corresponding image of fluorescent beads perfused in the hollow structures formed by the HUVECs (projection of 300 time points separated by 0.33 s, made using the Fiji plugin "Temporal-Color Code"). Images are taken after 6 days of culture

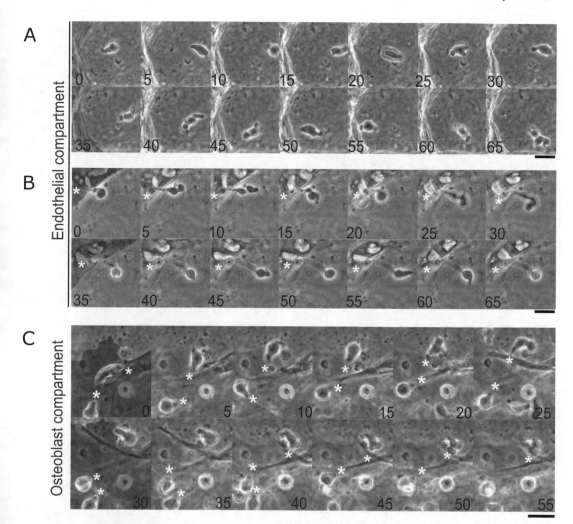

Fig. 3 Live-cell imaging of individual HSPCs in the BM-on-a-chip compartments. (**a**) Gallery of transmitted light images of a HSPC freely migrating in the hydrogel within the vascular compartment. Nearby endothelial cell appears salmon-colored in the left images of the gallery. Time points (in minutes) are indicated in the bottom right corners of each frame. Scale bar: 10 μm. (**b**) Gallery of transmitted light images of a HSPC interacting with an endothelial cell within the vascular compartment. The endothelial cell appears salmon-colored in the left images. The point of contact is highlighted with a white asterisk. Time points (in minutes) are indicated in the bottom left corners of each frame. Scale bar: 10 μm. (**c**) Gallery of transmitted light images of a HSPC interacting with an osteoblast within the endosteal compartment. The osteoblast is salmon-colored in the left images. 2 points of contact are highlighted with white asterisks. Time points (in minutes) are indicated in the bottom right corners of each frame. Scale bar: 10 μm

perfused fluorescent beads in the chip and observed their trajectory by fluorescence microscopy (Fig. 2c).

As illustrated in Fig. 3, individual HSPC behavior can indeed be investigated using live-cell imaging in each compartment of the BM-on-a-chip. Some HSPCs are found migrating freely in the hydrogel despite the proximity of stromal cells (Fig. 3a). Others can undergo heterotypic interactions with endothelial cells or osteoblasts, in the vascular and endosteal compartments respectively (Fig. 3b, c). In both cases, interacting HSPCs exhibit an elongated shape with a thin stalk at the point of contact with the stromal cell. Such interactions are stable in time (Fig. 3b, c). Our BM-on-a-chip appears therefore as a powerful system to (1) dissect the cellular and molecular mechanisms at play in these heterotypic interactions and (2) analyze the impact of such interactions on HSPCs fate.

Importantly, our model is compatible with chemical fixation, immunolabeling and high magnification imaging.

2   Materials

2.1   Microfabrication

1. Silicon wafer.
2. SU-8 3005 and SU-8 3050 negative photoresists (MicroChem Inc.).
3. Spin Coater.
4. Propylene glycol monomethyl ether acetate (PGMEA).
5. Hot plate.
6. Isopropyl alcohol.
7. Deionized (DI) water.
8. Acetone.
9. Polydimethylsiloxane (PDMS).
10. Silane: Trichloro(1H,1H,2H,2H-perfluoro-octyl)silane.
11. DMD-based photolithography device (e.g., PRIMO device from Alvéole, France).
12. UV lamp, 365 nm.
13. Lab oven.
14. Glass desiccator and vacuum pump.
15. Plasma cleaner.
16. Microfabrication tools: tweezers, scalpel, 1.5 mm and 5 mm punchers, Scotch tape.
17. Ultraclean glass coverslips 75 × 25 mm.

2.2   Hydrogels

1. Phosphate-buffered saline 1×, without calcium chloride and without magnesium chloride.
2. Phosphate-buffered saline 10× with calcium chloride and magnesium chloride.
3. 0.2 N Sodium hydroxide solution.
4. pH paper.

5. 0.4 μm sterile filters.

6. 5 mg/ml Rat tail Collagen-I solution.

7. 10 mg/ml Fibrinogen solution, (*see* **Note 1**).

8. 100 U/ml Thrombin solution (*see* **Note 2**).

2.3 Cells and Culture Media

1. Human umbilical vein endothelial cells (HUVECs) (*see* **Note 3**).

2. Human osteoblast cell line (hFOB).

3. Mitotically inactivated normal human lung fibroblast cells (NHLF) (*see* **Note 4**).

4. HSPCs, purified from cord blood as previously described in [16] (*see* **Note 5**). Cells are thawed and kept in suspension overnight at 37 °C, the day before the loading in the chip.

5. HUVEC medium containing EGM-2 MV Microvascular Endothelial Cell Growth Medium-2 BulletKit (Lonza CC-3203). Store at 4 °C and warm at 37 °C before use.

6. hFOB medium containing DMEM-F12 medium supplemented with 10% fetal bovine serum (FBS) and antibiotics/antimycotics (100 U/ml Penicillin, 100 μg/ml Streptomycin and 0.25 μg/ml Fungizone®). Store at 4 °C and warm at 37 °C before use.

7. NHLF cells medium containing FGM-2 Fibroblast Growth Medium-2 BulletKit (Lonza, CC-3132). Store at 4 °C and warm at 37 °C before use.

8. HSPC medium containing IMDM medium supplemented with 10% FBS, 100 ng/ml SCF, 10 ng/ml G-CSF, 20 ng/ml Il-3, and antibiotics/antimycotics (100 U/ml Penicillin, 100 μg/ml Streptomycin, and 0.25 μg/ml Fungizone®) (adapted from [17, 18]). Store at 4 °C and warm at 37 °C before use. In the chip, use the medium without cytokines.

3 Methods

3.1 Virtual Mask Design

The virtual mask is an 8-bit image (Fig. 2a) and can be drawn with an open-source software like Inkscape (https://inkscape.org/). The design of the chip might need to be optimized to meet your specific needs. The advantage of maskless photolithography is that you do not need to go through the entire process of the actual fabrication of a genuine photomask mask to test a new version of your PDMS chip, you just need to change the design of your virtual mask and expose a new layer of photoresist. This makes prototyping easy and fast.

1. The global size of the virtual mask (*see* Fig. 2a for the shape) is 11.5×14 mm (width $\times$ length).

2. The endosteal, vascular and NHLFs compartments must be at least 500 μm wide, and the HSPC compartment at least 300 μm wide.

3. Separate the channels with separation pillars: they measure 90×70 μm (length $\times$ width, *see* Fig. 2a for the shape) and are spaced by 70 μm (*see* **Note 6**).

4. Add some anchorage pillars (50 μm wide, 150 μm spacing) in the endosteal compartment (Fig. 2a, channel no. 2) to prevent the hydrogel from collapsing due to the constraints imposed by the cells (*see* **Note 7**).

3.2 Master Mold Fabrication

All the fabrication process (Fig. 1) must be performed in a clean room or at least in a laminar flow cabinet to prevent the presence of dust.

1. Grip layer (*see* **Note 8**): spin coat a 5 μm thick layer of SU-8 3005 resist onto the silicon wafer (refer to your manufacturer's protocol). You can also do it onto a glass slide (*see* **Note 9**).

2. Soft bake: ramp slowly from 65 °C to 95 °C for 2 min then bake it at 95 °C for 3 min.

3. Expose the whole SU-8 to UV light at 200 mJ/cm^2 using a UV lamp.

4. Post-Exposure bake: ramp slowly from 65 °C to 95 °C for 2 min then bake it at 95 °C for 2 min (this fully polymerized layer will make strong anchor for the structures on its top).

5. Spin coat a 75 μm thick layer of SU-8 3050 resist onto the grip layer of SU-8 (refer to your manufacturer's protocol).

6. Soft bake: ramp slowly from 65 °C to 95 °C for 5 min, then bake it at 95 °C for 35 min.

7. Spin coat a second 75 μm thick layer of SU-8 3050 resist onto the previous one to finally obtain a 150 μm thick layer of uncured SU-8.

8. Soft bake: ramp slowly from 65 °C to 95 °C for 5 min then bake it at 95 °C for 35 min.

9. UV exposure: place the SU-8 coated wafer onto the microscope holder with the SU-8 facing the objective (*see* **Note 10**) and make sure you focus on the SU-8 surface (*see* **Note 11**). Load your virtual mask image onto the dedicated software and project it with UV light (375 nm) using the DMD-based device, at 32 mJ/mm^2. Your system needs a "stitching mode" to create smooth junctions between the DMD images (*see* **Note 12**).

10. Post exposure bake: ramp slowly from 65 °C to 95 °C for 2 min then bake it at 95 °C for 8 min.

11. Development: this step must be performed under a chemical hood. Put some SU-8 developer (PGMEA) in a beaker, place the wafer (SU-8 facing up) in it and let it develop with gentle agitation for 30 min. After 20 min of development, change the developer with some fresh one for the remaining 10 min. At the end of the development time, only the exposed areas will remain. Using a wash bottle, wash the sample a first time with clean PGMEA, then with isopropyl alcohol until all the non-cured resist has been removed (*see* **Note 13**).

12. Dry it with air flow. You have obtained what is thereafter referred to as the master mold.

13. Hard bake: to solidify the master mold and make it more durable, bake it for 2 h at 150 °C.

14. Silanization: this surface treatment step is necessary to further prevent the PDMS from attaching to the master mold. In a glass desiccator, put the master mold. Place a centrifuge tube's cap in the desiccator and add 50 μl of silane in it. Apply a 10 mBar vacuum, wait for its stabilization then close hermetically the desiccator for 1 h (*see* **Note 14**).

15. Bake the silanized master mold for 2 h at 120 °C.

3.3 Microfluidic Chip Molding and Assembling

1. In a disposable weighing boat, prepare 10 g of fresh PDMS by mixing the silicone elastomer base and the curing agent at a 9:1 ratio (wt./wt.): 9 g of silicone and 1 g of curing agent. Mix vigorously.

2. Place the weighing boat in a glass desiccator and degas the PDMS mix until all air bubbles have been removed.

3. Place the master mold in a plastic dish and pour the PDMS mixture on it. The PDMS layer must be approximatively 3–4 mm thick. Degas in a glass desiccator to remove the remaining bubbles.

4. Bake in a lab oven at 70 °C for 1 h to cure the PDMS.

5. Carefully peel-off the PDMS from the master mold (*see* **Note 15**).

6. Cut the PDMS at the desired shape around the structures. Do not cut too close to the structures to prevent damages nor too far to be able to further bond it to the glass coverslip. You will obtain what is thereafter referred to as the PDMS chip.

7. Create the inlets of the channels: place the PDMS chip on a surface with the structures facing up and remove a cylinder of PDMS at the circular openings area using a 5 mm puncher for the medium reservoirs (Fig. 2a, channels 1 and 6) and a

1.5 mm puncher for the cell loading inlets (Fig. 2a, channels 2, 3, 4, and 5). The 1.5 mm holes are made to fill the channels using a 20 µl pipette tip.

8. The plasma bonding step enables to permanently bond the PDMS chip to a glass coverslip: place a clean glass coverslip and the PDMS chip (with the structures facing up) in a plasma reactor and plasma treat for 30 s to 3 min, depending on your plasma reactor. Place the PDMS chip on the treated surface of the glass coverslip. The bonding is covalent and very strong, you will have only one chance to place it correctly so do it very carefully. Once placed, press gently on the PDMS chip using tweezers to make sure all the chip is correctly bonded (*see* **Note 16**).

9. Your PDMS chip is now finished and is ready to be used the next day (*see* **Note 17**).

3.4 Hydrogel Loading with Cells

HUVECs, hFOBs, and HSPCs are all encapsulated in the same hydrogel, a mixture of collagen-I and fibrin (formed by the cleavage of fibrinogen by thrombin), hereafter referred to as Coll/fib hydrogel. NHLFs are encapsulated in fibrin only (*see* **Note 18**). Until cell loading, all the following steps need to be performed on ice to prevent gelation of collagen-I.

1. On ice, in a prechilled microtube, prepare 50 µl of 4 mg/ml collagen-I solution at neutral pH with 4.44 µl of 10× PBS, 4.37 µl of 0.2 N NaOH, 1.19 µl of 1× PBS, and 40 µl of 5 mg/ml rat tail collagen-I stock solution (*see* **Note 19**).

2. In another prechilled microtube, add the fibrinogen: mix 37.5 µl of the 4 mg/ml collagen-I solution prepared in **step 1** with 18.75 µl of 10 mg/ml fibrinogen and 18.75 µl of 1× PBS.

3. In another microtube, prepare a 20 µl solution of thrombin at 5 U/ml by mixing 1 µl of 100 U/ml thrombin stock solution in 19 µl of 1× PBS. Dispose 4 µl of this thrombin solution in 3 other prechilled microtubes.

4. Prepare the HUVECs (for the vascular compartment) and the hFOBs (for the endosteal compartment). Detach the cells, centrifuge, and count them, then put 2×10^5 HUVECs in a 1.5 ml centrifuge tube and 1.5×10^5 hFOBs in another 1.5 ml centrifuge tube (*see* **Note 20**). Use the dedicated medium for each cell line (*see* section Materials 2.2).

5. Centrifuge at $350 \times g$ for 5 min at room temperature.

6. Now focus only on the HUVECs to finish preparing them (**steps 7** to **10**).

7. Remove the supernatants using a P1000 pipette but let approximately 10 µl. After a few seconds (this will allow the remaining

medium to flow from the wall down to the bottom of the tube), remove the rest of the supernatant using a P20 pipette so that only the "dry" pellet remains.

8. Resuspend the HUVECs in 16 µl of the collagen-I + fibrinogen solution previously prepared. Mix gently while avoiding bubbles.

9. Using a P20 pipette set on 20 µl, take the HUVECs suspension and add it into one of the 3 prechilled microtubes (previously prepared in **step 3**) containing 4 µl of thrombin (*see* **Note 21**). The HUVECs are now at 10^7 cells/ml in the Coll/fib hydrogel.

10. Mix efficiently, quickly but gently, still while avoiding bubbles and load slowly the suspension in the inlet of channel no. 4 (Fig. 2a) using a P20 pipette (*see* **Note 22**). It should not be necessary to use all the 20 µl (*see* **Note 23**).

11. Repeat the **steps 7–10** with the hFOBs pellet (cells will be finally at 7.5×10^6 cells/ml) except that loading of the cell suspension will be in channel no. 2 (Fig. 2a).

12. Incubate 30 min at 37 °C in humid atmosphere to limit evaporation.

13. In the meantime, thaw the mitotically inactivated NHLFs, resuspend them in some HUVEC medium, centrifuge and count them, then put 1×10^5 cells in a centrifuge tube. Follow the same procedure to "dry" the pellet as in **step 7**.

14. The NHLFs are not encapsulated in Coll/Fib but in fibrin (fibrinogen + thrombin) hydrogel. In a microtube, prepare a 20 µl solution of 5 mg/ml fibrinogen by mixing 10 µl of the 10 mg/ml fibrinogen stock solution with 10 µl of PBS.

15. In the last microtube prepared in **step 3** containing 4 µl of 5 U/ml thrombin, add 4 µl of PBS 1× and mix to obtain an 8 µl solution of 2.5 U/ml thrombin.

16. Resuspend the NHLFs in 12 µl of the 5 mg/ml fibrinogen solution prepared in **step 14** and mix gently.

17. Using a P20 pipette set on 20 µl, add the 12 µl of NHLFs suspension in the 8 µl of 2.5 U/ml thrombin solution prepared in **step 15** and mix gently. Cells are now at 5×10^6 cells/ml in a hydrogel of 1 U/ml thrombin and 3 mg/ml fibrinogen.

18. Load them gently in channel no. 5 (Fig. 2a).

19. Incubate 5 min at 37 °C in humid atmosphere to limit the evaporation.

20. Add some HUVEC medium in one of the two reservoirs of channel no. 6 (Fig. 2a) and hFOB medium in one of the two reservoirs of channel no. 1 (Fig. 2a). Media should fill the

channels and form droplets in the opposite reservoirs (*see* **Note 24**).

21. Place your PDMS chip in a plastic petri dish and place it in your incubator at 37 °C (*see* **Note 25**). Do not change the medium for the next 3 days. However, if the medium level decreases in the inlets, add a few microliters of fresh one.

22. HSPCs are loaded 72 h after HUVECs and hFOBs to allow a proper auto-organization in the chip. Do not forget to thaw and resuspend the HSPCs in their dedicated medium (IMDM supplemented with FBS, SCF, G-CSF, and IL-3) the day before their loading. HSPCs are embedded in Coll/Fib hydrogel following exactly the same procedure as for the HUVEC and hFOB cells, except that 10^5 cells (final density of 5×10^6 cells/ml in the chip) will be loaded in channel no. 3 (Fig. 2a) (*see* **Note 26**).

23. After HSPC loading, incubate at 37 °C.

24. Change the media in channels no. 1 and no. 6 (Fig. 2a) with HSPC medium, without SCF, G-CSF and IL-3.

4 Notes

1. Dissolve the fibrinogen lyophilized powder in $1\times$ PBS at 37 °C under gentle agitation. Filter the solution using a 0.4 μm sterile filter and a syringe, aliquot and store at -20 °C.

2. Dissolve the thrombin lyophilized powder in a 0.1% BSA solution. Filter the solution using a 0.4 μm sterile filter and a syringe, aliquot and store at -20 °C.

3. Culture the HUVECs on 0.1% gelatin coated flasks. They must be harvested at 80% of confluency for the experiment. To do so, seed at 3.5×10^3 cells/cm^2 in a culture flask 72 h before the experiment. Use low-passage HUVECs for a better vascular network formation.

4. NHLFs secrete proangiogenic growth factors supporting endothelial cells morphogenesis [14]. In the chip, NHLFs tend to proliferate in the hydrogel having for consequence their escape from the hydrogel. Their proliferation is prevented by treating them with 10 μg/ml Mitomycin-c for 2 h. Freeze the inactivated NHLF in FGM-2 + 10% DMSO. Thaw them just before their loading in the PDMS chip.

5. HSPCs originate from human cord blood samples. Purify them using Ficoll™ density gradient, followed by MACS® enrichment using anti-CD34 antibody. Cryopreserve the HSPCs in FBS + 10% DMSO at 7×10^5 cell/ml.

6. These dimensions can be adapted but if the pillars are too close, the medium will not diffuse well into the compartments; if they

are too spaced, the hydrogels will leak from one channel to the other during their loading.

7. The size and spacing must be optimized so that the anchorage pillars stabilize the hydrogel without perturbing the its loading.

8. When peeling-off the PDMS layer from the SU-8 master mold, the SU-8 structures are submitted to high mechanical constraints. A first layer of SU-8 (5 μm thick) is spin coated and cured to prevent them from detaching from the substrate.

9. It can be quite complicated to control/monitor the UV exposure onto the SU-8 and to make sure the image is focused since the wafer is upside-down and not transparent. An alternative is to spin-coat the SU-8 on a glass slide so that it is easier to center the exposure on the SU-8 resist. Besides, a glass slide will fit on the microscope holder, which can be more complicated with a silicon wafer. Before spin coating a glass slide, wash it using acetone then isopropyl alcohol and bake it at 120 °C for 30 min.

10. Different objectives can be used to do maskless photolithography. In this case, since the desired structures are quite thick (150 μm), it is preferable to use an objective with a large depth of field (e.g., a 4× objective) so that the projected images are focused over a greater height (*see* **Note 11** to know how to focus).

11. Thanks to the autofluorescence of the SU-8, the UV image that is projected can be observed through the microscope to make sure of the focus. However, the area you will use for this purpose will be cured by the UV and thus cannot be used to project the chip design. To avoid wasting a too big area of your resist during this focusing step, define a "sacrificial" area. Make sure this sacrificial area is at least 1 cm far from the area you have chosen to be exposed to the virtual mask of the chip.

12. Stitching of DMD images. To make an image bigger than the size of a single DMD-image (a few hundreds of microns depending on the objective, for example around 2600 μm × 1600 μm for a 4× objective), the software must divide the virtual mask into subimages that will be sequentially projected. However, this process creates sharp junctions in the mold, which will lead the final PDMS chip to leak. To obtain smooth junctions between these subimages, PRIMO's dedicated software can add grey level gradients on the edges of each subimage, thus allowing the final structures to be smooth and the PDMS chip hermetically sealed onto the glass.

13. A visual control at the microscope must be performed after washing. If some photoresist remains in the non-exposed area, continue the development for a few minutes. Besides, a white film can remain after washing if the substrate has been underdeveloped. In this case, repeat the washing step: spray the

substrate with PGMEA to remove the film and then with isopropyl alcohol to remove the PGMEA.

14. Silanization can be performed in a plastic dish instead of a desiccator. Place the master mold in the dish, put the centrifuge tube's cap and add the 50 μl of silane in it. Seal the dish with parafilm and incubate for 1 h. The silanization can be done again after multiple uses of the master mold if the PDMS start to be hard to detach.

15. To peel-off the cured PDMS, use a scalpel and tweezers. First, cut the PDMS roughly around the structure's area. Remove the part of interest slowly using tweezers while making sure you do not degrade the mold. Afterward, cut the PDMS more precisely.

16. Some residual dust can remain on the PDMS chip and prevent efficient bonding. Before plasma treating it, use some 3 M Scotch tape to remove all the dust present on the surface that will be bonded to the glass. Use air flow or tape to remove the dust on the glass coverslip as well. Bond the PDMS chip to the glass directly after plasma treatment as the surface activation is stable for a few minutes only.

17. It is necessary to do the bonding the day before the experiment and to put the chip in the oven at 65 °C overnight so that the PDMS retrieves its hydrophobicity. Indeed, after the plasma treatment, the PDMS becomes hydrophilic, which can allow the hydrogel to go through the separation pillars and thus escape from its dedicated channel.

18. The volumes of most of the solutions prepared here are larger than necessary to account for volumetric errors in pipetting (e.g., Collagen-I solution is very viscous). Besides, this protocol is for one PDMS chip. Since some cells need to be thawed for the experiment, it can be more convenient to do the experiment on multiple PDMS chips in parallel. The protocol will thus need to be scaled up.

19. Collagen-I stock solution is at 17.5 mM in acetic acid (pH $\approx$ 3.8). For gelation, physiological pH and salt concentration are required and are tuned with the NaOH and the 10× PBS solutions, respectively. Always control the pH with the pH paper, 5 μl is sufficient.

20. Adapt the number of cells in each compartment if required.

21. The addition of the thrombin to the hydrogel mix makes it crosslink really fast. To make sure it remains liquid during its loading into the channels, add the cell suspension to the 4 μl thrombin solution at the last moment.

22. Cut the extremity of the 20 μl tip to enlarge its diameter. This will help to fit to the channel's inlet.

23. Once the solution reaches the outlet of the channel, stop applying pressure or the solution may leak in the other channels.

24. Due to the previous incubation times of the hydrogels in the absence of medium (for their polymerization), it is possible that they shrink a little bit. When loading the medium, some bubbles can thus form at the hydrogel/medium interface. If they are not too big, do not try to remove them, they will disappear after a few hours. If they are problematic or if the medium does not flow into the channel, gently apply pressure on the chip using a pipette tip to remove them.

25. Make sure that both inlets and outlets are filled with liquid and that no air bubble remains. Besides, to limit medium evaporation, create a humid atmosphere by placing caps of 15 ml conical tubes filled with sterile water in the petri dish.

26. In previous steps, nothing was loaded in the central channel (no. 3, (Fig. 2a)). Sometimes, this channel stays totally dry for days and sometimes it is full of medium. Based on our experience, it seems easier (and safer) to load HSPCs in the chip when the channel is filled with medium. If it is dry, fill the channel with PBS using a P20 pipette. Apply a Kimwipes on the opposite outlet in order to better "aspirate" the medium by capillarity.

Acknowledgments

This work was supported by the collaboration n°C29039 between Alvéole SAS and Laboratoire de Physiologie Cellulaire et Végétale (LPCV) (UMR 5168 CEA/CNRS/Université Grenoble Alpes (UGA)/UMR 1417 INRA).

References

1. Velten L, Haas SF, Raffel S et al (2017) Human haematopoietic stem cell lineage commitment is a continuous process. Nat Cell Biol 19:271–281. https://doi.org/10.1038/ncb3493

2. Morrison SJ, Scadden DT (2014) The bone marrow niche for haematopoietic stem cells. Nature 505:327–334. https://doi.org/10.1038/nature12984

3. Pinho S, Frenette PS (2019) Haematopoietic stem cell activity and interactions with the niche. Nat Rev Mol Cell Biol 20:303–320. https://doi.org/10.1038/s41580-019-0103-9

4. Sánchez-Aguilera A, Méndez-Ferrer S (2017) The hematopoietic stem-cell niche in health and leukemia. Cell Mol Life Sci 74:579–590. https://doi.org/10.1007/s00018-016-2306-y

5. Verovskaya EV, Dellorusso PV, Passegué E (2019) Losing sense of self and surroundings: hematopoietic stem cell aging and leukemic transformation. Trends Mol Med 25:494–515. https://doi.org/10.1016/j.molmed.2019.04.006

6. Christodoulou C, Spencer JA, S-CA Y et al (2020) Live-animal imaging of native haematopoietic stem and progenitor cells. Nature 578:278–283. https://doi.org/10.1038/s41586-020-1971-z

7. Guezguez B, Campbell CJV, Boyd AL et al (2013) Regional localization within the bone

marrow influences the functional capacity of human HSCs. Cell Stem Cell 13:175–189. https://doi.org/10.1016/j.stem.2013.06. 015

8. Höfer T, Busch K, Klapproth K et al (2016) Fate mapping and quantitation of hematopoiesis in vivo. Annu Rev Immunol 34:449–478. https://doi.org/10.1146/annurev-immunol-032414-112019

9. Zhao M, Perry JM, Marshall H et al (2014) Megakaryocytes maintain homeostatic quiescence and promote post-injury regeneration of hematopoietic stem cells. Nat Med 20:1321–1326. https://doi.org/10.1038/nm.3706

10. Chow A, Huggins M, Ahmed J et al (2013) CD169+ macrophages provide a niche promoting erythropoiesis under homeostasis and stress. Nat Med 19:429–436. https://doi.org/10.1038/nm.3057

11. Lutolf MP, Gilbert PM, Blau HM (2009) Designing materials to direct stem-cell fate. Nature 462:433–441. https://doi.org/10.1038/nature08602

12. Motazedian A, Bruveris FF, Kumar SV et al (2020) Multipotent RAG1+ progenitors emerge directly from haemogenic endothelium in human pluripotent stem cell-derived haematopoietic organoids. Nat Cell Biol 22:60–73. https://doi.org/10.1038/s41556-019-0445-8

13. Bessy T, Souquet B, Vianay B et al (2020) Hematopoietic progenitors polarize in contact with bone marrow stromal cells by engaging CXCR4 receptors. BioRxiv. https://doi.org/10.1101/2020.05.11.089292

14. Kim S, Lee H, Chung M et al (2013) Engineering of functional, perfusable 3D microvascular networks on a chip. Lab Chip 13:1489–1500. https://doi.org/10.1039/c3lc41320a

15. Nelson MR, Ghoshal D, Mejías JC et al (2019) A multi-niche microvascularized human bone-marrow-on-a-chip. bioRxiv 2019.12.15.876813. https://doi.org/10.1101/2019.12.15.876813

16. Biedzinski S, Faivre L, Vianay B et al (2019) Microtubules deform the nucleus and force chromatin reorganization during early differentiation of human hematopoietic stem cells. bioRxiv 763326. https://doi.org/10.1101/763326

17. Donaldson C, Denning-Kendall P, Bradley B et al (2001) The CD34+CD38neg population is significantly increased in haemopoietic cell expansion cultures in serum-free compared to serum-replete conditions: dissociation of phenotype and function. Bone Marrow Transplant 27:365–371. https://doi.org/10.1038/sj.bmt.1702810

18. Faivre L, Parietti V, Siñeriz F et al (2016) In vitro and in vivo evaluation of cord blood hematopoietic stem and progenitor cells amplified with glycosaminoglycan mimetic. Stem Cell Res Ther 7:3. https://doi.org/10.1186/s13287-015-0267-y

Part V

Sequencing-Based Analysis of Cell Heterogeneity

Chapter 21

In Vivo Tracking of Hematopoietic Stem and Progenitor Cell Ontogeny by Cellular Barcoding

Tamar Tak, Almut S. Eisele, and Leïla Perié

Abstract

Cellular barcoding is a powerful technique that allows for high-throughput mapping of the fate of single cells, notably hematopoietic stem and progenitor cells (HSPCs) after transplantation. Unique artificial DNA fragments, termed barcodes, are stably inserted into HSPCs using lentiviral transduction, making sure that each individual cell receives a single unique barcode. Barcoded HSPCs are transplanted into sublethally irradiated mice where they reconstitute the hematopoietic system through proliferation and differentiation. During this process, the barcode of each HSPC is inherited by all of its daughter cells and their subsequent mature hematopoietic cell progeny. After sorting mature hematopoietic cell subsets, their barcodes can be retrieved from genomic DNA through nested PCR and sequencing. Analysis of barcode sequencing results allows for determination of clonal relationships between the mature cells, that is, which cell types were produced by a single barcoded HSPC, as well as the heterogeneity of the initial HSPC population. Here, we give a detailed protocol of a complete HSPC cellular barcoding experiment, starting with barcode lentivirus production, isolation, transduction, and transplantation of HSPCs, isolation of target cells followed by PCR amplification and sequencing of DNA barcodes. Finally, we describe the basic filtering and analysis steps of barcode sequencing data to ensure high-quality results.

Key words Cellular barcoding, Hematopoiesis, Hematopoietic tree, Ontology, Lineage tracing, Hematopoietic clonality, Cell-family, Indexing library, Clone tagging, and Single-cell fate

1 Introduction

Cellular barcoding is a high-throughput single-cell fate mapping technique most extensively used in the field of hematopoiesis to analyze the fate of hematopoietic stem and progenitor cells (HSPCs) after bone marrow transplantation. The basis of the technique is the stable incorporation of unique artificial DNA sequences—termed barcodes—into the genome of single cells. During cell division, these barcodes are transmitted to daughter cells. The analysis of barcodes in the cell progeny thereby allows assessment of the clonal relationship between the progeny cells, that is, to determine which cells have been produced by each of the initially barcoded cells. Cellular barcoding was developed 20 years

Marion Espéli and Karl Balabanian (eds.), *Bone Marrow Environment: Methods and Protocols*, Methods in Molecular Biology, vol. 2308, https://doi.org/10.1007/978-1-0716-1425-9_21, © Springer Science+Business Media, LLC, part of Springer Nature 2021

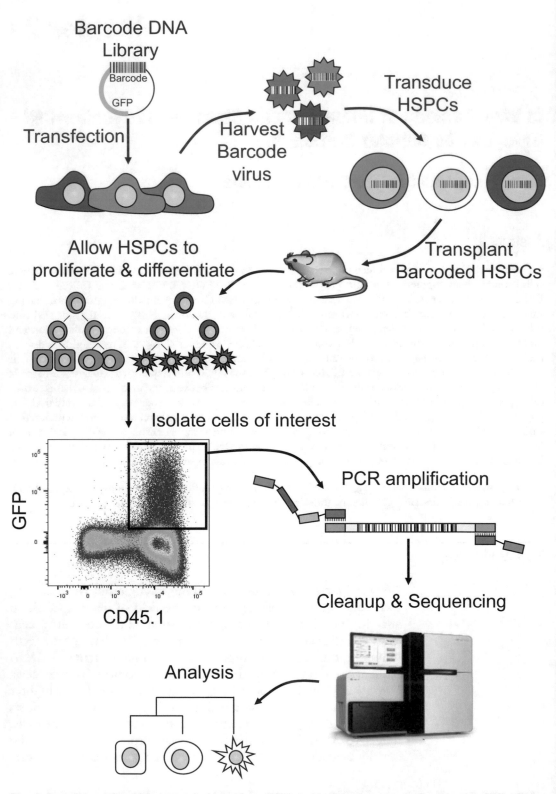

Fig. 1 Overview of the experimental steps of an HSPC barcoding experiment. The barcode DNA library, composed of 2,500 barcodes within the 3' UTR of a GFP cassette, is used to produce barcode virus. HSPCs are

ago [1, 2], but the technique for barcode detection has dramatically improved over time from sanger sequencing and microarrays to the use of nested PCR amplification and next-generation sequencing (NGS) [3]. This considerably increased the method's throughput, but also generated new challenges for the experimental setup.

We present here the detailed protocol of an HSPC cellular barcoding experiment using a library of around 2,500 semirandom barcodes of 98 base pairs (bp) [4], incorporated in the 3′UTR of a green fluorescent protein (GFP) expressed under the CMV promoter. This library has previously been applied to trace the fate of several types of HSPCs, namely, lymphoid-primed multipotent progenitors [5], hematopoietic stem cells (HSC) [6], and common myeloid progenitors [6], as well as dendritic cells [7] and breast cancer cells [8]. In the protocol, we describe all experimental steps starting from a pool of plasmids with different barcodes, called a barcode library (Fig. 1). Briefly, the protocol encompasses the production of barcode lentivirus from barcode plasmid DNA in the barcode library, the isolation of HSPCs from bone marrow of donor mice, the barcoding of HSPCs by lentiviral transduction and the transplantation of barcoded HSPCs into recipient mice. This is followed by the isolation of HSPC progeny and the processing of genomic DNA for barcode sequencing. A two-step PCR allows for barcode amplification using common barcode sequences (PCR1) and addition of an 8 bp sample index and Illumina sequencing primer sequences, including P5 and P7 Illumina flow cell attachment sequences (PCR2) (Fig. 2). After cleanup, amplified barcode DNA can be sequenced. Finally, we describe the analysis of the sequencing results, including retrieval of DNA barcode counts from sequencing files, quality control, and comparison of HSPC output patterns.

To allow reliable fate mapping, several critical points in the protocol need to be considered (Fig. 3). The first and most important step is to ensure that each HSPC receives a different and unique barcode. Errors in fate assignment arise when two HSPCs are transduced with the same barcode, as cells will seem to arise from the same HSPC whilst they are not. The chance of this occurring can be minimized by injecting a number of cells that is tenfold lower than the number of different barcodes present in the library. In addition, a control for this potential error should be

Fig. 1 (continued) isolated from donor mice (CD45.1$^+$) and transduced with barcode virus so that each HSPC incorporates a unique barcode in its genome. Barcoded HSPCs are transplanted into sub-lethally irradiated mice where they proliferate and differentiate to reconstitute the hematopoietic system and transmit their barcode to every daughter cell. To analyze barcodes in daughter cells, barcoded CD45.1$^+$ donor cells are isolated based on GFP expression. Barcodes from isolated cell populations are amplified and prepared for sequencing by PCR. PCR product is cleaned-up and sequenced on an Illumina sequencer. Analysis of sequencing results gives insights on the clonal relationship of cells, and thereby on the clonal output of transplanted HSPCs

a 1 barcode per HSPC

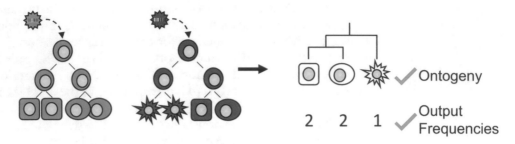

b 2 barcodes in 1 HSPC

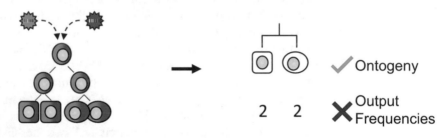

c 1 barcode in 2 HSPCs

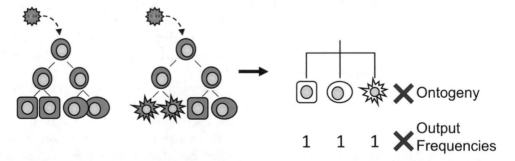

Fig. 2 Repeat use of barcodes during a cellular barcoding experiment leads to erroneous conclusions. (**a**) The integration of a unique barcode in each HSPC during a cellular barcoding experiment allows deduction of cell ontogeny and determination of the frequency of HSPCs with a certain clonal output. (**b**) The incorporation of two barcodes in one HSPC will lead to an overestimation of the frequency of its clonal output pattern. (**c**) The incorporation of one barcode in two HSPCs will lead to an erroneous estimation of both HSPC output patterns and the frequency of clonal output patterns

included in each experiment by injecting HSPCs from the same transduction batch in at least two recipient mice. The proportion of shared barcodes between mice is a proxy for the number of barcodes present in several cells. Secondly, integration of several different barcodes into a single HSPC leads to the same differentiation output being detected several times and, thereby, leads to an

PCR 1

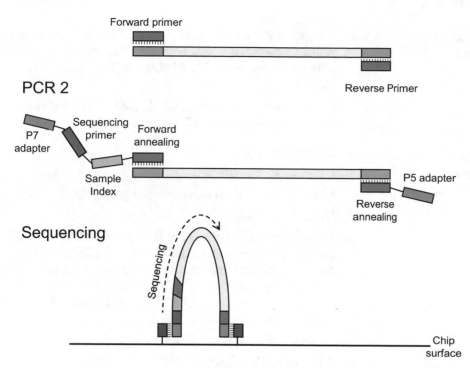

PCR 2

Sequencing

Fig. 3 A two-step PCR prepares for sequencing of barcodes. During PCR1 primers complementary to the common sequences upstream and downstream of all the barcodes are used to amplify barcode DNA. During PCR2, the common sequences are used to add a sample index of 8 bp, the Illumina sequencing primer sequence, and the P7 and P5 flow cell attachment sites to the barcode PCR product of PCR1. During sequencing the P5 and P7 flow cell attachment sites allow cluster generation. Sequencing 50 bp in single read mode will allow sequencing of the 8 bp sample index, the common sequence upstream of the barcode and the first 15 bp of the unique barcode sequence, all three of which will be used during sequencing analysis. The first 15 bp of the unique barcode sequence are sufficient to identify the approximately 2,500 barcodes in the barcode library

overestimation of the quantity of cells produced by this single HSPC. The chance of this error occurring is reduced by using a low (~10%) transduction efficiency but comes with the cost that not all HSPCs are barcoded. Lastly, PCR amplification and NGS detection of barcodes are error prone and can thereby lead to the false detection of barcodes that are not truly present. False barcodes can be filtered out of the data by comparing replicates split before the first PCR, the use of a reference list containing all barcodes known to be present in the barcode library and by controlling the expected sequencing depth. Taking these possible errors into account ensures high-quality results and allows for drawing valid conclusions.

2 Materials

2.1 Common Reagents

1. Phosphate buffered saline (PBS): 144 mg/l potassium phosphate (KH_2PO_4), 9 g/l sodium chloride (NaCl), 795 mg/l sodium phosphate ($Na_2HPO_4 \cdot 7H_2O$), with pH 7.4 (*see* **Note 1**).

2. RPMI medium containing RPMI1640, 10% fetal bovine serum (FBS), 100 U/ml penicillin–streptomycin.

2.2 Barcode Virus Production

1. HEK 293T cells.

2. HEK culture medium containing DMEM, GlutaMAX, 10% FBS, and 100 U/ml penicillin–streptomycin.

3. 10 cm petri dish.

4. Transfection mix containing 800 µl serum-free DMEM with 100 U/ml penicillin–streptomycin, add 25 µl FuGENE HD and incubate for 5 min at room temperature. Add 8 µg barcode library vector (*see* **Note 2**), 2.8 µg VSV-G vector, and 1.35 µg RSV-REV vector (*see* **Note 3**). Incubate for 15 min at room temperature before adding to cells.

5. Sodium butyrate ($C_4H_7NaO_2$).

6. 0.45 µm syringe filter.

7. Centrifugal filter units with a 100 kDa molecular weight cutoff (e.g., Amicon Ultra-15 Centrifugal Filter Units, Ultracel-100 regenerated cellulose membrane).

8. Opti-MEM culture medium.

2.3 Isolation of HSPCs

1. B6.SJL-*Ptprc^a Pepc^b*/BoyJ (also known as CD45.1 or Ly5.1) donor mice (*see* **Note 4**).

2. Scissors and tweezers.

3. 21G 50 mm needle and 2 ml syringe.

4. Reagent vials: 15 ml Falcon tubes, 50 ml Falcon tubes, 1.5 ml Eppendorf tubes.

5. 70 µm cell strainers.

6. RPMI medium.

7. CD117 MicroBeads UltraPure, mouse (Miltenyi).

8. MS or LS magnetic MACS column (1 column per 4 donor mice) (*see* **Note 5**).

9. MACS magnet (MiniMACS™ Separator or OctoMACS™ Separator) (*see* **Note 6**).

10. Insulin syringe.

11. Fluorochrome labeled antibodies for identification of HSPCs. The panel can be modified depending on the desired type of HSPCs (*see* **Note 7**).

(a) Anti-mouse CD117-APC, clone 2B8.

(b) Anti-mouse Sca-1-Pacific Blue, clone D7.

(c) Anti-mouse CD135-PE, clone A2F10 (*see* **Note 8**).

(d) Anti-mouse CD150-PE-Cy7, clone TC15-12F12.2.

(e) Anti-mouse CD34-Alexa Fluor 700, clone RAM34 (*see* **Note 9**).

(f) Anti-mouse CD48-APC-Cy7, clone HM48-1.

12. Cell sorter.

2.4 Transduction of HSPCs with Barcode Virus

1. Library of barcoding virus (*see* **Note 2**).

2. HSPC medium containing StemSpan Serum-Free Expansion Medium (SFEM), 50 ng/ml murine Stem Cell Factor (mSCF).

3. PBS.

4. RPMI medium.

5. U-bottom 96-well plate.

6. Plate centrifuge.

2.5 Transplantation of Transduced HSPCs

1. Wild-type C57BL/6 recipient mice expressing the Ptprc[b] allele (also known as CD45.2 of Ly5.2) (*see* **Note 4**).

2. Gamma or X-ray irradiator.

3. Insulin syringes.

4. Hank's Balanced Salt Solution (HBSS).

2.6 Isolation of Target Populations

1. Fluorescently labelled antibodies to identify target cell populations.

2. Cell sorter.

3. 1.5 ml Eppendorf tubes.

4. Deep 96-well PCR plate (*see* **Note 10**).

5. PCR plate seals (*see* **Note 11**).

6. 96-well PCR cycler.

7. Cell lysis buffer containing Viagen Direct PCR Lysis Reagent; 0.5 mg/ml Proteinase K. Prepare fresh.

2.7 PCR Amplification of Barcode DNA

1. DNA Decontamination Reagent.

2. PCR reaction mix 1 for a single sample containing 125.5 μl MilliQ, 20 μl 10× PCR buffer (from Invitrogen Taq DNA Polymerase, recombinant kit), 8 μl 50 mM $MgCl_2$, 4 μl of 40 mM dNTP mix (10 mM each);1 μl 100 μM forward primer tgctgccgtcaactagaaca; 1 μl 100 μM reverse primer gatctcgaatcaggcgctta, and 0.5 μl Taq DNA polymerase.

3. PCR reaction mix 2 for each duplicate containing 18 μl MilliQ, 3 μl 10× PCR buffer (from Invitrogen Taq DNA Polymerase,

recombinant kit), 1.2 μl 50 mM $MgCl_2$, 0.6 μl of 40 mM dNTP mix (10 mM each), 0.15 μl 100 μM reverse primer caagcagaagacggcatacgagatgatctcgaatcaggcgctta, and 0.075 μl Taq DNA polymerase.

4. Index primers (*see* **Note 12**) 3.0 mM in MilliQ.

5. PCR plate seals (*see* **Note 11**).

6. Disposable reagent reservoir for multichannel pipette.

7. PCR cycler.

2.8 Sequencing

1. Low-bind 1.5 ml Eppendorf tubes.

2. Magnetic beads for cleanup and DNA size selection (e.g., Agencourt AMPure XP beads).

3. 70% ethanol. Prepare fresh from absolute ethanol and MilliQ.

4. MilliQ.

5. Strong magnet that fits 1.5 ml Eppendorf tubes (e.g., DynaMag-2 magnet).

6. Vortex.

7. Automated electrophoresis machine for assessing DNA quality suitable for analyzing DNA fragments between 100 and 1000 bp (e.g., a BioAnalyzer device from Agilent with DNA 1000 kit).

8. Phix Illumina phage genome library.

2.9 Data Processing and Analysis

1. XCALIBR for Perl; eXtracting Counting and LInking to Barcode References https://github.com/NKI-GCF/xcalibr.

2. Reference list of all barcodes present in the library.

3. List of all sample indices used in the experiment.

3 Methods

Perform all procedures with live virus and animal materials according to the relevant safety regulations. The experiment should be planned over several rooms to prevent barcode contamination. Perform all steps prior to the PCR steps in a room where no amplified PCR products are handled and ensure to minimize any risk of contaminating your samples. Accordingly, cells and all non-amplified DNA product should be handled in a different room (cells with barcode room) than where the PCR reagents are handled (pre-PCR room) as well as amplified PCR products (post-PCR room). Similarly, all amplified PCR products should be handled in a third, separate room (post-PCR room). Store all reagents and materials in the room they will be used and do not move materials or equipment between rooms. All procedures are performed on ice and protected from light, unless stated otherwise.

3.1   Barcode Virus Production

1. Seed 1–1.2 million HEK 293T in 8 ml HEK culture medium in a 10 cm petri dish.

2. Allow cells to grow for 24 h by placing the cells at 37 °C in an incubator under 5% CO_2 atmosphere.

3. At the day of transfection, change culture medium with 8 ml fresh HEK culture medium.

4. Prepare transfection mix.

5. Add transfection mix dropwise to HEK 293T cells, mix by swirling the plate and incubate for 24 h at 37 °C in an incubator with 5% CO_2 (*see* **Note 13**).

6. Add 8 ml fresh HEK culture medium supplemented with 10 mM sodium butyrate.

7. Harvest cell supernatant and remove dead cells and cellular debris by passing the solution over a 0.45 μm syringe filter.

8. Transfer up to 15 ml of filtered cell supernatant on top of filter unit. Spin down at 800 × *g* for 10 min (*see* **Note 14**).

9. Harvest supernatant on top of the column. If desired, adjust volume using Opti-MEM culture medium.

10. Aliquot virus and store at −80 °C.

11. For each new batch of virus, test transduction efficiency before using it in an experiment (*see* **Note 15**).

3.2   Isolation of HSPCs

1. Remove femur, tibia and iliac crest from B6.SJL-*Ptprc^a Pepc^b*/ BoyJ mice as demonstrated by Amend et al. [9] using scissors and tweezers.

2. Flush bones with RPMI medium to remove cells using a 21G 50 mm needle, 2 ml syringe and tweezer into a 15 ml Falcon tube.

3. Pipette solution up and down several times to generate a single-cell suspension.

4. Pass solution over a 70 μm cell strainer to remove cell clumps and bone fragments.

5. Spin down at 300 × *g* for 5 min and resuspend in 25 μl/mouse CD117 MicroBeads.

6. Add 125 μl RPMI medium per mouse and incubate for 5 min.

7. Add 1:100 anti-CD117 APC and incubate for another 20 min (*see* **Note 16**).

8. Wash cells by adding 2 ml of RPMI medium and spinning down at 300 × *g* for 5 min.

9. Whilst spinning down, prepare MACS columns: put columns in magnet and rinse with 500 μl RPMI medium.

10. Resuspend cells in 500 μl RPMI medium per MS column using an insulin syringe (*see* **Note 17**) and put cell suspension directly onto the column using the insulin syringe and allow all liquid to pass through the column.

11. Wash column 3× with 500 μl RPMI medium.

12. Collect CD117$^+$ cells by removing column from the magnet, placing the tip of the column in a new, clean tube, adding 1 ml RPMI medium and flushing the column by firmly pushing the plunger into the column.

13. Spin down CD117$^+$ cells at 300 × *g* and resuspend in RPMI medium.

14. Leave the CD117$^+$ cells in a closed 15 ml Falcon tube in the fridge overnight. When staining for CD34, resuspend cells in 1.7 ml RPMI medium, add 17 μl anti-CD34 antibody and incubate overnight (*see* **Note 9**).

15. The following day, proceed to staining for the other markers: wash away medium by spinning down cells at 300 × *g* and resuspending in fresh RPMI medium.

16. Add fluorescent labelled antibodies and incubate for at least 40 min (*see* **Note 8**).

17. Remove medium with antibodies by spinning down at 300 × *g* for 5 min and resuspend cells in RPMI medium at a concentration of 20 million cells per ml.

18. Sort the desired cell populations into Eppendorf tubes precoated with 100% FCS and filled with 1 ml of PBS (*see* **Note 18**).

3.3 Transduction of HSPCs with Barcode Virus and Transplantation

1. Spin down Eppendorf tubes containing HSPCs at 300 × *g* for 5 min.

2. Carefully remove any supernatant (*see* **Note 19**), resuspend HSPCs in 100 μl PBS and put up to 10^5 cells in one well of a 96-well U-bottom plate. Rinse Eppendorf tube with another 100 μl PBS and add to the well of the 96-well plate.

3. Spin down plate at 300 × *g* and resuspend cells in 100 μl HSPC medium. Add virus to the cells (*see* **Notes 15** and **20**).

4. Put plate in a plate centrifuge and spin inoculate for 1.5 h at 800 × *g* with low acceleration and deceleration.

5. Place plate in an incubator with 5% CO_2 at 37 °C for another 4.5 h.

6. During the incubation condition recipient mice with 6.0 Gy sublethal irradiation using an X-ray generator or a gamma source (*see* **Note 21**).

7. Resuspend cells and place in a new Eppendorf tube containing 1 ml of RPMI medium (*see* **Note 19**). Rinse the well of the 96-well plate with 100 μl RPMI medium to make sure you do not lose any cells.

8. Spin down cells at $300 \times g$ and resuspend in 110 μl HBSS per recipient mouse.

9. Take approximately 10% of the cells for a transduction control. Spin down and resuspend in SFEM not SFEM but HSPC medium. Culture for 3 days in an incubator with 5% CO_2.

10. Intravenously inject 100 μl of transduced HSPCs into each recipient mouse.

11. After 3 days, take the cultured HSPCs from the incubator and wash once with PBS.

12. Determine the transduction efficiency by measuring the percentage GFP^+ cells on an FACS analyzer (*see* **Note 22**).

3.4 Isolation of Target Populations

As the cell population to be isolated depends on the research question, we will not discuss protocols for isolating specific cell types. Before isolating cells, HSPCs should have enough time to reconstitute the hematopoietic system. For terminally differentiated cells, Boyer et al. provide good estimations of the time required to produce each cell type from different types of progenitor cells [10]. If necessary, fixing cell in 1% paraformaldehyde (PFA) for 30 min after FACS staining is compatible with the downstream PCR reactions. Before you start, clean working surfaces and pipettes with DNA decontaminating reagent. During all procedures ensure you minimize chances for barcode cross-contamination of samples. The isolation procedure should be performed in the room where only nonamplified barcode material is processed. Remember that each mouse should be handled separately to keep the identity of the barcodes.

1. After obtaining a single cell suspension and performing FACS staining, sort $GFP^+/CD45.1^+$ target cells, whilst making sure to include GFP^{low} cells (*see* **Note 23**). Note the number of sorted GFP^+ cells for each population as well as the remaining volume after sort if the entire tube cannot be sorted. This will allow for estimation of the total number of cells.

2. Transfer sorted cells to a 96-well plate (*see* **Note 10**), filling only every other row (*see* **Note 24**) and leaving wells empty to include negative controls (*see* **Note 25**).

3. Spin down plate with sorted cells at $500 \times g$ for 5 min. Remove all supernatant without disturbing the cell pellet and resuspend cells in 40 μl cell lysis buffer, also add cell lysis buffer to one of the negative controls.

4. Seal the plate (*see* **Note 11**) and place in a PCR cycler.

5. Lyse cells in a PCR cycler using the following program: 55 °C 120 min, 85 °C 30 min, 95 °C 5 min and 4 °C indefinitely to ensure complete lysis of the cells and Proteinase K inactivation. Samples remain stable if kept at 4 °C overnight.

6. Samples can now be frozen and stored at −20 °C until further processing.

3.5 PCR Amplification of Barcode DNA

During all procedures, prevent cross-contamination and include negative controls for each step (*see* **Note 25**). Store buffers and equipment in a dedicated pre-PCR room. Prepare buffers in a PCR hood which has been cleaned with special DNA decontaminating reagents and, if possible, sterilized under UV light. Tubes containing amplified barcodes and reagents or equipment that have been in rooms where amplified barcodes are processed should not be allowed to enter this room. Spin down plates each time before removing the plate seal to prevent cross-contamination by fluid on the seal.

1. Plan which samples to combine for a single sequencing run and plan which unique indices to use for samples and controls. If possible, sequence all samples from a single experiment together. Determine the number of expected barcode positive cells in each sequencing run by adding up the numbers you noted down during isolation. Ensure your sequencer has at least 50–100 clusters per expected barcode positive cell.

2. Thaw samples if necessary and add 160 μl of PCR1 mix to each well using a multichannel pipette. Mix carefully by pipetting the sample up and down at least five times.

3. Pipette 100 μl of the sample into a neighboring empty well to create duplicates for the PCR reaction (*see* **Note 26**).

4. Run PCR reaction 1 in a PCR cycler performing a pseudo-hot start (*see* **Note 27**): 95 °C 5 min followed by 30 cycles of 95 °C 15 s, 57.2 °C 15 s, and 72 °C 15 s followed by a final elongation step of 72 °C 10 min and a last step keeping the sample at 4 °C indefinitely.

5. Run 5 μl of the product on a 2% agarose gel to check whether the PCR reaction has been completed successfully. The expected size of the PCR product is 150 bp.

6. Aliquot 23 μl of PCR reaction mix 2 in a new 96-well plate.

7. Add 5 μl of index primers to each well. Use one unique index primer for each duplicate and make sure to note which index is used for which sample as this will allow you to identify your samples in your sequencing results.

8. Transfer plate to post-PCR room and add 5 μl of PCR product from PCR reaction 1 to each well.

9. Run PCR reaction 2, again using a pseudo hot-start (*see* **Note 27**): 5 min 94 °C followed by 30 cycles of 5 s 94 °C, 5 s 57.2 °C, and 5 s 72 °C followed by a final elongation step of 10 min 72 °C and a final step keeping the sample at 4 °C indefinitely.

10. Since amplified barcodes from each sample can now be identified by their index incorporated during PCR 2, samples to be sequenced in a single sequencing run can now be pooled (as planned above) making sure not to combine samples with the same index. Pipette 5 µl of each sample into a clean reagent reservoir.

11. Run a gel to assess whether the PCR reaction was successful for each sample using 5 µl of the remaining PCR product. Expected size of the PCR product is 231 bp.

12. Transfer pooled samples to a clean low-bind Eppendorf tube. The sample can be frozen at −20 °C or immediately processed for sequencing.

3.6 Sequencing

1. Place beads at room temperature 30 min before starting the procedure.

2. Transfer 50 µl of the pooled barcode amplicons into a clean low-bind Eppendorf tube.

3. Add 90 µl beads and mix thoroughly by vortexing for 30 s. Allow beads to bind DNA for at least 5 min.

4. Place the Eppendorf tube in the magnet and wait 2 min for the beads to clear from the solution. You should clearly see the beads close to the magnet.

5. Remove solution from the beads and discard.

6. Add 500 µl of 70% ethanol whilst keeping the tube in the magnet and without disturbing the beads.

7. Wait 30 s and remove solution.

8. Wash again by repeating **steps 6** and **7**.

9. Remove as much ethanol as possible, you can quickly centrifuge the tube, place it in the magnet again and remove the remaining ethanol.

10. Allow the tube to air-dry for 10–20 min to allow any residual ethanol to evaporate. Do not allow the beads to completely dry (the ring will appear to be cracked) as this will negatively influence the elution efficiency.

11. Elute the PCR product from the beads by adding 40 µl MilliQ, vortex for 30 s and leave at room temperature for 2 min.

12. Place the tube in the magnet and put solution in a new tube.

13. Measure DNA quality and concentration using an Agilent DNA 1000 Kit on an Agilent Bioanalyzer or a similar device.

14. Take 5 nM of DNA and add 0.5 nM Phix Illumina phage genome library to prevent the sequencer from stopping due to a too low library diversity.

15. Run sample on an Illumina sequencer with at least 50–100 clusters per barcode positive cell in single-read mode, sequencing 50 bp.

3.7 Data Processing and Analysis

Both the PCR reactions and the sequencing can induce sequencing errors, which can lead to barcodes being detected that were not really present. Therefore, we perform several filtering steps before analyzing the data to ensure that the samples are of good quality. Most steps of the analysis can be performed in multiple software. Therefore, we will only describe the steps themselves. An example of the code written in R can be found here: https://github.com/lperie/data-from-Peri-L-Cell-2016. Note that thresholds for filtering steps may differ depending on the number of barcoded cells in each sample and on the efficiency of your PCR or sequencing. We recommend running the analysis using different thresholds to ensure that changing the threshold does not change the conclusions of your experiment. During each filtering step, note which samples are removed.

1. Use Xcalibr to count how often each of the DNA barcodes is present in each sample. The software needs to be supplied with a reference file listing all barcodes in the lentiviral library (*see* **Note 28**) and another reference file listing all the indices used in PCR reaction 2. It will retrieve barcodes with a perfect match to the reference library without allowing indels or mismatches. Subsequently, it will assign retrieved barcodes to a sample based on the sample index, again without allowing mismatches and requiring a perfect match to an index from the supplied reference file.

2. In the resulting file, match the sample indices to sample names.

3. Determine for each sample the percentage of reads that could not be matched to a barcode. If you are using Xcalibr, the number of reads that does not match a barcode is provided in the row "nohit_rows". Note down which samples have over 60% nonmatched barcodes, as these contain a large number of sequencing errors and are likely low-quality samples.

4. Ensure that the negative controls are truly negative or contain only nonmatched DNA. Remove the negative controls from further analysis.

5. Gather duplicates of each sample and sum the number of reads from both samples. Remove samples with less than 1000 reads in total (*see* **Note 29**).

6. Normalize barcode reads to sum up to 10,000 reads per sample (*see* **Note 30**) to remove variation due to differences in GFP$^+$ cell numbers in each sample.

7. For each sample, plot the normalized reads for all barcodes in the two duplicates after hyperbolic arcsine transformation. A high-quality sample will show barcodes on or close to a 45° line (Fig. 4).

8. For each sample, determine the Pearson correlation between the two replicates and filter out samples with a low correlation (<0.8).

9. For each sample, remove barcodes that are present in only one duplicate. These barcodes are likely to be sequencing errors.

10. For each sample, sum reads of the two duplicates.

11. Perform hierarchical clustering on samples from all mice together using the complete linkage method on a Euclidean distance matrix of the data. The data should be log or hyperbolic arcsine transformed to allow better visualization. Plot the data as a heatmap in which each row represents a barcode and each column represents a sample. Ensure that only very few barcodes are present in more than one animal (*see* **Note 31**).

12. Perform hierarchical clustering on samples from each mouse separately. If barcodes are present in more than one sample, that means that cells from these samples originate from the same progenitor cell.

13. Perform statistical analysis, for example by performing a permutation test or determining the bootstrapping probability of clusters of interest.

4   Notes

1. We prefer to buy premade solutions to remove any risks of barcode contamination during preparation of buffers. For all reagents, use at least reagent grade ingredients. Handle and store all reagents in a room where no amplified barcode PCR products are handled.

2. The barcode library for which this protocol was optimized consists of 98 bp semirandom DNA fragments in the 3′ UTR of a GFP cassette as described by Schepers et al. [4]. Alternatively, DNA barcoding libraries can be bought ready-made. The barcode can also be incorporated into the UTR of a

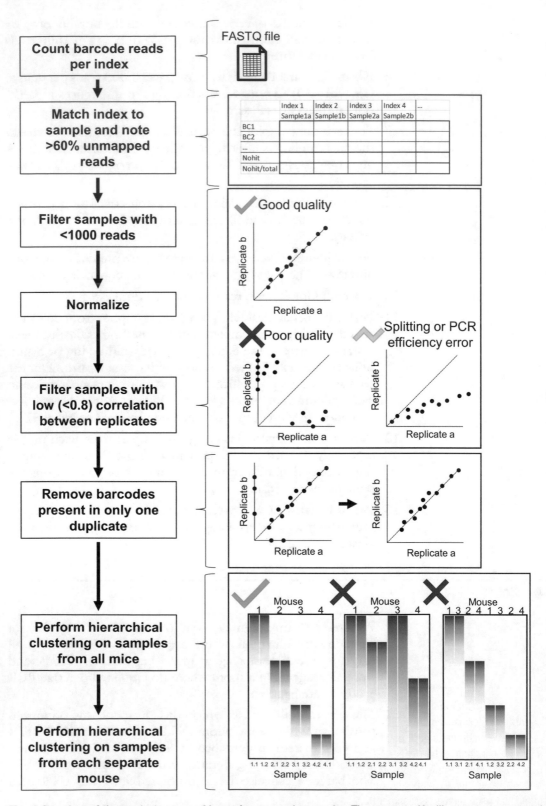

Fig. 4 Overview of the analysis steps of barcode sequencing results. The program Xcalibr generates a count matrix of reads per barcode sequence for different sample indices. These have to be matched back to the

different fluorescent protein or in a novel protein which can be identified by labelling with a fluorescently labeled antibody (e.g., CD45.1/CD45.2).

3. For safety reasons, the components of the virus are split among several plasmids. Therefore, the transfer plasmid containing the barcode library DNA needs to be cotransfected with plasmids expressing envelope (VSV-G) and packaging (RSV-REV) enzymes.

4. For experiments transplanting HSCs, we use cells from one donor mouse to transplant one recipient. For multipotent progenitors 4 (MPP4), we use one donor for two recipients. You may need to adapt this ratio for other cell types. Consider simultaneously performing experiments with different types of HSPCs to reduce the number of mice required for your experiment. CD45.1 and CD45.2 mice can be used as donor/recipient interchangeably.

5. MS columns are suited for isolation of up to 10^7 labelled cells. We recommend using one MS column for the cells of up to two mice. We achieved higher cell recovery by using two MS columns than by using one LS column. When using LS columns, adjust volumes in Subheading 3.2, **steps 9–11** to 1 ml.

6. In order to keep the cells cold while in the column, you can prechill the magnet by putting it at 4 °C for at least 1 h before use.

7. We have carefully titrated all antibodies and use all at a 1:100 dilution, except for the Pacific Blue-conjugated anti-Sca-1 antibody (1:200). Optimal antibody concentrations may vary depending on product lot or manufacturer and experimental setting.

8. Flt3 is dimly expressed, so a bright fluorochrome should be chosen for this marker. PE and APC have proven to work well. Furthermore, when including this antibody, we recommend staining for at least 40 min.

9. This antibody binds only weakly to its target. We stain with anti-CD34 Alexa Fluor 700 overnight. For APC conjugated anti-CD34 we have successfully stained cells by incubating for 90 min.

Fig. 4 (continued) input samples. Samples with a high number of unmapped reads should be noted. Read counts are normalized to allow comparison between samples. Samples with a low number of reads are omitted. The comparison of duplicates from split samples helps to distinguish true barcodes from erroneous and low-confidence barcodes. Samples with a low correlation between duplicates are filtered out and barcodes detected only in one of the two PCR duplicates are removed. Hierarchical clustering of samples from all mice allows estimation of repeat use of barcodes, while hierarchical clustering of samples from individual mice allows assessment of HSPC clonal output patterns and their frequency

10. To prevent cross-contaminations between wells during pipetting steps of the protocol we advise to use deep well 96-well PCR plates with a capacity of at least 250 μl/well.

11. To avoid cross-contaminations between wells during removal of the PCR plate seal, ensure the seal can be easily detached from the PCR plate.

12. Index primers add an index sequence of 8 bp to the PCR product. We use indices designed based on Faircloth and Glenn et al. [11] such that sequences differ by at least two bases, and homopolymers of more than 2 bp, hairpins, and complementary regions with the rest of the primer sequence are absent.

13. To avoid a skewed barcode distribution in the barcode library, it is essential to keep culture time short and not to extend the 48 h culture time.

14. Alternatively, you can concentrate virus using an ultracentrifuge by centrifuging for 90 min at 31,000 × g at 4 °C.

15. To reduce the chance of a single cell being transduced with two different barcodes, we aim for a transduction efficiency of approximately 10%. When working with cells with a lower engraftment efficiency than HSCs, going up to 20–30% transduction efficiency is acceptable. As the transduction efficiency varies for each batch of virus, the exact amount of virus being added to the cells needs to be carefully titrated.

16. When making single-stained controls for compensations, take some unstained sample before adding APC-conjugated anti-CD117 antibody.

17. In our experience insulin syringes allow for better separation of cell doublets or bigger cell aggregates than using pipette tips. If necessary, you can add an additional filtering step with a 70 μm cell-strainer.

18. Tusi et al. have extensively tested different settings and nozzle sizes optimal for sorting HSPCs [12]. In short, sorting with low pressure and a large nozzle increases HSPC viability. However, this could lead to a prohibitively long sorting time. In our hands, sorting using a 70 μm nozzle at a rate of 15,000 cells/s yielded high numbers of viable HSPCs.

19. The FCS in RPMI medium inactivates the lentivirus. Therefore, we sort our cells in PBS and ensure to remove all FCS before adding virus to the cells.

20. Although up to 100,000 cells can be transduced with 10% transduction efficiency, keep in mind that the use of 10,000 barcoded cells exceeds the capacity of the library of approximately 2,500 barcodes. The number of cells injected per mouse

should be greatly lower than the diversity of the library. Therefore, it is recommended to keep cell numbers low. We recommend to inject 100–400 barcoded cells per mouse.

21. We aim for a 50% transplantation efficiency, which is achieved on our X-ray generator with a dose of 6 Gy. Due to differences in the energy output between machines and especially between X-ray generators and gamma sources we recommend testing what radiation dose corresponds to a 50% transplantation efficiency on your own machine.

22. Depending on the safety regulations in your facility you may need to inactivate the virus before analyzing GFP expression. We incubate cells with 1% PFA in PBS for 30 min at room temperature.

23. Large amounts of DNA can negatively affect the efficiency of the PCR reaction. On the other hand, you do not want to miss any GFP^{low} cells, so it is better to include some negative cells (Fig. 1).

24. Filling only every other row of wells in a 96-well plate reduces the chance of contaminating neighboring wells when splitting samples for PCR **step 1**. If you have too many samples/too few PCR cyclers to fill only half of each PCR plate you can of course completely fill the plate. However, take extra care during the splitting step to minimize the chance of cross-contamination.

25. Due to the sensitivity of the assay, contamination can happen even when taking precautions. Therefore, always include negative controls at every step of the sample processing to ensure you find and identify this contamination as soon as it happens. We include three negative controls, the first including the cell lysis and onward, the second including PCR reaction 1 and onward, and the third including only PCR reaction 2.

26. This step has a high risk for cross-contamination. Use a multichannel pipette and prevent moving used pipette tips over the plate as much as possible. For example, change the setting of the pipette from 160 µl to 100 µl, while keeping the pipette above the wells you are currently pipetting into.

27. A pseudo hot-start is performed by allowing the PCR machine to reach 95 °C before inserting the sample.

28. The barcode reference list contains all barcodes present in the library. It is established before the first use by duplicate sequencing of the barcode plasmid library after PCR amplification as described in the protocol. Resulting read numbers are normalized between replicates, sequences present in both replicates are sorted by frequency and the cumulative read number is determined. Barcodes contributing less than 92.5% of the cumulative read number are discarded. Barcodes over

1.5-fold more prevalent in one replicate as well as the most abundant barcodes are discarded. For the described library the reference list encompasses 2,609 barcodes.

29. The number of 1,000 reads corresponds to a total of only 10–20 barcoded cells in your sample. If you had a very low number of GFP$^+$ cells during the sort, you can reduce this number, but you should be aware that this may affect the quality of the data. Preferably, test multiple values for filtering and compare the results. Different filtering thresholds should not lead to different conclusions, unless the threshold is set very high or very low.

30. The value of 10,000 is chosen arbitrarily.

31. Considering that we adapt the number of cells transduced and transplanted per mouse to the library size, the number of barcodes present in more than one mouse should be low. If many barcodes are shared between mice, this could indicate that HSPCs have undergone division after transduction and before transplantation. Alternatively, it could indicate cross-contamination.

References

1. Walsh C, Cepko CL (1992) Widespread dispersion of neuronal clones across functional regions of the cerebral cortex. Science 255 (5043):434–440. https://doi.org/10.1126/science.1734520

2. Golden JA, Fields-Berry SC, Cepko CL (1995) Construction and characterization of a highly complex retroviral library for lineage analysis. Proc Natl Acad Sci U S A 92:5704–5708

3. Lu R, Neff NF, Quake SR et al (2011) Tracking single hematopoietic stem cells in vivo using high-throughput sequencing in conjunction with viral genetic barcoding. Nat Biotechnol 29:928–933. https://doi.org/10.1038/nbt.1977

4. Schepers K, Swart E, van Heijst JWJ et al (2008) Dissecting T cell lineage relationships by cellular barcoding. J Exp Med 205:2309–2318. https://doi.org/10.1084/jem.20072462

5. Naik SH, Perié L, Swart E et al (2013) Diverse and heritable lineage imprinting of early haematopoietic progenitors. Nature 496:229–232. https://doi.org/10.1038/nature12013

6. Perié L, Duffy KR, Kok L et al (2015) The branching point in Erythro-myeloid differentiation. Cell 163:1655–1662. https://doi.org/10.1016/j.cell.2015.11.059

7. Lin DS, Kan A, Gao J et al (2018) DiSNE movie visualization and assessment of clonal kinetics reveal multiple trajectories of dendritic cell development. Cell Rep 22:2557–2566. https://doi.org/10.1016/j.celrep.2018.02.046

8. Merino D, Weber TS, Serrano A et al (2019) Barcoding reveals complex clonal behavior in patient-derived xenografts of metastatic triple negative breast cancer. Nat Commun 10:766. https://doi.org/10.1038/s41467-019-08595-2

9. Amend SR, Valkenburg KC, Pienta KJ (2016) Murine hind limb long bone dissection and bone marrow isolation. J Vis Exp 110:53936. https://doi.org/10.3791/53936

10. Boyer SW, Rajendiran S, Beaudin AE et al (2019) Clonal and quantitative in vivo assessment of hematopoietic stem cell differentiation reveals strong erythroid potential of multipotent cells. Stem Cell Rep 12:801–815. https://doi.org/10.1016/j.stemcr.2019.02.007

11. Faircloth BC, Glenn TC (2012) Not all sequence tags are created equal: designing and validating sequence identification tags robust to Indels. PLoS One 7:e42543. https://doi.org/10.1371/journal.pone.0042543

12. Khoramian Tusi B, Socolovsky M (2018) High-throughput single-cell fate potential assay of murine hematopoietic progenitors in vitro. Exp Hematol 60:21–29.e3. https://doi.org/10.1016/j.exphem.2018.01.005

Chapter 22

Single-Cell Analysis of Hematopoietic Stem Cells

Katherine H. M. Sturgess, Fernando J. Calero-Nieto, Berthold Göttgens, and Nicola K. Wilson

Abstract

The study of hematopoiesis has been revolutionized in recent years by the application of single-cell RNA sequencing technologies. The technique coupled with rapidly developing bioinformatic analysis has provided great insight into the cell type compositions of many populations previously defined by their cell surface phenotype. Moreover, transcriptomic information enables the identification of individual molecules and pathways which define novel cell populations and their transitions including cell lineage decisions. Combining single-cell transcriptional profiling with molecular perturbations allows functional analysis of individual factors in gene regulatory networks and better understanding of the earliest stages of malignant transformation. In this chapter we describe a comprehensive protocol for scRNA-Seq analysis of the mouse bone marrow, using both plate-based (low throughput) and droplet-based (high throughput) methods. The protocol includes instructions for sample preparation, an antibody panel for flow cytometric purification of hematopoietic progenitors with index sorting for plate-based analysis or in bulk for droplet-based methods. The plate-based protocol described in this chapter is a combination of the Smart-Seq2 and mcSCRB-Seq protocols, optimized in our laboratory. It utilizes off-the-shelf reagents for cDNA preparation, is amenable to automation using a liquid handler, and takes 4 days from preparation of the cells for sorting to producing a sequencing-ready library. The droplet-based method (using for instance the $10\times$ Genomics platform) relies on the manufacturer's user guide and commercial reagents, and takes 3 days from isolation of the cells to the production of a library ready for sequencing.

Key words Hematopoiesis, scRNA-Seq, Transcriptome, Gene regulatory networks, mcSCRB-Seq, Smart-Seq2, mcSmart-Seq2, Hematopoietic stem cell, Hematopoietic stem/progenitor cell, Bone marrow

1 Introduction

Hematopoiesis is characterized by a hierarchical differentiation process, wherein stem cells and progenitors choose either to self-renew, or enter differentiation pathways toward more than 10 distinct cell lineages. Traditional descriptions of this differentiation hierarchy have depended on immunophenotyping, using combinations of several cell-surface markers to delineate cellular populations with defined differentiation and self-renewal potentials. Sorting

Marion Espéli and Karl Balabanian (eds.), *Bone Marrow Environment: Methods and Protocols*, Methods in Molecular Biology, vol. 2308, https://doi.org/10.1007/978-1-0716-1425-9_22, © Springer Science+Business Media, LLC, part of Springer Nature 2021

strategies have been refined over decades with the identification of new combinations of cell surface markers, but the purified populations remain functionally heterogeneous. For instance, only a proportion of "phenotypic hematopoietic stem cells" display long term multilineage reconstitution properties. The application of single-cell RNA sequencing (scRNA-Seq) is transforming our ability to classify the hematopoietic system in detail. Such analyses permit interrogation of molecular profiles at an unprecedented resolution. The transcriptome-wide coverage allows unbiased correlation of self-renewal and differentiation properties with specific gene signatures, providing insight into mechanisms controlling cell state [1–4]. Computational analysis of thousands of transcriptomes can identify rare novel populations which would be lost in bulk analysis, while also allowing for an overview at tissue or whole organ level [3, 5]. Transcriptomic analysis of hematopoiesis has demonstrated that most populations defined by cell-surface phenotype are indeed heterogeneous and revealed that lineage decisions are made earlier than previously anticipated [2, 6–9].

Starting from the premise that scRNA-Seq captures a snapshot of cells at various stages of a differentiation trajectory, computational methods can arrange single-cell profiles into coherent landscapes progressing from a multipotent state through unilineage states to mature populations. This, in turn, enables identification of putative branchpoints, at which cells "choose" a specific differentiation potential [10–12]. Using this approach, entry points into distinct hematopoietic lineages have been identified, resulting in a comprehensive single-cell landscape of the hematopoietic compartment within mouse bone marrow [8, 13, 14]. Navigating this landscape allows for observation of transitions and inference of cellular origins, while unbiased capture of the entire transcriptome uncovers previously inaccessible and novel molecular pathways and gene combinations defining distinct cell transitions. A comparative analysis of specific cell populations facilitates identification of new cell surface markers for prospective cell isolation and correlative functional analysis [2]. The combination of single-cell transcriptomics with genetic perturbations allows for an insight into the earliest stages of leukemogenesis, and molecular cooperation which may underly disease states [13, 15, 16]. Using high-throughput droplet-based methods, hundreds of thousands of single cells can be analyzed in a time- and cost-efficient manner, generating sufficient resolution to draw conclusions about even very rare populations [17, 18]. High-throughput techniques (e.g., InDrops [19], Drop-Seq [20], 10× Genomics [21], and SPLiT-seq [22]) allow for profiling of over 10,000 cells per experiment. While these methods tend to detect fewer genes per cell, the higher throughput permits analysis of a vast diversity of cell-types at a much higher scale, allowing refinement of cellular populations. Conversely, lower-throughput plate-based techniques (e.g., Smart-Seq2 [23],

Smart-seq3 [24], mcSCRB-Seq [25], Quartz-Seq2 [26], CEL-Seq2 [27], MARS-seq [28], and RamDA-seq [29]) are very effective to profile fewer cells per experiment (in the range of 100 s) but with a higher sensitivity for the detection of genes per cell, offer the advantage of index sorting data (a feature which enables the recording of every measured parameter by the sorter) which can be integrated with transcriptomic data. Some plate based methods additionally offer the advantage of full-length transcript coverage.

Modifications of these techniques have permitted multimodal measurements, such as transcriptomics alongside genomic or epigenomic profiling in a single cell in so-called multiomics protocols [30]. One of the most clinically relevant applications thus far has been the combined analysis of single-cell gene expression and mutation status. Giustaccini et al. modified the Smart-Seq2 protocol to simultaneously genotype for the BCR-ABL fusion gene and capture transcriptomes of over 2000 stem cells from chronic myeloid leukemia patients [31]. This allowed for characterization of a subgroup of BCR-ABL mutant stem cells which persist through therapy, and also transcriptionally aberrant wild type HSCs present in CML patients whose presence predict the response to therapy. TARGET-Seq, a further refinement of this protocol from the Mead group, allows parallel targeted mutation analysis from cDNA and gDNA in addition to scRNA-Seq [32]. Applied to over 4500 hematopoietic stem/progenitor cells (HSPCs) from patients with myeloproliferative disorders, this technique reveals clonal heterogeneity correlated with aberrant transcriptional profiles, and can resolve the order of acquisition of mutations.

A further study of acute myeloid leukemia (AML) has employed nanowell technology and nanopore long read sequencing to integrate targeted genotyping with transcriptional profiling [33]. While genotyping data remained relatively sparse, integrating these with single-cell transcriptomes from over 38,000 cells in identified six malignant AML phenotypes which shared features with healthy HSPCs. Intratumoral diversity can be dissected by this technique as demonstrated by the detection of different differentiation states associated with FLT3-ITD and FLT3-TKI mutations in the same patient.

Nam et al. [34] modified the 10x Genomics scRNA-Seq protocol by spiking in primers for specific mutant transcripts associated with myeloproliferative disorders, which allowed the authors to demonstrate the progenitor-specific fitness effects of driver mutations and correlate these with patient phenotypes. The application of single-cell transcriptomics and mutational profiling techniques to cancer biopsy samples illuminates the molecular programs underlying carcinogenesis and clonal evolution. The ability to separate tumor cells from their nonmutant counterparts also extends insights into the tumor microenvironment and non–cell-autonomous effects of driver mutations.

Here we present comprehensive experimental protocols for plate-based scRNA-Seq (mcSmart-Seq2) and droplet-based scRNA-Seq (using the 10x Genomics platform). These techniques are complementary and can be combined in a single experiment if sample quantity permits. A droplet-based workflow has several advantages; its higher throughput provides an overview of organs at tissue level, the high cell number yield confers greater power to detect rare cell populations, and finally the use of unique molecular identifiers reduces amplification-related noise. However, the method provides only minimal splicing data. By default, the $10\times$ Genomics platform does not allow genotyping of mutant cells in mixed populations, but several labs have tried to enable this by adding further processing steps [34, 35]. By contrast, plate-based techniques are more labor-intensive, expensive and the number of cells which can be profiled is more limited. Nevertheless, the method provides full length transcript sequencing, which requires a greater read depth but confers increased sensitivity and permits more even coverage with the ability to distinguish RNA isoforms. The plate-based approach is generally better suited when looking at a specific population of cells or when cell numbers are a limiting factor.

The plate-based scRNA-Seq protocol that we describe in this chapter is a combination of two previously published protocols from Picelli et al. [23] and Bagnoli et al. [25]. This protocol has also been successfully used for small pools of cells (up to 100) without any modification to the reagents; only reducing the number of PCR cycles for the amplification of the cDNA library will be required, to account for larger starting material. The plate-based method described in this chapter captures polyadenylated transcripts and the signal detected encompasses the full length cDNAs. As with all polyA capture methods, the obtained sequencing reads are $3'$ biased, but it is possible to analyze all the exons of highly expressed genes, different spliced variants and even the spliced vs unspliced transcripts (although the latter is a small proportion of the detected material). This method takes advantage of template-switching during the reverse transcriptase reaction and therefore the choice of reverse transcriptase (RT) is essential as not all RT enzymes are able to add untemplated nucleotides onto the end of the cDNA molecule.

While this protocol is plate based and therefore ultimately more expensive per cell than the droplet methods, it allows the retention of additional metadata (such as index sorted data) which the droplet methods do not. Of note, droplet methods can now be combined with oligo-labeled antibodies therefore allowing surface phenotype to be measured in parallel with the transcriptome [36]. Two major advantages of this plate-based protocol are that all the cDNA amplification steps use off-the-shelf reagents and the entire protocol takes around 4 days, from preparation of the cells for sorting to producing a final library ready to be sequenced. While we present a manual version of the protocol, automation could be implemented at several stages (Subheadings 3.3.5 and 3.3.8). Automation at the cleanup stages will ensure that repetitive pipetting steps are performed uniformly, and miniaturization of the library preparation reaction offers large savings in commercial reagents.

2   Materials

2.1   Bone Marrow Harvest

1. PBS (calcium and magnesium free) supplemented with 2% heat-inactivated FCS.
2. Pestle and mortar.
3. 15 ml centrifuge tube.
4. 50 μM filter.
5. p1000 pipette.
6. p200 pipette.
7. p20 pipette.
8. p2 pipette.
9. p1000 filter tips.
10. p200 filter tips.
11. p20 filter tips.
12. p2 filter tips.
13. Pipette boy.
14. Stripettes.
15. Ammonium chloride.
16. 5 ml round bottom polystyrene tube.
17. 5 ml round bottom polypropylene tube.
18. Hematopoietic progenitor enrichment kit (For example Easy-Sep™ Mouse Hematopoietic Progenitor Cell Enrichment Kit, Stemcell Technologies).
19. Magnet (this will depend on the progenitor enrichment kit used—e.g., EasySep™ Magnet, Stemcell Technologies).
20. Antibodies (*see* Subheading 2.2).
21. 1.5 ml Eppendorf Tubes.
22. FACs sorter.

2.2   Antibodies

1. For dilutions of the individual antibodies used see Table 2 and **Note 1**.
2. Mouse Hematopoietic Progenitor Isolation Cocktail PN (19856C) part of the EasySep™ Mouse Hematopoietic Progenitor Cell Enrichment Kit (Stemcell Technologies).

Table 2
Detailed description of antibodies used in this protocol

Tube	Fluorophore	Antibody	Clone	Vol (μl)/ 100 μl
Lineage	–	Mouse hematopoietic progenitor isolation cocktail[a]	–	1
EPCR	PE	EPCR	RMEPCR1560	0.3
CD48	APC	CD48	HM48-1	0.5
CD150	PE-Cy7	CD150	TC15-12F12.2	0.5
c-kit	APC-Cy7	c-kit	2B8	0.5
Sca 1	BV421	Sca 1	D7	1
CD45	FITC	CD45	30-F11	0.5
Viability	7AAD		–	0.1
			TOTAL	4.3

[a]Mouse Hematopoietic Progenitor Isolation Cocktail PN (19856C) part of the EasySep™ Mouse Hematopoietic Progenitor Cell Enrichment Kit (Stemcell Technologies)

3. EPCR PE, Clone RMEPCR1560.

4. CD48 APC, Clone HM48-1.

5. CD150 PE-Cy7, Clone TC15-12F12.2.

6. c-kit APC-Cy7, Clone 2B8.

7. Sca1 BV421, Clone D7.

8. CD45 FITC, Clone 30-F11.

9. BV510 Streptavidin.

10. 7AAD.

2.3 scRNA-Seq

1. qPCR machine.

2. Spectro fluorometer or fluorescence microplate reader.

3. Bioanalyzer or TapeStation (e.g., from Agilent).

4. Temperature controlled Microcentrifuge.

5. Temperature controlled benchtop centrifuge (15 ml tube).

6. 96-well PCR plate (e.g., nonskirted Starlab or skirted Framestar 96 4titude).

7. Adhesive film lid (*see* **Note 2**).

8. Optical caps for 96-well plate (qPCR).

9. 8-strip 0.2 ml PCR tubes with cap.

10. 15 ml centrifuge tubes.

11. Index Plate Fixture (Illumina).

12. Microplate, 384-well, PS, Flat-bottom, Fluotrac, medium binding, black (Greiner Bio-One).

13. Magnetic stand (e.g., Life Technologies DynaMag-96 side).

14. Magnetic stand (e.g., Life Technologies DynaMag-2).

15. 8-channel multichannel pipette 1–10 µl (pre-amp).

16. 8-channel multichannel pipette 1–10 µl (post-amp).

17. 8-channel multichannel pipette 5–50 µl (post-amp).

18. 8-channel multichannel pipette 3–300 µl (post-amp).

19. 12-channel multichannel pipette 1–10 µl (post-amp).

20. Eppendorf Multipette® E3 (Eppendorf)—optional.

21. p1000 pipette.

22. p200 pipette.

23. p20 pipette.

24. P2 pipette.

25. p1000 filter tips low retention.

26. p200 filter tips low retention.

27. p20 filter tips low retention.

28. p2 filter tips low retention.

29. Index Adapter Replacement Caps (Illumina).

30. 1× PBS (Calcium and Magnesium free) supplemented with 0.04% BSA solution (400 µg/ml).

31. 10% Triton X-100.

32. DEPC treated Distilled water.

33. Distilled water—PCR grade.

34. RNase Inhibitor (e.g., SUPERase-In, Ambion).

35. dNTP mix (10 mM each).

36. ERCC RNA Spike-In Mix (ThermoFisher) (*see* **Note 3**).

37. Maxima H minus Reverse Transcriptase (200 U/µl) (ThermoFisher).

38. Terra PCR Direct Polymerase Mix (250 U) (Takara Bio).

39. AMPure XP beads (Beckman Coulter).

40. Elution buffer (EB) solution: 10 mM Tris–Cl, pH 8.5.

41. 80% Ethanol (*see* **Note 4**).

42. Bioanalyzer High Sensitivity DNA reagents or TapeStation High Sensitivity DNA Reagents (Agilent).

43. Quant-iT™ PicoGreen double stranded DNA assay kit (ThermoFisher).

44. 1 × TE.

45. Nextera XT DNA sample preparation kit 96 samples (Illumina).

46. Nextera XT Index Kit v2 (96 Indexes, 384 samples) (Illumina).

47. KAPA Library Quantification Kit (Roche).

48. KAPA dilution buffer: 10 mM Tris–HCl, pH 8.0–8.5 + 0.05% Tween® 20.

2.4 Oligonucleotides

1. Oligos—for details of sequence for individual oligos *see* Table 1. Oligo-dT30VN and IS PCR oligo were ordered from Integrated DNA Technologies, Inc. (IDT). LNA-modified TSO oligo was ordered from Qiagen. All oligos were ordered with HPLC purification. Avoid repeated freezing-thawing cycles for all oligos.

2. Resuspend oligonucleotides in 1× TE as described in Table 1.

3. Oligos can be stored for up to 6 months at −80 °C.

3 Methods

3.1 Sample Preparation

1. Cool down the centrifuge before use (4–8 °C).

2. Ensure PBS + 2% FCS and Ammonium chloride are precooled.

3.1.1 Harvest Bone Marrow

1. Sacrifice the mouse/mice according to local rules.

2. Dissect femur, tibia, and pelvic bones ensuring to remove all tissue.

3. Harvest cells from femurs, tibia, and pelvic bones of a mouse by crushing bones with a pestle and mortar in 3 ml cold PBS + 2% FCS (*see* **Note 5**).

4. Place 50 µM filter onto a 15 ml centrifuge tube (*see* **Note 6**).

5. Ensure complete resuspension of the crushed bones by pipetting the PBS + 2% FCS up and down using a p1000 pipette.

6. Using a p1000 pipette collect cell suspension and filter into a 15 ml centrifuge tube.

7. Add 3 ml of cold PBS + 2% FCS to mortar and pipette up and down with a p1000 pipette. Filter solution into 15 ml centrifuge tube, combining all bone marrow.

8. Repeat **step 7** (total of two washes).

9. Top up to final volume of 10 ml in cold PBS + 2% FCS.

10. Centrifuge for 5 min at 300 × *g* at 4 °C.

11. Discard supernatant and resuspend in 3 ml of cold PBS + 2% FCS.

Table 1
Oligos used for scRNA-Seq protocol

Oligo	Sequence (5′ – 3′)	Concentration for stock (μM)	Resuspend in	Storage (°C)	Time in storage (months)
TSO	AAGCAGTGGTATCAACGCAGAGTACATrGrG +G	100	1 x TE	−80	6
Oligo-dT30VN	AAGCAGTGGTATCAACGCAGAGTAC(T30) VN	100	1 x TE	−80	6
ISPCR oligo	AAGCAGTGGTATCAACGCAGAGT	100	1 x TE	−20	6

12. Lyse the red blood cells by adding 5 ml of cold Ammonium chloride.

13. Shake well or vortex, then incubate on ice for 5 min, invert the tube again or vortex and then continue to incubate on ice for a further 5 min (10 min total) (*see* **Note 7**).

14. Top up to 15 ml with cold PBS + 2% FCS (to dilute the Ammonium Chloride) (add ~7 ml).

15. Centrifuge for 5 min at $300 \times g$ at 4 °C.

16. Discard supernatant and resuspend the pellet in 500 µl of cold PBS + 2% FCS.

17. Transfer all the bone marrow sample to a 5 ml round bottom polystyrene tube and place on ice.

18. Keep the centrifuge tube as remaining cells in the centrifuge tube will be used for antibody single stain controls (remaining volume and volume/cells stuck to walls of tube).

3.1.2 Single Antibody Staining Controls

1. The volume in which to resuspend the controls will depend on the number of different antibodies used for FACS; add 100 µl cold PBS + 2% FCS per control to the 15 ml centrifuge tube which contained the red cell lysed bone marrow (retained in **step 18** of Subheading 3.1.1) (*see* **Note 8**).

2. The controls are not lineage depleted.

3. Add 1 ml cold PBS + 2% FCS for the 10 control tubes to the 15 ml centrifuge tube which contained all the red cell lysed bone marrow (retained in **step 18** of Subheading 3.1.1).

4. Place a 50 µM filter into 15 ml centrifuge tube.

5. Filter samples into 15 ml centrifuge tube (*see* **Note 9**).

6. Label control tubes (unstained, all individual antibodies as in Table 2 and all stain control).

7. Aliquot 100 µl into each of the 10 tubes (unstained, single stains, all stain control).

8. Add antibodies as described (*see* Table 2), for all stain control add 4.3 µl of Antibody Master Mix (*see* **Note 10**).

9. Mix and incubate on ice, in the dark for 30 min.

10. After incubation, wash by adding 1 ml of cold PBS + 2% FCS to each tube (except the viability stain or unstained sample as these contain no antibodies).

11. Centrifuge samples at $300 \times g$ for 5 min at 4 °C.

12. Remove supernatants, being careful not to disturb the cell pellet. The cell pellet can be almost invisible so a small volume of supernatant can be left behind.

13. Resuspend the cell pellet by flicking the tubes.

14. Keep "Lineage" single stain and "All stain" control tubes to one side.

15. Resuspend remaining controls in 500 µl cold PBS + 2% FCS.

16. Place on ice and in darkness.

17. To "Lineage" single stain and "All stain" control add 100 µl cold PBS + 2% FCS.

18. Add 0.5 µl Streptavidin conjugated antibody (BV510 Streptavidin).

19. Mix and incubate on ice for 15 min.

20. Add 1 ml of cold PBS + 2% FCS and centrifuge at 300 × *g* for 5 min at 4 °C.

21. Remove supernatants, being careful not to disturb the cell pellet. The cell pellet can be almost invisible so a small volume of supernatant can be left behind.

22. Resuspend the cell pellet by flicking the tubes.

23. Resuspend in 500 µl of cold PBS + 2% FCS.

24. Add 0.5 µl 7AAD to viability single stain and "All stain" control tubes (*see* **Note 11**).

25. These control samples are now ready to be used to set voltages and gates on the flow cytometry sorter.

3.1.3 Lineage Depletion

1. Lineage depletion protocol (*see* **Note 12**).

2. Use EasySep™ Mouse Hematopoietic Progenitor Cell Enrichment Kit (Stemcell Technologies).

3. Retrieve samples from ice (**step 18** of Subheading 3.1.1).

4. Make samples up to an equal volume if processing more than one mouse (normally 600 µl).

5. Add 1:100 of EasySep™ Mouse Hematopoietic Progenitor Isolation Cocktail. Mix and incubate on ice for 15 min.

6. Vortex EasySep™ Streptavidin RapidSpheres™.

7. Ensure the particles are in a uniform suspension with no visible aggregates.

8. Add 1:25 EasySep™ Streptavidin RapidSpheres™.

9. Mix and incubate on ice for 10 min.

10. Add cold PBS + 2% FCS to bring volume up to 2.5 ml, pipette up and down, and place the tube (without lid) into EasySep™ magnet.

11. Incubate for 3 min at room temperature.

12. Pick up magnet, bracing the tube against the side of the magnet using index finger, and in one continuous motion pour into a new 5 ml round bottom tube (*see* **Note 13**). Do not shake or blot.

Table 3
Detailed description of antibodies used to prepare the master mix for this protocol

Fluorophore	Antibody	Clone	Vol (μl)/ 100 μl	Master mix (500 μl) (*see* Note 14)
–	Mouse hematopoietic progenitor isolation cocktail[a]	–	1	5
PE	EPCR	RMEPCR1560	0.3	1.5
APC	CD48	HM48-1	0.5	2.5
PE-Cy7	CD150	TC15-12F12.2	0.5	2.5
APC-Cy7	c-kit	2B8	0.5	2.5
BV421	Sca 1	D7	1	5
FITC	CD45	30-F11	0.5	2.5
7AAD		–	0.1	
		TOTAL	4.3	

[a]Mouse Hematopoietic Progenitor Isolation Cocktail PN (19856C) part of the EasySep™ Mouse Hematopoietic Progenitor Cell Enrichment Kit (Stemcell Technologies)

13. The supernatant contains the enriched bone marrow cell suspension (lineage depleted).

14. Centrifuge for 5 min at 300 × *g* at 4 °C.

15. Resuspend in 300 μl of cold PBS + 2% FCS.

3.1.4 Specific Antibody Staining of Samples

1. Add antibody master mix to sample (*see* Table 3), mix and incubate on ice, in the dark for 30 min.

2. After incubation, wash by adding 3 ml of cold PBS + 2% FCS.

3. Centrifuge samples at 300 × *g* for 5 min at 4 °C.

4. Remove supernatant, being careful not to disturb the cell pellet.

5. Resuspend the cell pellet by flicking the tube.

6. Resuspend sample in 300 μl cold PBS + 2% FCS.

7. Add 1.5 μl Streptavidin conjugated antibody (BV510 Streptavidin).

8. Mix and incubate on ice for 15 min.

9. Add 3 ml of cold PBS + 2% FCS and centrifuge at 300 × *g* for 5 min at 4 °C.

10. Remove supernatant, being careful not to disturb the cell pellet.

11. Resuspend the cell pellet by flicking the tube.

12. Resuspend in 500 µl of cold PBS + 2% FCS.

13. Place a 50 µM filter onto a new 5 ml round bottom tube.

14. Filter antibody stained enriched bone marrow sample into 5 ml round bottom tube.

15. Wash the original 5 ml round bottom tube with an additional 500 µl cold PBS + 2% FCS by pipetting up and down.

16. Filter supernatant into tube containing first supernatant of antibody stained enriched bone marrow sample.

17. Add 1 µl 7AAD to sample (*see* **Note 11**).

18. The sample is now ready to be sorted.

19. Specifications for each FACS sorter will be specific to the individual machine, please speak to your local flow cytometry expert for details.

20. Ensure to use index sorting mode on the sorting machine to allow retrospective review of the surface phenotype of each individual single cell sorted into a 96-well plate [37].

21. Each plate should be sealed with an adhesive film lid and clearly labeled. If cell quantities permit, cells prepared as above can be FACS sorted in bulk for droplet-based scRNA-Seq 10× Genomics (Subheading 3.2) in parallel to index sorting individual cells for plate-based scRNA-Seq (Subheading 3.3).

3.2   Droplet-Based scRNA-Seq Using the 10x Genomics Platform

1. Droplet-based scRNA-Seq (*see* **Note 15**).

2. Collect sorted cells for this experiment into 1.5 ml Eppendorf tube(s) containing 200 µl cold PBS + 2% FCS.

3. FACS sort cells in bulk (*see* **Note 16**).

4. Harvest cells by centrifuging at $300 \times g$ for 7 min at 4 °C.

5. Remove supernatant, taking care not to touch the cell pellet (normally not visible).

6. Flick the tube to resuspend the cells.

7. Resuspend the cells in 1× PBS + 0.04% BSA solution (400 µg/ml) (*see* **Note 17**).

8. Process the samples following the 10x Genomics user guide.

3.3   Plate Based scRNA-Seq Using Molecular Crowding Smart-Seq2 (mcSmart-Seq2)

Note that the protocol is adapted from Smart-Seq2 [23] and mcSCRB-Seq [25] (*see* **Note 18**). It is essential that all steps are performed at 4 °C, unless otherwise stated. Cool down the centrifuge before use (4–8 °C). All quantities in master mixes are in µl (*see* **Note 19**). Use a thermal cycler with a preheated lid set to 105 °C for all incubations. Care should be taken to ensure that the same batch(s) of reagents are used for the entirety of a standalone experiment to reduce any potential batch effects.

3.3.1 Single-Cell Sorting

1. This step should be performed in preamplification conditions.
2. Prepare lysis buffer, with reagents maintained on ice during preparation.

	×1 well	× 100 wells
RNase Inhibitor (SUPERase-in) 20 U/µl	0.115	11.5
10% Triton X-100	0.046	4.6
DEPC-treated or RNase-free H$_2$O	2.139	213.9
	2.3	230

3. Aliquot 2.3 µl of lysis buffer into each well of a 96-well PCR plate (*see* **Notes 20** and **21**).
4. Cover plates with adhesive film lid.
5. Centrifuge at 700 × *g* for 1 min in a precooled centrifuge.
6. Keep plates on ice or in refrigerator until required (*see* **Note 22**).
7. Isolate single cells into 2.3 µl of lysis buffer by FACS (*see* **Notes 23–25**).
8. Cover the plates with an adhesive film lid.
9. Ensure that each plate is labeled clearly.
10. Vortex well (*see* **Note 26**).
11. Centrifuge the plates in a precooled centrifuge at 700 × *g* for 2 min.
12. Store at −80 °C for up to 6 months (*see* **Note 27**).

3.3.2 Cell Lysis and Hybridization

1. This step should be performed in preamplification conditions.
2. Prepare the annealing mixture, with reagents maintained on ice during preparation (*see* **Note 28**).

	×1 well	×100 wells
ERCC (dilution 1:300,000)[a]	0.1	10
Oligo-dT (100 µM)	0.02	2
dNTP (10 mM)	1	100
DEPC-treated or RNase-free H$_2$0	0.88	88
	2	200

[a]Dilution of ERCCs is batch specific; start with 1:300,000 (*see* **Note 2**)

3. Add 2 µl of annealing mix per well (Vf = 4.3 µl) (*see* **Notes 29** and **30**).
4. Cover the plates with a new adhesive film lid.

5. Centrifuge at 700 × g briefly (ensure that all the reagents are at the bottom of the well).

6. Incubate the samples at 72 °C for 3 min and immediately place the samples on ice.

7. Centrifuge at 700 × g briefly.

3.3.3 Reverse Transcription

1. This step should be performed in preamplification conditions.

2. Prepare the reverse transcription mix, with reagents maintained on ice during preparation.

	×1 well	×100 wells
Maxima H Minus (200 U/μl)	0.1	10
RNase inhibitor (20 U/μl)	0.25	25
5× Maxima RT buffer	2	200
TSO (100 μM)	0.2	20
PEG 8000 (40% v/v)	1.875	187.5
DEPC-treated or RNase-free H_2O	1.275	127.5
	5.7	570

3. Remove and discard the cover.

4. Add 5.7 μl of reverse transcription mix per well (Vf = 10 μl) (*see* **Note 30**).

5. Cover the plates with an adhesive film lid.

6. Centrifuge at 700 × g briefly (ensure that all the reagents are at the bottom of the well).

7. Run in the thermocycler with preheated lid.

Step	Temperature (°C)	Time	Cycle
RT and template switching	42	90 min	1
Enzyme inactivation	70	15 min	1
Storage	4	Hold	1

8. Centrifuge at 700 × g for 1 min.
   Samples can be frozen at this stage, although it is preferable to continue to PCR stage.

3.3.4 PCR Preamplification

1. This step should be performed in preamplification conditions.

2. If samples have been frozen, defrost on ice and centrifuge at 700 × g for 1 min prior to PCR preamplification.

3. Prepare the PCR mix, with reagents maintained on ice during preparation.

	×1 well	×100 wells
Terra PCR direct polymerase (1.25 U/μl)	1	100
2× Terra PCR direct buffer	25	2500
IS PCR primer (10 μM)	1	100
PCR grade dH₂O	13	1300
	40	4000

For preparation of IS PCR primer *see* **Note 31**.

4. Remove and discard the adhesive film lid.

5. Add 40 μl PCR mix per well (Vf = 50 μl) (*see* **Note 30**).

6. Cover the plates with a new adhesive film lid.

7. Centrifuge at 700 × *g* briefly.

8. Perform the PCR in a thermocycler (with preheated lid) following the program:

Step	Temperature (°C)	Time	Cycle(s)
Denature	98	3 min	1
Denature	98	15 s	19
Anneal	65	30 s	
Extend	68	4 min	
Extend	72	10 min	1
Storage	4	Hold	1

Number of PCR cycles (*see* **Note 32**).

9. Centrifuge at 700 × *g* for 1 min.

PCR products can be stored at −20 °C for 6 months or −80 °C for longer.

3.3.5 PCR Purification

1. This step should be performed in postamplification conditions.

2. Pipetting of PCR purification steps (*see* **Note 33**).

3. Equilibrate AMPure XP beads to room temperature for at least 15 min.

4. If samples have been frozen, defrost on ice and centrifuge at 700 × *g* for 1 min prior to PCR purification.

5. Vortex room temperature equilibrated AMPure XP beads until all beads are resuspended (approximately 30 s).

6. Remove and discard the adhesive film lid.

7. Add 29 μl of AMPure XP beads to each sample (for 0.6× cleanup) (*see* **Note 34**).

8. Mix by pipetting up and down ten times or until the solution appears homogeneous.

9. Incubate the mixture for 8 min at room temperature to allow the DNA to bind to the beads.

10. Place the 96-well plate on the magnetic stand for 5 min (check that the solution is clear and that all the beads are against the wall of the well where the tube is in contact with the magnet).

11. Carefully remove the supernatant without disturbing the beads.

12. Wash the beads with 200 μl of freshly prepared 80% ethanol (vol/vol) (*see* **Notes 4** and **35**).

13. Incubate the samples for 30 s on the magnetic stand.

14. Remove the ethanol.

15. Repeat wash **steps 12–14**.

16. Leave the plate at room temperature for 5 min to remove any trace of ethanol and allow the beads to dry.

17. Avoid over drying of the beads as this makes resuspension difficult and leads to loss of material.

18. Add 26 μl of elution buffer (EB) and pipette up and down ten times to resuspend the beads (*see* **Note 36**).

19. Incubate the 96-well plate off the magnet at room temperature for 2 min.

20. Place the 96-well plate on the magnet and leave it for 2 min at room temperature (*see* **Note 37**).

21. Transfer 25 μl of the supernatant without disturbing the beads to a fresh 96-well plate (*see* **Notes 38** and **39**).

22. Cover the plates with a new adhesive film lid.

23. Samples can be placed at −20 °C for longer term storage or 4 °C overnight.

3.3.6 Quality Check the cDNA Library

1. This step should be performed in postamplification conditions.

2. If samples have been frozen, ensure that they are fully defrosted and then centrifuge at $700 \times g$ for 1 min prior quality check.

3. Check the size distribution of the amplified cDNA library on an Agilent Bioanalyzer high-sensitivity DNA chip (*see* **Note 40**). The library should be free of products smaller than 500 base pairs (bp) (primer dimers, etc.) and while will have a distribution of fragment sizes, should show a peak around 1.5–2 kb (Fig. 1).

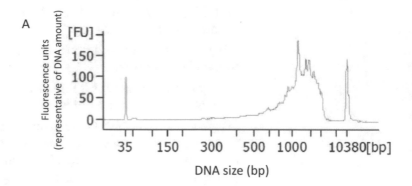

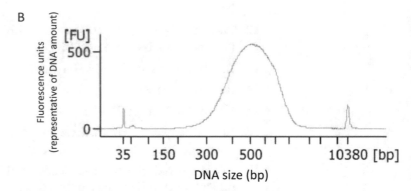

Fig. 1 Examples of Bioanalyzer electropherograms of cleaned libraries. (**a**) A representative electropherogram from the Bioanalyzer of a cleaned cDNA library, showing typical size distribution (bp). The example shows a single-cell. (**b**) A representative electropherogram from the Bioanalyzer of a clean indexed library, showing a peak of the double-side size selected library at around 500 bp. The example shows a pool of 96 cells

3.3.7 cDNA
Quantification

1. This step should be performed in postamplification conditions.

2. Use Quant-iT™ PicoGreen™ dsDNA Assay Kit (*see* **Note 41**).

3. Prepare "sample" plate - label a 96-well plate and add 28 μl of 1 × TE to each well.

4. To each well of the "sample" plate add 2 μl of the purified cDNA library (*see* **Note 42**).

5. Seal plate with adhesive film lid.

6. Prepare standard curve.

7. Dilute Lamdba DNA stock solution in 1 × TE to obtain 2000 pg/μl working solution (100 μl required per quantification plate).

8. Add 30 μl of the 2000 pg/μl working stock of Lamdba DNA solution to Std 1 and 2 of the standard curve (no 1 × TE will be added to Std 1) (*see* **Note 43**).

Table 4
Preparation of standard curve for PicoGreen assay

	Well	Concentration (pg/μl)	Volume of 1 × TE (μl)	Volume of diluent (μl)
Std 1	A1	2000	–	–
Std 2	B1	1000	30	30
Std 3	C1	500	30	30
Std 4	D1	250	30	30
Std 5	E1	125	30	30
Std 6	F1	62.5	30	30
Std 7	G1	31.25	30	30
Std 8	H1	15.625	30	30
Std 9	A2	7.813	30	30
Std 10	B2	3.906	30	30[a]
Blank	C2	–	30	–
Blank	D2	–	30	–
Blank	E2	–	30	–
Blank	F2	–	30	–
Blank	G2	–	30	–
Blank	H2	–	30	–

[a]Remove and discard 30 μl from this well after mixing

9. Prepare "standards" plate - label a 96-well plate and prepare serial dilutions of the Lamdba DNA working stock using 1 × TE as indicated in Table 4.

10. At each dilution step mix well by pipetting up and down ten times.

11. Prepare Quant-iT™ PicoGreen™ working solution (*see* **Note 44**).

12. For one quantification plate add 31.7 μl of PicoGreen™ stock to 6304.3 μl of 1 × TE.

13. Pipette 30 μl of PicoGreen™ working solution into each well of the standard curve plate and the sample plate (*see* **Note 45**).

14. Seal all 96-well plates with adhesive film lids.

15. Vortex to ensure mixing and then centrifuge the plates at 700 × *g* for 1 min.

16. Transfer 25 μl of all samples and standards into a 384-well (PS, Flat-bottom, Fluotrac, medium binding, black) plate in duplicate as shown in Fig. 2.

A

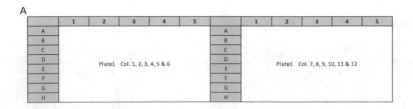

B

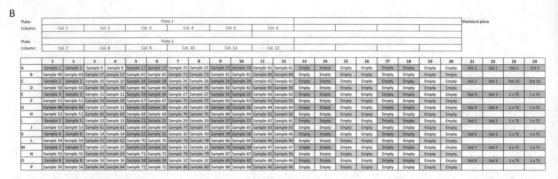

Fig. 2 Layout of PicoGreen Quantification plate. (**a**) Layout of the "sample dilution" plates. (**b**) Location of each diluted sample and the standard curve within the 384-well PicoGreen quantification plate

17. Cover the plates with an adhesive film lid.

18. Incubate the 384-well plate for 2–5 min at room temperature and protect from light.

19. Centrifuge the plate at $700 \times g$ for 1 min.

20. Measure the fluorescence intensity in a 384-well fluorometer (excitation ~480 nm, emission ~520 nm).

21. Use 100-150 pg of DNA per sample for library preparation (*see* **Note 46**).

3.3.8 Library Preparation

1. This step should be performed in post-amplification conditions.

2. Library preparation for multiple plates (*see* **Note 47**).

3. Equilibrate Tagment DNA Buffer and NT Buffer to room temperature.

4. Prepare the Tagmentation mix, with the Amplicon Tagment Mix maintained on ice during preparation:

	×1 well	×106 wells
Tagment DNA buffer	2.5	265
Amplicon Tagment mix	1.25	132.5
	3.75	397.5

5. Pipette master mix up and down gently to mix all components.

6. Aliquot 49 μl of Tagmentation mix into each tube of an 8-strip PCR tube.

7. Take a clean 96-well plate.

8. Add 3.75 μl of Tagmentation mix per well into a 96-well plate using an 8-channel pipette.

9. Add 1.25 μl of sample (Vf = 5 μl) (*see* **Note 48**).

10. Seal the plate with an adhesive film lid and centrifuge at 700 × *g* briefly.

11. Run the following program in a thermocycler:

Step	Temperature (°C)	Time	Cycle
Tagmentation	55	10 min	1
Storage	10	Hold	1

12. While tagmentation is taking place, aliquot NT buffer into an eight-strip PCR tube (16.5 μl per tube).

13. Defrost the Indexes (*see* **Note 49**), vortex, and centrifuge briefly.

14. Place the index tubes into the Index Plate Fixture, taking care to place them in the correct order and location.

15. Remove and discard the adhesive film lid.

16. Using a multichannel pipette, add 1.25 μl of NT buffer per well as soon as sample reaches 10 °C to neutralize the samples (Vf = 6.25 μl).

17. Seal the plate with an adhesive film lid and briefly centrifuge the plate at 700 × *g*.

18. Place the plate onto the Index Plate Fixture.

19. Remove and discard the adhesive film lid.

20. Add 1.25 μl of Index Primer 1 (N701-N707, N710-N712, N714, N715) to the corresponding well using a 12-channel pipette (Vf = 7.5 μl) (*see* **Note 50**) *see* Fig. 3 for example layout of Indexes.

21. Add 1.25 μl of Index Primer 2 (S502, S503, S505-S508, S510, S511) to the corresponding well using an eight-channel pipette (Vf = 8.75 μl).

22. Aliquot 49.5 μl of Nextera PCR Master Mix (NPM) into an eight-strip PCR tube.

23. Add 3.75 μl of NPM to each well using an 8-channel pipette (Vf = 12.5 μl).

24. Seal the plate with an adhesive film lid and centrifuge at 700 × *g* briefly.

		N701	N702	N703	N704	N705	N706	N707	N710	N711	N712	N714	N715
		1	2	3	4	5	6	7	8	9	10	11	12
S502	A	N701-S502	N702-S502	N703-S502	N704-S502	N705-S502	N706-S502	N707-S502	N710-S502	N711-S502	N712-S502	N714-S502	N715-S502
S503	B	N701-S503	N702-S503	N703-S503	N704-S503	N705-S503	N706-S503	N707-S503	N710-S503	N711-S503	N712-S503	N714-S503	N715-S503
S505	C	N701-S505	N702-S505	N703-S505	N704-S505	N705-S505	N706-S505	N707-S505	N710-S505	N711-S505	N712-S505	N714-S505	N715-S505
S506	D	N701-S506	N702-S506	N703-S506	N704-S506	N705-S506	N706-S506	N707-S506	N710-S506	N711-S506	N712-S506	N714-S506	N715-S506
S507	E	N701-S507	N702-S507	N703-S507	N704-S507	N705-S507	N706-S507	N707-S507	N710-S507	N711-S507	N712-S507	N714-S507	N715-S507
S508	F	N701-S508	N702-S508	N703-S508	N704-S508	N705-S508	N706-S508	N707-S508	N710-S508	N711-S508	N712-S508	N714-S508	N715-S508
S510	G	N701-S510	N702-S510	N703-S510	N704-S510	N705-S510	N706-S510	N707-S510	N710-S510	N711-S510	N712-S510	N714-S510	N715-S510
S511	H	N701-S511	N702-S511	N703-S511	N704-S511	N705-S511	N706-S511	N707-S511	N710-S511	N711-S511	N712-S511	N714-S511	N715-S511

Fig. 3 Layout of Index Combinations. The combinations of different indexes in each well of a 96-well plate. Shown in this Figure is the Nextera XT Index Kit v2 Set A

25. Perform the PCR in a thermocycler (with preheated lid) following the program:

Step	Temperature (°C)	Time	Cycle(s)
Extend	72	3 min	1
Denature	95	30 s	1
Denature	95	10 s	12
Anneal	55	30 s	
Extend	72	60 s	
Extend	72	5 min	1
Storage	10	Hold	1

26. Centrifuge at $700 \times g$ for 1 min and proceed with cleaning.

3.3.9 Library Pooling and Clean up

1. Equilibrate AMPure XP beads to room temperature for at least 15 min.

2. Samples will be cleaned up using a double-sided size selection ($0.5 \times$ and $0.7 \times$) (*see* **Note 51**).

3. This step should be performed in postamplification conditions.

4. Vortex AMPure XP beads until all beads are resuspended (approximately 30 s).

5. At this step, the library of each cell is specifically labeled and can be pooled. For this, remove 2 µl from each sample. The final library should be in a 1.5 ml low bind tube (*see* **Note 52**).

6. Measure the final volume of the pooled sample (should be ~192 µl).

7. Add the required amount of AMPure XP beads to the sample ($0.5 \times$) (~96 µl).

8. Mix by pipetting up and down 5–10 times or until the solution appears homogeneous.

9. Incubate for 5–8 min at room temperature to allow the DNA to bind to the beads.

10. Place the 1.5 ml low bind tube on the magnetic stand for 2 min.

11. Do not discard supernatants.

12. Carefully transfer the supernatant to a fresh 1.5 ml low bind tube without disturbing the beads.

13. Add AMPure XP beads to the sample to obtain the desired ratio (0.7×) (~38.4 μl).

14. Mix by pipetting up and down 5–10 times or until the solution appears homogeneous.

15. Incubate for 5–8 min at room temperature to allow the DNA to bind to the beads.

16. Place the 1.5 ml low bind tube on the magnetic stand for 2 min.

17. Discard the supernatant without disturbing the beads.

18. Add 1 ml of freshly prepared 80% ethanol (vol/vol) to the bead pellet.

19. Incubate for 30 s on the magnetic stand.

20. Remove the ethanol.

21. Repeat wash **steps 18–20**.

22. Leave the 1.5 ml low bind tube on the magnetic stand at room temperature for 5–7 min to remove any trace of ethanol but avoid over drying the beads as this makes resuspension difficult (*see* **Note 37**).

23. Resuspend the beads by adding 37.5 μl of EB (Qiagen) and mix ten times or until the solution appears homogeneous.

24. Incubate off the magnet for 2 min at room temperature.

25. Place the 1.5 ml low bind tube on the magnet and incubate for 2 min or until the solution appears clear.

26. Transfer 35 μl of supernatant to a fresh 1.5 ml low bind tube.

27. Samples can be stored at −20 °C for up to 3 months.

3.3.10 Quality Check and Quantification of the Indexed Library

1. This step should be performed in postamplification conditions.

2. If samples have been frozen, ensure that they are fully defrosted, vortex briefly, and centrifuge at $700 \times g$ briefly prior to quantification.

3. Check the size distribution of the final library on an Agilent Bioanalyzer high-sensitivity DNA chip (*see* **Note 40**). Due to the double-sided cleaning a discrete peak should be visible with an average size between 450 and 600 bp (Fig. 1). Determine

the average size of the library by defining a range of 42 bp–3561 bp.

4. Quantify pooled indexed library using KAPA Library Quantification Kit (Roche) according to the manufacturer's instructions.

5. Dilute 1 μl of the sample (product from Subheading 3.3.9) with KAPA dilution buffer (*see* **Note 53**); run the dilutions in triplicate.

6. Prepare enough Master Mix for two DNA dilutions per sample (run in triplicate), Standard Curve and No Template Controls (NTC) (water and KAPA dilution buffer) (run in duplicate).

7. Prepare the Master Mix 1× (volumes shown here are for two dilutions of one sample).

	×1 well	×18 wells
SYBR fast master mix + primer	12	216
Water	4	72
	16	288

8. Mix and briefly centrifuge the Master Mix.

9. To a 96-well plate, add 4 μl of DNA Standards, diluted samples and NTC to appropriate wells.

10. Briefly centrifuge the plate at $700 \times g$.

11. Dispense 16 μl of the Master Mix into each well and pipette up and down several times. Fresh pipette tips must be used for each well to prevent well-to-well contamination.

12. Seal the plate with optical caps.

13. Briefly centrifuge the plate at $700 \times g$.

14. Transfer to qPCR machine.

15. Run the following program on the qPCR machine (*see* **Note 54**):

Step	Temperature (°C)	Time	Cycle(s)
Initial denaturation	95	5 min	1
Denaturation	95	30 s	35
Annealing/extension/data acquisition	60	45 s	
Dissociation curve analysis			

16. Follow instructions in the manufacturer's protocol for data analysis.

17. Use the average size determined in **step 2** to calculate the final concentration.

3.3.11 Pooling of Multiple Samples for Next Generation Sequencing

1. This step should be performed in post-amplification conditions.

2. If multiple 96-well plates (already pooled in Subheading 3.3.9 Library pooling and clean up) are to be pooled and sequenced together, care must be taken to ensure that each pooled 96-well plate is evenly represented.

3. Distribute the number of moles of each library to be added to the sequencing pool according to the desired ratios, taking into consideration the number of cells in each pool. In that way, if 4 pooled libraries, each one containing 96 cells are to be pooled together, add equimolar quantities. 460 fmole in a volume of 20 µl corresponds to a final concentration of 23 nM. Typically, we would pool ~115 fmole of each library, which assuming a similar concentration of each single-cell library corresponds to ~1.2 fmole of each single-cell library.

4. Using these guidelines, the ratios can be adjusted to the number of cells, availability of material and desired ratio of sequencing.

3.3.12 Sequencing

1. We would routinely perform single end 50 bp sequencing on scRNA-Seq libraries which reduces the expenses of sequencing without compromising the results.

2. The amount of sequencing required will depend on the number of single cells which were processed and of the desired depth of sequencing.

3. A standard lane of Illumina HiSeq 4000 would render approximately 400 Million reads. We find that around 0.5 million mappable reads per cell should be sufficient and we typically obtain between 50% and 60% mappability with this protocol.

4. We typically sequence 4 plates of 96 single cells (384 cells in total) in one lane of HiSeq 4000.

4 Notes

1. The dilutions for individual antibodies can be found in Table 2. Each dilution has been carefully titrated within our laboratory. It is important to remember that the precise optimal antibody concentration may vary depending on the supplier, the product lot and the experimental setting. We would recommend that antibody panels should be optimized by the end user.

2. We use optical adhesive film lids from Starlab, the lids do not need to be optical, but it is important to test the stickiness of

the lids. The lids need to be reasonably easy to remove, otherwise removal of the lid can lead to the splashing of reagents and mixing of wells. But the lids also need to be able to withhold storage at −80 °C and the extremes of the PCR conditions throughout the protocol. We recommend testing lids before proceeding.

3. Make dilutions of ERCC RNA spike-in mix in water with RNase inhibitor (1 U/μl). 1:300,000 dilution. Make a serial dilution. First prepare 1:10 dilution (this can be stored at −80 °C). Then make a dilution 1:100 (1 μl + 99 μl of water + RNase inhibitor) and a further dilution 1:300 (4 μl + 1196 μl of water + RNase inhibitor) to obtain a final dilution 1:300,000. Make aliquots and store at −80 °C—these aliquots can be used for multiple plates of the same experiment but should not be stored long term.

4. 80% Ethanol (vol/vol) needs to be made fresh every time, as ethanol absorbs moisture and therefore this will lead to changes in the final concentration if stored.

5. Bones can also be flushed using a 23G needle, but we have found that the number of cells harvested from the bone marrow is more reproducible when using a pestle and mortar. Also, the use of needles is an additional biohazard to consider. If bones are crushed then the inclusion of a CD45 antibody is necessary within the stem cell antibody staining panel to ensure that all cells which are sorted are hematopoietic and not contaminating stromal cells.

6. Any 50 μM filter can be used. We prefer Partec filters as they fit snuggly into a 15 ml centrifuge tube and the nylon mesh inside the filter is at an angle to optimize recovery of cells.

7. Do not leave red cell lysis step longer than 15 min in total otherwise all granulocytes can be removed from the sample. While red cell lysis is taking place, begin to label the tubes for the antibody single stains and prepare the antibody master mix—*see* Tables 2 and 3.

8. We do not include fluorescence minus one (FMO) controls in this protocol as we no longer routinely run them for these experiments. If establishing this protocol for the first time, FMO controls should be included in the control staining panel. FMOs are particularly important when setting up multicolor FACs panels, as they will facilitate the identification of a positive vs negative populations and will help to determine where the gates should be set.

9. When preparing single stains place the undepleted bone marrow cells onto ice and begin with the lineage depletion step of the protocol (Subheading 3.1.3) and then return to add

control antibodies while the cells are incubating with the Mouse Hematopoietic Progenitor Isolation Cocktail.

10. For the "All stain" control and actual samples which will be sorted, it is better practice to prepare a master mix; this means that there is less chance that an antibody will be missed from the panel and the pipetting is generally more accurate as the volume to be pipetted is larger. Details of the antibody master mix are in Table 3.

11. The viability stain should only be added to the controls and samples just before proceeding to the flow cytometry facility as the signal can increase over time.

12. The lineage depletion performed in this protocol is very mild, this was undertaken to ensure that some of the more mature lineages which still expressed the markers used in the antibody staining panel would be included in the final analysis. If different populations are required, we would recommend following the manufacturer's instructions for the lineage depletion as this will greatly save on sorting time.

13. At this stage the cells should be transferred into a tube suitable for the available flow cytometry sorter. In our case we use 5 ml round bottom polypropylene tube.

14. The master mix volume is 500 µl, this will allow for a single all stain control (100 µl), an individual sample (300 µl) and 100 µl excess. The additional/excess volume can be altered. The volume in which the sample is stained can also be adjusted to a larger volume, but we would not recommend staining the whole material obtained from one mouse in less than 300 µl.

15. We have always used the 10x Genomics platform to be able to investigate larger numbers of cells to be able to characterize the entire hematopoietic compartment of the mouse bone marrow. The antibody panel we use contains multiple additional antibodies which are used to be able to ensure that the different populations are present and at expected ratios within the bone marrow. We have always used the 3′ reagents (referred to as single-cell gene expression) from 10× Genomics but 5′ reagents are also available (single-cell immune profiling).

16. The number of cells required will depend on the population which has been sorted. If samples are precious, we would sort an exact number of cells and rely on the number specified by the sorter. If cells are plentiful, a higher number of cells can be sorted and counted after being resuspended in 1x PBS + 0.04% BSA solution (400 µg/ml). We find that bone marrow samples do not require filtering before being used in the 10× Genomics protocol, as after sorting they tend to be a homogeneous single-cell suspension. There are considerations to deciding on the cell number to use in the 10× Genomics protocol. As

the number of cells increases, the number of droplets that will contain doublets will also increase. For many of our experiments characterizing the Lineage negative c-kit+ compartment of the mouse bone marrow, we have always loaded ~16,000 cells which will produce an estimated doublet rate of ~8%. For more details, we recommend reading the 10× Genomics protocols/user manuals.

17. The volume in which to resuspend the cells for the 10× Genomics protocol will depend of the desired concentration of the cells. This will be determined by the total number of cells obtained from the sorting steps and by the desired number of cells to be used in the droplet protocol. Please refer to the 10× Genomics manual. Once the concentration has been chosen, we normally resuspend the cells in half of the required volume. We then count the cells in an improved Neubauer counting chamber and adjust the final volume to achieve the desired concentration. For experiments using Chromium Next GEM Single Cell 3′ Reagent Kits v3.1, we would resuspend the cells at a concentration of ~370 cells/μl.

18. mcSmart-Seq2 experiments should be performed in RNase, RNA, and DNA-free areas to minimize contamination. If possible, pre- and postamplification work should be carried out in different laboratories. These experiments rely on the amplification of the RNA from single cells, so tiny amounts of contaminating material could drastically affect the output of the experiments. Be extremely careful in the handling of the samples if human material is processed to avoid contamination introduced by material from the operator. If separate laboratory space is not possible then the use of a UV PCR cabinet and separate equipment (particularly pipettes) is highly advisable to minimize the risk of contamination. If using UV PCR hood, all equipment should be sterilized by exposing to UV light before and after use.

19. All master mixes are enough for a 96-well plate, including an excess enough for 100 wells (~4% excess).

20. Aliquoting of the lysis buffer can be done using a multichannel pipette. We would normally aliquot the lysis buffer into eight-strip PCR tubes on a chilled metal block and then use a multichannel pipette to aliquot the lysis buffer into all the wells of the 96-well plate. Care must be taken to ensure that each tip contains 2.3 μl, this should be verified visually for each pipetting step. The same pipette tips can be used for the entire 96-well plate as there are no cells present at this time.

21. The type of 96-well PCR plate will depend on the type of PCR machine which is available and the sorting options. We favor

using nonskirted PCR plates, but this required our sorting facility to create a holder which could fit the 96-well plate in place while sorting [37]. Fully skirted PCR plates can generally be utilized in the FACS sorters as they have the same footprint as a 6-well culture plate.

22. Lysis plates should be kept cold until required, this can be either on ice or in the refrigerator. We do not recommend preparing the lysis plates the day before the experiment and storing in the refrigerator overnight as considerable evaporation occurs. We have found that it is very difficult to recuperate the volume back into the wells. It is much preferable to prepare the lysis plates fresh on the day of the experiment. The lysis buffer can be prepared in advance and stored at 4 °C for 6 months or −20 °C.

23. Use index sorting when possible, even if only one or two parameters are used during the sorting, as this will keep a record of the number of single cells that were sorted into wells. We would also recommend stringent settings for the sorting procedure, it is preferable to have no cell in the well than two cells. It is very important to avoid refilling the wells which are reported empty, as we have seen that even though no event may be recorded by the index sorter the scRNA-Seq may show data for this well [37].

24. Care should be taken when planning the layout of lysis plates. It is advisable to have self-contained experiments within each plate. Sorting different populations or conditions onto different plates can lead to batch effects, if two different conditions or populations are to be compared then they should be on the same plate. If multiple plates with similar content are to be used, it is also advisable to rotate the samples within the plate positions so that there are no plate effects when comparing across multiple plates (Fig. 4).

25. We would also recommend that test plates are sorted for each sorting session. The test plates serve several purposes as they allow the investigator to: (1) check that cells were successfully sorted on the day, (2) confirm that cells are of good quality, (3) test the quality of the batch of reagents to be used during the processing, and (4) optimize the number of PCR cycles to be used for the cDNA amplification (Subheading 3.3.4). Each test plate normally comprises of single columns of single cells and one well of ten cells (Fig. 5). A test should be sorted for each of the different cell types to be investigated within an experiment. The test plates are only processed up to Subheading 3.3.6, once a successful cDNA trace can be seen the full 96-well plates can be processed. Several columns can be sorted on an individual 96-well plate (if using nonskirted plates) as

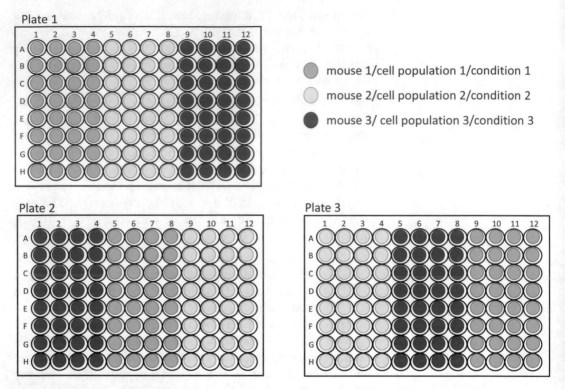

Fig. 4 Planning of lysis plates. If comparisons need to be drawn from different variables then the layout of the sorting/lysis plates must be taken into careful consideration. Sorting one variable onto a single plate can lead to batch effects. Care must be taken to ensure that all the comparable variables are distributed across all plates, and that the correct number of total cells are sorted per variable

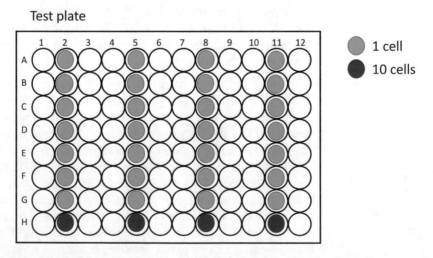

Fig. 5 Test plate. A test plate is essential to ensure that cells have been successfully sorted, to allow the batches of reagents to be tested and to optimize the number of PCR cycles for the cDNA amplification. Several columns can be sorted on an individual 96-well plate (if using nonskirted plates) as these can be cut into separate columns

these can be cut into separate columns and processed separately.

26. The vortexing of the sorted cells in the lysis buffer is controversial since the solution contains detergent which creates foam. We find that it does not seem to be an issue in such a small volume. The vortexing should help with the lysis of the cells, and specific protocols have shown it to be essential [38].

27. When freezing the lysis plates, we would recommend placing them onto a precooled metal shelf within the −80 °C freezer to accelerate the freezing process. Nonskirted 96-well plates tend to bend once stored at −80 °C and if not properly sealed the adhesive lid can become unstuck. It is good practice that once frozen, the plates should be stored together in zip lock plastic bags to keep an experiment together, as individual plates can very easily be lost in the freezer. It is essential to avoid freeze-thawing of the plates as this will lead to degradation of the material.

28. We would normally prepare the master mixes for both the annealing and reverse transcription steps at the same time. This means that as soon as the cell lysis/annealing step has finished the reverse transcription mix can be added to the wells.

29. For the aliquoting of all reagents for reverse transcription and cDNA amplification in the protocol we suggest the use of a Multipette® (Eppendorf), as this saves the use of multiple pipette tips and repeated pipetting actions. The volume of retention within the combitips is also very small, which means that large excesses of reagents do not need to be used and the pipetting steps are more accurate.

30. Care should be taken when adding reagents to wells. Each well now contains a single cell and so touching inside of the wells could lead to cross-contamination. Reagents should be added to the wells dropwise to edge of well, touch very lightly if at all, capillary action should attract the drop to the side of the well without needing to physically need to touch the inside of the well. We do not vortex the plate after the addition of reagents.

31. The IS PCR primer should be prepared fresh. If multiple plates are going to be processed as a part of the same experiment then one dilution can be made and aliquoted, this should then be stored at −20 °C. Avoid freeze–thaw cycles.

32. The number of PCR cycles depends on the starting amount of RNA. Cells which are quiescent will contain less RNA than rapidly cycling cells, for example hematopoietic stem cells contain less RNA than more committed progenitor cells (i.e., megakaryocyte–erythroid progenitors or granulocyte–monocyte progenitors). Typically, use 19 cycles for single hematopoietic cells. Try to maintain the PCR cycles at the minimum

possible to minimize amplification bias and PCR duplicates. Not a lot of cDNA is required for library preparation. If multiple cell types are been compared within one experiment then the number of cycles of PCR should be fixed and remain constant for all cells from that experiment. This will allow for bioinformatic comparison across the populations.

33. We would normally perform the PCR purification using a liquid handling robot. While this is not essential, if many plates are being processed it ensures consistency between the plates and minimizes pipetting. Care also must be taken to ensure that the beads do not over dry following the ethanol wash, as this can lead to a loss of material.

34. We have found that AMPure XP beads–sample ratio which works optimally for the PCR purification is 0.6×. We suggest to add 29 μl as there is normally a slight loss of volume from the wells, we find that on average there is approximately 48 μl remaining in the wells. Care must be taken when pipetting the AMPure beads, as inaccuracy or droplets of beads on the side of tips will lead to a change in the ratio of beads to sample, and will alter the fragment sizes which are captured.

35. If manually washing an entire 96-well plate of AMPure XP beads, wash one column at a time to avoid over-drying of the beads which will reduce elution of the DNA.

36. When resuspending AMPure beads ensure that all the sample is at the bottom of the well and no drops remain on the walls of the tube.

37. Care must be taken when drying AMPure beads. This protocol recommends leaving the beads for 2 min, but time may vary with the temperature of the laboratory, the amount of beads, and plasticware used. Upon the removal of ethanol, the beads will appear shiny, as they dry they will lose this shiny appearance and a crack will begin to appear. If multiple cracks begin to appear this is a sign that the beads have begun to over dry and then elution efficiency may be reduced.

38. When transferring the eluted PCR material to a new 96-well plate maintain the well location, or keep clear records of which wells have moved to new plate locations. This will be essential to be able to track the individual cells in the sequencing files and be able to combine the scRNA-Seq to the metadata of each individual cell (flow cytometry index data, etc.).

39. When removing eluted material a small volume of EB will be left behind; this is to ensure that no beads are transferred with the cleaned cDNA library.

40. The size of the amplified cDNA library can also be checked on an Agilent TapeStation High Sensitivity DNA tape. We find

that the trace of the amplified cDNA material is more accurate on the Bioanalyzer as the material can be larger than 5000 bp. It is not essential to accurately assign the size of the material, but we recommend using the Bioanalyzer unless it is performed by an experienced user of this technique.

41. For cDNA quantification we recommend following the manufacturer's protocol and specifications of the analyzers used to read the assay.

42. If multiple plates are being processed it is not feasible that all wells can be quantified. As a result we would recommend to quantify as many samples as possible across a cross section of the 96-well plate (typically 40 which is the equivalent to 5 columns). This will also depend on the design of your experiment and the layout of the individual 96-well plates. To characterize the hematopoietic stem and progenitor compartments of the mouse bone marrow we tend to have multiple different cell types on one 96-well plate. Attention should be paid to ensuring that similar numbers of cells of all cell types present in the plate are included in the quantification process. Multiple quantification plates can be run, but this is dependent on the researcher and the time and cost that you wish to dedicate to this. It is however very important to quantify the samples accurately since it will have a strong impact in the efficiency and quality of the final library.

43. If multiple PicoGreen™ quantification plates are being set up at once, the standard curve preparation should be multiplied as necessary, these should all be made up in the same "standards" plate to ensure comparison across the plates. The standards should then be transferred to adjacent wells to allow the separate addition of PicoGreen™ working solution.

44. When preparing Quant-iT™ PicoGreen™ working solution use a plastic container rather than glass as reagents may absorb to glass surfaces and ensure to protect the working solution from light by covering with foil or placing it in the dark.

45. If multiple quantification plates are being analyzed the standards should have been transferred to adjacent wells. Add the PicoGreen™ working solution to one standard curve at a time.

46. For library preparation of single cells, use within the range of 100–150 pg DNA per sample for tagmentation. The amount of DNA has been optimized for the amount of Tagmentase recommended in this protocol. The use of too much starting material will produce under-tagmented products that will eventually result in suboptimal material for sequencing. Prepare dilutions of DNA samples in EB buffer, the dilutions should be made fresh as low concentrations of DNA are not

stable for extended periods of time and run the risk of molecules sticking to the plastic of the plates.

47. If preparing libraries for multiple plates, process one plate at a time. It is very difficult to control the time that the Tagmentase is incubated with the DNA if multiple plates are processed at once.

48. 100 to 150 pg of cDNA should be used as input for tagmentation. Dilution of cDNA plates in EB buffer is normally required so that concentration of the majority of samples is within range. Prepare dilutions of DNA samples in EB buffer, the dilutions should be made fresh as low concentrations of DNA are not stable for extended periods of time and run the risk of molecules sticking to the plastic of the plates.

49. There are four index options available from Illumina at the time of writing. The use of the four individual kits allows the combining of four 96-well plates. This can be useful, if a smaller amount of sequencing is required as multiple plates can be pooled into one single lane of sequencing. Care has to be taken to ensure that indexes are not repeated and that the order of the indexes is maintained, as it will be the indexes which will allow the researcher to link back to any metadata that exists for the individual cells. When adding indexes to individual wells it is essential to change tips between each row and each column. We normally discard all the lids from the index tubes and replace them with new lids so as not to cause any cross contamination between the different index tubes.

50. The indexes given as an example in the method and Fig. 3 are from the Nextera XT Index Kit v2 Set A (Illumina). Nextera XT Index Kit v2 Set B, C, and D are also available.

51. We have found that a single side clean up using the AMPure beads can leave some larger untagmented fragments within the sample that lead to a reduced performance of the sample in sequencing. We find that the indicated double-sided size clean up works best. The AMPure beads bind DNA depending on the ratio of AMPure beads used [39] which is determined by the volume of the sample to be cleaned (e.g., volume of AMPure beads/volume of DNA sample). Care must be taken when performing a double-sided size selection, as the supernatant from the first addition of AMPure beads must not be discarded and is transferred to a new tube. In this step, the larger fragments (bound to the beads) will be discarded. During the second clean, the beads are kept and the smallest fragments are discarded in the supernatant. The second addition of AMPure beads is accumulative, and therefore in our protocol to clean a 100 μl reaction with ratios 0.5 and 0.7,

50 μl of beads will be added in the first instance followed by the addition of 20 μl of beads (50 μl + 20 μl/100 μl = 0.7×).

52. To reduce the number of samples which need to be cleaned, prepare a pool of the individual indexed libraries. Assume at this stage that the concentration of each individual library is equal. If preferred, samples can be cleaned up individually or in small pools prior to pooling all libraries, quantifying each library individually (or small pool) and pooling as desired. As only 1/5 of the each single-cell library will be pooled, it is always possible to repeat the process in the future using different ratios of each library. We would normally use an 8-channel multichannel pipette and transfer 2 μl from each well into an eight-tube strip, using a new pipette tip for each well to avoid cross-contamination. Once all 12 columns have been pooled into the eight-tube strip, use a p200 pipette to pool the 8-tubes into one 1.5 ml low bind tube.

53. When quantifying pooled indexed libraries, dilutions and Standard Curves used will be qPCR machine dependent. We typically prepare dilutions 1:200,000 and 1:2,000,000 in KAPA dilution buffer, but it may vary according to the expected concentration of the libraries. Due to the dynamic range of our qPCR machine we only use Standards 3 to 6 for the Standard Curve. It is essential that the used dilutions are contained within the range of the standard curve. It is important to run control wells on the quantification plate using the KAPA dilution buffer and water used to prepare the dilutions and master mixes, to ensure that there is no contamination within the reagents.

54. When quantifying the library, please see additional notes within the manufacturer's protocol for longer fragment lengths and details of specific protocols for individual qPCR machines.

Acknowledgments

We would like to acknowledge core support from Wellcome and MRC to the Wellcome-MRC Cambridge Stem Cell Institute (grant number 203151/Z/16/Z).

References

1. Moignard V, Macaulay IC, Swiers G et al (2013) Characterization of transcriptional networks in blood stem and progenitor cells using high-throughput single-cell gene expression analysis. Nat Cell Biol 15:363–372. https://doi.org/10.1038/ncb2709

2. Wilson NK, Kent DG, Buettner F et al (2015) Combined single-cell functional and gene expression analysis resolves heterogeneity within stem cell populations. Cell Stem Cell 16:712–724. https://doi.org/10.1016/j.stem.2015.04.004

3. Psaila B, Barkas N, Iskander D et al (2016) Single-cell profiling of human megakaryocyte-erythroid progenitors identifies distinct megakaryocyte and erythroid differentiation pathways. Genome Biol 17:83. https://doi.org/10.1186/s13059-016-0939-7

4. Giladi A, Paul F, Herzog Y et al (2018) Single-cell characterization of haematopoietic progenitors and their trajectories in homeostasis and perturbed haematopoiesis. Nat Cell Biol 20:836–846. https://doi.org/10.1038/s41556-018-0121-4

5. Miyawaki K, Iwasaki H, Jiromaru T et al (2017) Identification of unipotent megakaryocyte progenitors in human hematopoiesis. Blood 129:3332–3343. https://doi.org/10.1182/blood-2016-09-741611

6. Pietras EM, Reynaud D, Kang Y-A et al (2015) Functionally distinct subsets of lineage-biased multipotent progenitors control blood production in normal and regenerative conditions. Cell Stem Cell 17:35–46. https://doi.org/10.1016/j.stem.2015.05.003

7. Nestorowa S, Hamey FK, Pijuan Sala B et al (2016) A single-cell resolution map of mouse hematopoietic stem and progenitor cell differentiation. Blood 128:e20–e31. https://doi.org/10.1182/blood-2016-05-716480

8. Rodriguez-Fraticelli AE, Wolock SL, Weinreb CS et al (2018) Clonal analysis of lineage fate in native haematopoiesis. Nature 553:212–216. https://doi.org/10.1038/nature25168

9. Laurenti E, Göttgens B (2018) From haematopoietic stem cells to complex differentiation landscapes. Nature 553:418–426. https://doi.org/10.1038/nature25022

10. Haghverdi L, Büttner M, Wolf FA et al (2016) Diffusion pseudotime robustly reconstructs lineage branching. Nat Methods 13:845–848. https://doi.org/10.1038/nmeth.3971

11. Qiu X, Mao Q, Tang Y et al (2017) Reversed graph embedding resolves complex single-cell trajectories. Nat Methods 14:979–982. https://doi.org/10.1038/nmeth.4402

12. Setty M, Tadmor MD, Reich-Zeliger S et al (2016) Wishbone identifies bifurcating developmental trajectories from single-cell data. Nat Biotechnol 34:637–645. https://doi.org/10.1038/nbt.3569

13. Dahlin JS, Hamey FK, Pijuan-Sala B et al (2018) A single-cell hematopoietic landscape resolves 8 lineage trajectories and defects in Kit mutant mice. Blood 131:e1–e11. https://doi.org/10.1182/blood-2017-12-821413

14. Tusi BK, Wolock SL, Weinreb C et al (2018) Population snapshots predict early haematopoietic and erythroid hierarchies. Nature 555:54–60. https://doi.org/10.1038/nature25741

15. Ostrander EL, Kramer AC, Mallaney C et al (2020) Divergent effects of Dnmt3a and Tet2 mutations on hematopoietic progenitor cell fitness. Stem Cell Rep 14:551–560. https://doi.org/10.1016/j.stemcr.2020.02.011

16. Izzo F, Lee SC, Poran A et al (2020) DNA methylation disruption reshapes the hematopoietic differentiation landscape. Nat Genet 52:378–387. https://doi.org/10.1038/s41588-020-0595-4

17. Svensson V, Vento-Tormo R, Teichmann SA (2018) Exponential scaling of single-cell RNA-seq in the past decade. Nat Protoc 13:599–604. https://doi.org/10.1038/nprot.2017.149

18. Hay SB, Ferchen K, Chetal K et al (2018) the human cell atlas bone marrow single-cell interactive web portal. Exp Hematol 68:51–61. https://doi.org/10.1016/j.exphem.2018.09.004

19. Zilionis R, Nainys J, Veres A et al (2017) Single-cell barcoding and sequencing using droplet microfluidics. Nat Protoc 12:44–73. https://doi.org/10.1038/nprot.2016.154

20. Macosko EZ, Basu A, Satija R et al (2015) Highly parallel genome-wide expression profiling of individual cells using Nanoliter droplets. Cell 161:1202–1214. https://doi.org/10.1016/j.cell.2015.05.002

21. Zheng GXY, Terry JM, Belgrader P et al (2017) Massively parallel digital transcriptional profiling of single cells. Nat Commun 8:14049. https://doi.org/10.1038/ncomms14049

22. Rosenberg AB, Roco CM, Muscat RA et al (2018) Single-cell profiling of the developing mouse brain and spinal cord with split-pool barcoding. Science 360:176–182. https://doi.org/10.1126/science.aam8999

23. Picelli S, Faridani OR, Björklund ÅK et al (2014) Full-length RNA-seq from single cells using Smart-seq2. Nat Protoc 9:171–181. https://doi.org/10.1038/nprot.2014.006

24. Hagemann-Jensen M, Ziegenhain C, Chen P et al (2020) Single-cell RNA counting at allele and isoform resolution using Smart-seq3. Nat Biotechnol 38:708–714. https://doi.org/10.1038/s41587-020-0497-0

25. Bagnoli JW, Ziegenhain C, Janjic A et al (2018) Sensitive and powerful single-cell RNA sequencing using mcSCRB-seq. Nat Commun 9:1–8. https://doi.org/10.1038/s41467-018-05347-6

26. Sasagawa Y, Danno H, Takada H et al (2018) Quartz-Seq2: a high-throughput single-cell RNA-sequencing method that effectively uses limited sequence reads. Genome Biol 19:29. https://doi.org/10.1186/s13059-018-1407-3

27. Hashimshony T, Senderovich N, Avital G et al (2016) CEL-Seq2: sensitive highly-multiplexed single-cell RNA-Seq. Genome Biol 17:77. https://doi.org/10.1186/s13059-016-0938-8

28. Jaitin DA, Kenigsberg E, Keren-Shaul H et al (2014) Massively parallel single-cell RNA-seq for marker-free decomposition of tissues into cell types. Science 343:776–779. https://doi.org/10.1126/science.1247651

29. Hayashi T, Ozaki H, Sasagawa Y et al (2018) Single-cell full-length total RNA sequencing uncovers dynamics of recursive splicing and enhancer RNAs. Nat Commun 9:619. https://doi.org/10.1038/s41467-018-02866-0

30. Zhu C, Preissl S, Ren B (2020) Single-cell multimodal omics: the power of many. Nat Methods 17:11–14. https://doi.org/10.1038/s41592-019-0691-5

31. Giustacchini A, Thongjuea S, Barkas N et al (2017) Single-cell transcriptomics uncovers distinct molecular signatures of stem cells in chronic myeloid leukemia. Nat Med 23:692–702. https://doi.org/10.1038/nm.4336

32. Rodriguez-Meira A, Buck G, Clark S-A et al (2019) Unravelling intratumoral heterogeneity through high-sensitivity single-cell mutational analysis and parallel RNA sequencing. Mol Cell 73:1292–1305.e8. https://doi.org/10.1016/j.molcel.2019.01.009

33. van Galen P, Hovestadt V, Wadsworth MH II et al (2019) Single-cell RNA Seq reveals AML hierarchies relevant to disease progression and immunity. Cell 176:1265–1281.e24. https://doi.org/10.1016/j.cell.2019.01.031

34. Nam AS, Kim K-T, Chaligne R et al (2019) Somatic mutations and cell identity linked by Genotyping of Transcriptomes. Nature 571:355–360. https://doi.org/10.1038/s41586-019-1367-0

35. Saikia M, Burnham P, Keshavjee SH et al (2019) Simultaneous multiplexed amplicon sequencing and transcriptome profiling in single cells. Nat Methods 16:59–62. https://doi.org/10.1038/s41592-018-0259-9

36. Stoeckius M, Hafemeister C, Stephenson W et al (2017) Simultaneous epitope and transcriptome measurement in single cells. Nat Methods 14:865–868. https://doi.org/10.1038/nmeth.4380

37. Schulte R, Wilson NK, Prick JCM et al (2015) Index sorting resolves heterogeneous murine hematopoietic stem cell populations. Exp Hematol 43:803–811. https://doi.org/10.1016/j.exphem.2015.05.006

38. Teles J, Enver T, Pina C (2014) Single-cell PCR profiling of gene expression in hematopoiesis. In: Bunting KD, Qu C-K (eds) Hematopoietic stem cell protocols. Springer, New York, NY, pp 21–42

39. Quail MA, Swerdlow H, Turner DJ (2009) Improved protocols for the Illumina genome analyzer sequencing system. Curr Protoc Hum Genet 62:18.2.1–18.2.27. https://doi.org/10.1002/0471142905.hg1802s62

INDEX

Marion Espéli and Karl Balabanian (eds.), *Bone Marrow Environment: Methods and Protocols*, Methods in Molecular Biology,
vol. 2308, https://doi.org/10.1007/978-1-0716-1425-9, © Springer Science+Business Media, LLC, part of Springer Nature 2021

Printed in the United States
by Baker & Taylor Publisher Services